OPNsense Beginner to Professional

Protect networks and build next-generation firewalls easily with OPNsense

Julio Cesar Bueno de Camargo

BIRMINGHAM—MUMBAI

OPNsense Beginner to Professional

Group Product Manager: Vijin Boricha
Publishing Product Manager: Mohd Riyan Khan
Senior Editor: Arun Nadar
Content Development Editor: Nihar Kapadia
Technical Editor: Nithik Cheruvakodan
Copy Editor: Safis Editing
Project Coordinator: Ashwin Kharwa
Proofreader: Safis Editing
Indexer: Hemangini Bari
Production Designer: Prashant Ghare
Marketing Coordinator: Hemangi Lotlikar

First published: June 2022
Production reference: 1180522

Published by Packt Publishing Ltd.
Livery Place
35 Livery Street
Birmingham
B3 2PB, UK.

ISBN 978-1-80181-687-8

www.packt.com

To the memory of my father, Nivaldo B. de Camargo, my life's sensei. To my wife, who always encourages and supports me with her unconditional love and dedication, and our daughter, Maria Catarina, our little angel! To all friends (especially the Cloudfence team!) and family who supported me during this journey! To God, for keeping me standing to face life's challenges.

– Julio Cesar B. de Camargo

Contributors

About the author

Julio Cesar Bueno de Camargo is a cybersecurity professional with 15+ years of experience working with open source software. He started with Conectiva Linux and later became the official instructor helping dozens of students. As an aviation enthusiast and airplane pilot, he strives to bring all the aviation best practices to his professional routine.

Julio started working with OPNsense in 2016, contributing to the project with code, official forum moderation, articles, a Udemy course, and promotions in Brazil and Europe. He founded CloudFence in 2018, a cybersecurity start-up and a Luso-Brazilian-managed security services firm with an open source DNA. As its CTO, Julio aims to spread open source security as a service to companies from different parts of the world.

First, I want to thank my wife, the love of my life, Maria Eugenia, she always makes me believe! To our baby, Maria Catarina, daddy loves you! Also, I want to thank tech reviewers Franz Fabian and Nicolas Goralski ... and all the Packt team! To the OPNsense community that maintains this fantastic open source security project! To my teammates at Cloudfence. Last but not least, my friends and family were essential in motivating me to write this book!

About the reviewers

Fabian Franz BSc is a senior software developer (mainly working in the Jakarta EE / Spring and JavaScript environment nowadays) who has a strong security background. He was a pupil at HTL Dornbirn when his team won the Austrian Cyber Security Challenge national final. He continued studying for a bachelor's degree in **Sichere Informationssysteme** (*secure information systems*) at the University of Applied Sciences Upper Austria – Campus Hagenberg.

While Fabian was studying at the university, at some point, he read an article about a new firewall appliance on the internet. He was interested and started contributing quite quickly. Since then, he has written a few OPNsense plugins (such as the os-nginx plugin, covered in this book).

> *I have to thank all my teachers, who always taught me more than needed, which helped me to learn more to become an expert in specific areas. I also have to thank all those developers of open source software / free software out there whose software I use daily and which makes a firewall distribution such as OPNsense or a Linux distribution possible. A special thanks also to Ad and Franco for all the help I got when I learned how OPNsense works.*

Nicolas Goralski is a Linux system engineer with over 24 years of experience in various companies.

Nicolas has a global vision of IT architectures to better understand business projects and propose the best economic and technical approach to achieve customer satisfaction.

Motivated by the leitmotiv **Keep It Simple, Stupid (KISS)** and what can be automated must be automated, Nicolas uses open source technology as much as he can to fulfill his duty.

To my wife, Isabelle, who has believed in me for years and who supports me, I love you.

To my kids, who try to understand what I do, and the fact that I'm not always playing on my computer.

To my family and friends, thank you for everything you have given me.

To Packt Publishing, thank you for giving me the opportunity to review this book.

To all the network packets that I've blocked, it's not personal, it's business.

Table of Contents

3
Configuring an OPNsense Network

4
System Configuration

Section 2: Securing the Network

5
Firewall

6
Network Address Translation (NAT)

7
Traffic Shaping

8

Virtual Private Networking

9

Multi-WAN – Failover and Load Balancing

10

Reporting

Section 3: Going beyond the Firewall

11
Deploying DHCP in OPNsense

12
DNS Services

13
Web Proxy

14

Captive Portal

15

Network Intrusion (Detection and Prevention) Systems

16

Next-Generation Firewall with Zenarmor

17

Firewall High Availability

18

Website Protection with OPNsense

19

Command-Line Interface

20

API – Application Programming Interface

Index

Other Books You May Enjoy

Preface

OPNsense is one of the most powerful open source firewalls and routing platforms available. With OPNsense, you can now protect networks using features that were only available to closed source commercial firewalls before.

This book is a practical guide to building a comprehensive network defense strategy using OPNsense. You'll start with the basics, understanding how to install, configure, and protect network resources using native features and additional OPNsense plugins. Next, you'll explore real-world examples to gain in-depth knowledge about firewalls and network defense. You'll then focus on boosting your network defense, preventing cyberthreats, and improving your knowledge of firewalling using this open source security platform.

By the end of this OPNsense book, you'll be able to install, configure, and manage the OPNsense firewall by making the most of its features.

Who this book is for

This OPNsense firewall book is for system administrators, network administrators, network security professionals, and enthusiasts who wish to build and manage an enterprise-grade firewall using OPNsense.

What this book covers

Chapter 1, *An OPNsense Overview*, will introduce you to the OPNsense project and tell you about its history, license, fork motivations, and where you can find help if you need it. We will learn a little bit about FreeBSD and its fork, HardenedBSD, and explore OPNsense features and the common deployment scenarios you can use them in.

Chapter 2, *Installing OPNsense*, will teach you how to choose the right OPNsense version for your project, download it, and do the initial configuration. We will also see how to expand OPNsense features with plugin installations and briefly discuss FreeBSD's packages.

Chapter 3, Configuring an OPNsense Network, discusses networking configuration and concepts in OPNsense. We will dive into each network interface type and see some examples of how to use each one and learn about the different types of virtual IP addresses. At the end of the chapter, we will tackle some of the common problems with networking and how to solve them.

Chapter 4, System Configuration, provides steps on how to configure OPNsense common and advanced settings, managing users, groups, and certificates, how to add external authentication, and how to perform backups and restores.

Chapter 5, Firewall, starts with firewalling concepts and the features available on OPNsense. We will learn how to manage rules, change firewalling settings when necessary, and troubleshoot common issues using diagnostic tools and logs.

Chapter 6, Network Address Translation (NAT), explores the different types of **Network Address Translation** (**NAT**), such as port forwarding, outbounds, and one-to-ones, and how to use each one. We will also briefly discuss IPv6 network prefix translation and how to troubleshoot NAT common problems.

Chapter 7, Traffic Shaping, provides an overview of traffic shaping and how to use it on OPNsense to prioritize and limit network bandwidth. We will learn about pipes and queues, how to combine them to create rules, and how to monitor them.

Chapter 8, Virtual Private Networking, will dive into the **Virtual Private Network** (**VPN**) world. We will explore the different types of deployments and technologies available on OPNsense, and learn how to troubleshoot some common issues and monitor VPN tunnels.

Chapter 9, Multi-WAN – Failover and Load Balancing, explores some multi-**Wide Area Network** (**WAN**) strategies such as load balancing and failover. We will learn how to create gateway groups and policy-based rules using them. We also will see some caveats while using multi-WAN on OPNsense and how to solve the most common issues with it.

Chapter 10, Reporting, will teach you how to correctly read graphs, which is a very important part of managing a firewall. We will explore the available graphs and how to use them to identify possible unexpected behaviors in a network or see a firewall's health.

Chapter 11, Deploying DHCP in OPNsense, discusses one of the possible firewall duties – providing IP addresses to network hosts. We will learn about the **Dynamic Host Configuration Protocol** (**DHCP**) concepts used by OPNsense and how to use them to perform dynamic IP address leasing.

Chapter 12, *DNS Services*, covers DNS resolvers, what the available options are on OPNsense core, and the features available in each one. We will also take a brief look at dynamic DNS and explore some available DNS plugins to see how to troubleshoot common issues with DNS resolving.

Chapter 13, *Web Proxy*, shows how to configure and understand the different options to deploy a web proxy, one of the top features of a firewall solution. With it, you will be able to extend the control capabilities of OPNsense to another level.

Chapter 14, *Captive Portal*, shows how to configure and use a captive portal with OPNsense, and covers the most common deployments and issues and how to solve them.

Chapter 15, *Network Intrusion (Detection and Prevention) System*, explores IDS/IPS concepts, Suricata and Netmap implementations on OPNsense, and how to use them to alert or block threats on a network.

Chapter 16, *Next-Generation Firewall with Zenarmor*, Zenarmorexplores the ZenarmorZenarmor plugin, which broke the commercial-only next-generation firewall barrier and brought to the open source world this wonderful feature. We will examine its features and how to install and use it to apply a layer 7 control in a network.

Chapter 17, *Firewall High Availability*, shows how to configure high availability by connecting two firewalls to sync configuration, connect states, and preserve network connectivity if something goes wrong with one of our firewalls.

Chapter 18, *Website Protection with OPNsense*, delves into the NGINX plugin, with which OPNsense became a strong full-featured **Web Application Firewall** (**WAF**), helping you to protect your network and web servers.

Chapter 19, *Command Line Interface*, explores the shell command-line interface and some of the most relevant FreeBSD commands to manage the operating system, networking, and firewalling. We also will learn how to customize some parts of the system and use commands to improve information extraction from logs.

Chapter 20, *API – Application Programming Interface*, explores the APIs on OPNsense, how they work, and how to use them, with some scripting examples.

To get the most out of this book

A basic understanding of how a firewall works will be helpful to make the most of this book.

Software/Hardware used	Version
OPNsense	21.x (required)
Ubuntu	18.04
VirtualBox	6.0

Download the color images

We also provide a PDF file that has color images of the screenshots and diagrams used in this book. You can download it here: `https://static.packt-cdn.com/downloads/9781801816878_ColorImages.pdf`.

Conventions used

There are a number of text conventions used throughout this book.

`Code in text`: Indicates code words in text, database table names, folder names, filenames, file extensions, pathnames, dummy URLs, user input, and Twitter handles. Here is an example: "For the WAN interface, type em0."

A block of code is set as follows:

```
end value: 1000
current states number: 750
start value: 500
(1000 - 750) / (1000 - 500) = 0,5
```

When we wish to draw your attention to a particular part of a code block, the relevant lines or items are set in bold:

```
opnsense@ubuntu:~$ traceroute 8.8.8.8
traceroute to 8.8.8.8 (8.8.8.8), 30 hops max, 60 byte packets
 1  _gateway (192.168.56.3)  2.277 ms  4.733 ms  4.707 ms
 2  10.0.2.2 (10.0.2.2)  4.685 ms  4.548 ms  4.512 ms
 3  * * *
 4  192.168.15.1 (192.168.15.1)  13.798 ms  14.349 ms  14.316
ms
```

Any command-line input or output is written as follows:

```
$ bzip2 -d <filename>.bz2
```

Bold: Indicates a new term, an important word, or words that you see onscreen. For instance, words in menus or dialog boxes appear in **bold**. Here is an example: "With VirtualBox installed and running, click on the **New** button."

> **Tips or Important Notes**
> Appear like this.

Get in touch

Feedback from our readers is always welcome.

General feedback: If you have questions about any aspect of this book, email us at customercare@packtpub.com and mention the book title in the subject of your message.

Errata: Although we have taken every care to ensure the accuracy of our content, mistakes do happen. If you have found a mistake in this book, we would be grateful if you would report this to us. Please visit www.packtpub.com/support/errata and fill in the form.

Piracy: If you come across any illegal copies of our works in any form on the internet, we would be grateful if you would provide us with the location address or website name. Please contact us at copyright@packt.com with a link to the material.

If you are interested in becoming an author: If there is a topic that you have expertise in and you are interested in either writing or contributing to a book, please visit authors.packtpub.com.

Share Your Thoughts

Once you've read *OPNsense Beginner to Professional*, we'd love to hear your thoughts! Scan the QR code below to go straight to the Amazon review page for this book and share your feedback.

https://packt.link/r/1-801-81687-5

Your review is important to us and the tech community and will help us make sure we're delivering excellent quality content.

Section 1: Initial Configuration

In this part, we will explore OPNsense's characteristics and features, install and configure it, take a look at some networking concepts, configuration, troubleshooting, and see how to configure and manage the system.

This part of the book comprises the following chapters:

- *Chapter 1, An OPNsense Overview*
- *Chapter 2, Installing OPNsense*
- *Chapter 3, Configuring an OPNsense Network*
- *Chapter 4, System Configuration*

1
An OPNsense Overview

This chapter will introduce you to the OPNsense project, tell you about its history, license, and fork motivations, and where you can find help if you need it. Before installing and configuring your own OPNsense installation, it is essential to learn about some concepts and how the OPNsense project was started. We will also learn about FreeBSD, and its fork, HardenedBSD, exploring OPNsense features and the typical deployments scenarios that we can use it.

In this chapter, we're going to cover the following main topics:

- About the OPNsense project
- Rock-solid FreeBSD – HardenedBSD
- Why OPNsense?
- Features and common deployments
- Getting help

About the OPNsense project

To introduce you to the OPNsense project, I'll first need to tell a bit of my story and how I fell in love with it.

Project history

To tell the OPNsense story, we need to go back to 2003, when the initial release of m0n0wall was released. The main goal of this project was to have FreeBSD-based firewall software with an easy-to-use web interface (based on PHP) that worked on embedded PCs and old hardware with a good performance but that was just focused on Layer 3 and Layer 4 firewalling. m0n0wall was a good achievement. Still, picky network and security admins were claiming for other features such as web proxying, intrusion detection and prevention systems, and some other features that commercial firewalls were delivering as a default **Unified Threat Management Solution** (**UTM**). So, in 2004 a new project began, a m0n0wall fork, with its first public released in 2006. The fork's name? pfSense, and, as the name suggests, it used **Packet Filter** (**PF**) as a firewall-based system instead of the ipfilter (another FreeBSD packet filter)of its predecessor. For a long time, pfSense was a unique open source firewall solution, with a big active community and constant improvements. Many network and security administrators that only accepted Linux-based firewalls (yes, I was one of them too!) started to migrate to this FreeBSD-based firewall. These two projects coexisted until 2015, when m0n0wall was discontinued. There were signs of discontent back then; part of the pfSense community was not happy with some things such as changes in licenses and the direction the project was heading in.

Back in 2014, a brave group of developers decided to fork from pfSense and m0n0wall and started the OPNsense project. The first official release was in January 2015, inheriting a lot of code from its predecessors. Still, with a very ambitious plan to change how a lot of things were being done, OPNsense quickly rose as a pfSense alternative and received an important recommendation from the m0n0wall founder, Manuel Kasper, encouraging users from his project to migrate to OPNsense. It was the start of one of the best open source firewall projects.

A new project with a lot of improvements on old code

The following are some of the key features that OPNsense came with:

- OPNsense came with many new concepts and features that the community could claim credit for, such as a **Model View Controller** (**MVC**)-based web interface, a fixed release cycle, and a genuinely open source aspiration. The release cycle is done in two major versions each year, one in January and another in July (the community version) – for example, in 2021, the first version was 21.1 (January 2021), and the second one was 21.7 (July 2021), with a predictable and well-written roadmap. For the business edition, the releases are launched in April and October. The business editions are targeted at businesses and enterprises, containing the improvements delivered to the community version users first.

- As a **Chief Technology Officer** (**CTO**) with dozens of managed OPNsense-based firewalls, it is strategic to use firewall firmware with a predictable roadmap and release life cycle. This way, we can plan things with companies whose business depends on our managed firewalls.

Talking about versions, we need to introduce you to the flavor available:

- **OpenSSL**: The default one.

If you don't have any reason to choose LibreSSL, I'll advise you to pick the default one, OpenSSL. We will talk more about versions and installation media in the next chapter.

Talking about improvements, we must speak of the project architecture, starting with the frontend, the Phalcon PHP framework. This framework is used to implement webGUI and its APIs (another considerable improvement compared with its predecessors). It will do the work to render and control all that you can see and do using your web browser to manage your OPNsense.

The OPNsense framework also contains a backend, which is a Python-based service, also known as `configd`. This backend service will be in charge of controlling services, generating daemons and service config files from Jinja2 templates, and applying these configurations to an operating system.

With this architecture, OPNsense has a significant advantage – a secure way to manage and apply configurations to an operating system without executing `root` commands directly from the PHP web interface (as pfSense did, for example), reducing the risk of a flaw in webGUI compromising the whole firewall system.

So, now that we know how OPNsense evolved and its benefits, let's take a look at the operating system that serves as the base to this incredible firewall platform – FreeBSD's fork, HardenedBSD. It's essential to understand how the whole system and its components work to become a good OPNsense administrator. Let's go!

Rock-solid FreeBSD – HardenedBSD

Before exploring OPNsense, let's look at its kernel: the almighty FreeBSD, the operating system that has *the power to serve*!

FreeBSD

First, FreeBSD isn't Linux! If you are a long-time FreeBSD user, don't be mad with me, as you probably have heard this statement before! If not and you thought that FreeBSD and Linux were the same, no problem! That is a common mistake people make. Let's first find out what FreeBSD is in a short introduction.

FreeBSD is a free and open source operating system. It is a Unix-like system but different from Linux; it is a complete operating system, including the kernel, drivers, and other user utilities and applications. Linux only includes the kernel and drivers, and everything else is built as the distribution or just *distro*. The FreeBSD project has an outstanding security reputation, and it has a dedicated team taking care of this at the code level. This has built a well-known reputation for a very secure operating system!

If you have never installed FreeBSD before, you must be thinking that you will use it for the first time on OPNsense, right? Not necessarily! You have probably used FreeBSD a lot already. Don't believe me? One of the strong points of the FreeBSD project is its licensing model, which is considered permissive. Add to that the system liability, robust security, and good hardware support and you have a lot of reasons to choose FreeBSD as your new product base operating system. Many companies have – Apple's macOS, iOS, and other OSes are based on FreeBSD code, Sony PlayStations 3 and 4 use it, and many network appliances make use of it too! If you have an iPhone, that also runs a FreeBSD-based operating system!

Now that we are introduced to OPNsense's operating system, let's explore why we should consider it for our network firewall.

Why OPNsense?

To give a short answer – because it is the best genuinely open source firewall project currently available!

Not satisfied with this answer? Fair enough! Let's go to the long one!

My personal experience

Back in 2009, I was a pfSense user and enthusiast when I decided to move from Linux-based firewalls to FreeBSD. Back then, I created a managed firewall service for small and medium businesses using Linux firewall distros. At that time, I was using IPCop and SmoothWall Express, two Linux-based distros. I started this project with IPCop, but I decided to move to Smoothwall Express later, from which IPCop was forked. Both were good options to run in an embedded firewall appliance with limited resources. As a long-time Linux user, I was very comfortable using something based on this operating system. I am from a time when we built firewalls using `ipchains` and, later, `iptables` scripts, without any GUI help. I know – I'm getting old! Changes were happening and the service grew, with customers demanding bigger firewall appliances with high availability capability. At this point, Linux had a problem; there was no good support for this feature, so I needed to go back to the drawing board.

From research, I have found two possible alternatives – OpenBSD and FreeBSD. The first one was a well-known security-focused operating system with a strong reputation, which looked like a great option to run as a firewall solution. However, there were some hardware compatibility limitations and a lack of a GUI to manage it. It was a crucial need that customers were asking for. I couldn't find any firewall GUI option available back then, and developing a new one as a one-person development team was not a suitable option. It was time to look at the FreeBSD choices.

By doing a quick Google search, I found a good option – m0n0wall. So, I downloaded it and started the tests! It impressed me! It was a light operating system with a good WebGUI running on a solid operating system. But at the very beginning, I found a problem – reading the official documentation, specifically on the topic of what m0n0wall is not, I discovered that it wasn't a proxy server, an **Intrusion Detection System (IDS)**, or an **Intrusion Prevention System (IPS)**, and losing these features was not an option! So, it was back to square one!

Googling again, I found a m0n0wall fork – a project called pfSense. It looked like a promising one, with everything I needed to implement at that time – WebGUI, high availability, and some plugins such as Squid/SquidGuard for proxying and web filtering, Snort for IDS/IPS, pf as the packet filter. Bingo! I had found what I was looking for! My first impression when testing on the PC engine-based hardware was disappointing! The performance was too low compared with the Linux options and even with m0n0wall. There were too many features for a hardware platform with few resources. Another problem to solve was that even the more powerful hardware I was testing on used a compact flash disk as mass storage, and enabling a service such as Squid for proxying and web filtering with a lot of disk writing meant that the **Compact Flash** (**CF**) card got damaged very quickly. But there was another option – installing the NanoBSD flavor of pfSense made things easier for the embedded hardware to run more smoothly, and with some coding, I was able to bypass the CF card limitations. So, I ran this solution as a firewall to many customers until 2015. That was the year I became a little bit frustrated with how the project was being driven; a lot of ideas were popping into my head, but the project community was not the same anymore. It was time for a fresh start. I even talked with some project contributors about starting a new fork, but the idea was not accepted by most of them. It was time to go back to Google to look for a good open source firewall project.

On one of these rainy days, when your creativity is low and you have no intention to code or think of a new solution, I decided to google for some open source firewall alternatives, and the results surprised me! I found something, a project called OPNsense, and it was a pfSense fork! At that moment, the sun shined on my keyboard, and I thought, that's it!! The more I read about the project history, the more convinced I became that I had found the perfect solution backing!

So, I started the tests, and it was fascinating! It was pretty easy to convert pfSense's XML configuration file to OPNsense, so my first labs were based on production systems, and the results were very satisfying. It was time to migrate the first customers to this new firewall platform. However, it had a limitation – web filtering. We had developed a custom SquidGuard plugin for pfSense, but it wasn't easy to convert it, as the webGUI was different. Fortunately, our first customers didn't need this feature. The first migration to OPNsense was a success! So, it was decided that OPNsense was the new option for our firewall appliances! Later, I developed a SquidGuard-based plugin for OPNsense, and we were able to migrate all the customers. Since then, I have moved from this company to a new one, CloudFence, and we have decided to support the OPNsense project as much as possible.

Some of the key features that led me to dive into the OPNsense project were: genuine open source motivations with the freedom that an actual open-sourced project must have – an **MVC**-based webGUI without direct root access to the operating system (except for legacy components), IDS and IPS with `netmap` support, excellent available plugins, two-factor authentication for VPN, and cloud backups, to mention just a few! The list is increasing with each new version; later, we will see each feature in detail, so don't worry about it for now!

Okay, so this book isn't a novel, but I had to tell my personal history with OPNsense because choosing an open source project as a contributor and user isn't like buying a product. It is about personal life decisions; it's about what you want to support and are passionate about, and that project was right for me. I tried to help it! This is a little bit of my personal story with OPNsense; I hope it can inspire you somehow! The whole story might need an entire book, and it isn't the purpose of this one to tell stories but rather to improve your OPNsense skills, so let's move on to OPNsense's features!

Features and common deployments

Let's dive into OPNsense's core features and the most common scenarios to deploy it.

Core features

What are the core features? The OPNsense core features all come with the default OPNsense installation, without any additional plugins.

The core features are as follows:

- **802.1Q virtual LAN (VLAN) support**: The IEEE's 802.1Q, also known as **Dot1q**, is a network standard for supporting VLANs. This allows us to set a lot of different networks, with logical divisions or broadcast domains, using a VLAN-capable network switch. This is very useful when we need to define different networks using a single physical network interface, with which we can separate packets from different networks sources, using VLAN tagging. We will explore this in more detail in *Chapter 3, Configuring an OPNsense Network*.

- **Stateful inspection firewall**: OpenBSD's PF firewall, or just pf, was ported to FreeBSD in version 5.3. This packet filter is very flexible and easy to use. As someone who comes from the Linux world, I must admit it's easier to understand than `iptables`. The OPNsense webGUI generates the pf rules that are used for packet filtering and **Network Address Translation (NATs)**. If, like me, you're a curious person and have access to a running OPNsense firewall, you can sneak a peak at the `/tmp/rules.debug` file to see some of the pf rules. But be warned – don't touch anything there yet! In *Chapter 5*, *Firewall*, we will dive into the world of firewalls. If you are not running OPNsense yet, don't worry! In the next chapter, we'll install and configure it.

- **Traffic shaper**: OPNsense uses another firewall component, `ipfw`, the native packet filtering for FreeBSD, to classify and prioritize packets for the traffic shaping. With a traffic shaper, you'll be able to limit and reserve bandwidth and prioritize **Quality of Service (QoS)** traffic.

- **DHCP server and relay**: The **Dynamic Host Configuration Protocol (DHCP)**, as the name suggests, is a protocol to lease IP addresses to hosts in a network. OPNsense has both server and relay capabilities; the most common one, the DHCP server, is used to set an address pool configured dynamically to hosts on the network. The second one is used when hosts can't access the DHCP server directly, such as if the DHCP server relies on another network segment.

- **DNS forwarder**: The **Domain Name System (DNS)** is the base of our modern internet; without it, we would need to know every website IP address to access it. The DNS server and forwarder do the job of resolving domains to IP addresses. OPNsense has more than one native service to do this job, Unbound and Dnsmasq; both are resolvers. To enable a DNS server, such as Bind, you will need to install the Bind plugin. We will talk about plugins in the next chapter. The default option in the core OPNsense installation is the Unbound service.

- **Dynamic DNS (DDNS)**: OPNsense has a dynamic DNS client to update its hostname to an external DNS service. This is often used to access OPNsense externally while using a dynamic IP address; in this way, every time the ISP's IP address changes, the DDNS client will update the hostname in the external DNS service. Otherwise, you will only be able to access OPNsense by externally finding a way to discover which IP address your OPNsense machine is using at the moment.

> **Important Note**
> Dynamic DNS is a plugin installed by default on OPNsense.

- **Intrusion Prevention System (IPS)**: Also known as an IDS, an IPS, or an IDPS, this is one of the most significant improvements in OPNsense compared to pfSense. The service used for pfSense is Suricata. Unlike Snort, which was used in the pfSense versions back in 2015 (when OPNsense was forked from it), it runs multithreaded. The OPNsense team implemented support for netmap, a network framework for high-speed packet processing. Moreover, the OPNsense project has Proofpoint support, which allows its users to use a high-quality ruleset. Instead of blocking a source or destination IP after matching a rule (as pfSense's IPS implementation used to work), OPNsense now just blocks the connection that corresponds with a rule, if in IPS mode; otherwise, it is in IDS mode, it will just alert. How does this improve IPS filtering? Suppose that one host in your LAN matches with an IPS rule, and it's a false positive if the system blocks the source IP. The host will stop communicating with the internet, and the user will be calling your boss, complaining about you and the firewall. However, if the IPS blocks the single connection that is matching with a rule, maybe the user will notice that a single application or website has stopped working, which has to be better, right?

- **Forwarding caching proxy**: This is also known as a web proxy. This service is native to OPNsense; it can be used to cache websites components like JavaScript, **Cascading Style Sheets (CSS)**, images, fonts, and so on. You can also use it to control access to the internet using authentication, block websites with blocklists, make some basic access control lists, and intercept HTTPS/SSL traffic in transparent mode. In *Chapter 13, Web Proxy*, we will talk about it in detail and present some alternatives maintained by third-party developers such as CloudFence, such as installing and using those alternative plugins. What really matters is the possibility of doing high-quality web filtering using OPNsense, and it is indeed very much possible because of this open source beauty!

- **Virtual Private Network (VPN)**: The VPN options available in the OPNsense core are IPSec and OpenVPN. Both can be used as site-to-site and client-to-site (also known as roadwarrior) setups to connect a user securely over the internet.

  Captive portal: Talking about guest networks and controlling users to join a network, this also applies to the captive portal in OPNsense. This feature can be used with the web proxy to authorize users to use the internet and has widespread usage in hotels, airports, shopping centers, and so on.

- **Built-in reporting and monitoring tools**: These features can help a lot in troubleshooting scenarios. There are real-time and historical graphs, with a friendly user interface, packet capture (also known as `tcpdump`), and Netflow, and the list is increasing with each new version.

Here are some of the other great OPNsense features:

- QoS
- **Two-Factor Authentication (2FA)**
- OpenVPN
- IPSec
- High availability (CARP)
- A captive portal
- Proxy
- A web filter
- IDPS
- Netflow

There are many other features that can be added in OPNsense through plugins, and we will see in detail each core function and some plugins later in this book.

> **Note**
> You can obtain a full list of features at `https://opnsense.org/about/features//`.

Common deployments

OPNsense is very powerful and versatile and can be used in many ways. I'll try to cover the most common deployments, as follows:

- **Network router**: We can use OPNsense as a network router. It even has an option to completely turn off packet filtering, which improves the network throughput a lot, becoming just a network router without firewalling and NAT functions. Without additional plugins, the routing capabilities are minimal, straightforward, and serve well in a small network. Using OPNsense as a simple network router is the same as buying a Cirrus aircraft to fly in the same airdrome forever; you know that you can fly for hundreds of miles but prefer to stay just a mile away from the same runway. As a private pilot, I can't think of anything better than this comparison.

- **Firewall with WAN failover**: This is one of the most common deployments – OPNsense as a perimeter or an internal firewall. You can even use it as a cloud firewall, combined with some plugins such as the ZeroTier VPN, which we will explore in detail in *Chapter 8, Virtual Private Networking*. A firewall without additions will cover network Layer 3 and 4 packet filtering, and only that! In this scenario, it will probably be used to block untrusted packets from an external network. It can also be used to port forwarding (NAT and PAT) and block outgoing packets that aren't allowed to leave the LAN. When more than one WAN is available, it is possible to enable failover and outbound load balance to ensure good availability of internet access.

- **I[DP]S**: Whether combined or not with the firewall function, OPNsense can be used as a great network IDS or IPS, alerting and blocking (with the IPS turned on) packets from the monitored networks. The Suricata implementation in OPNsense is very well rounded, and with suitable hardware, you can achieve a few gigabits per second of throughput.

- **Guest network wireless gateway (a guest network)**: With a captive portal enabled, you have a lot of control over a guest network. You can combine a firewall, WAN failover, an IPS, and a web proxy with the captive portal to build a robust solution.

- **VPN server**: OPNsense has excellent support for **Certificate Authority** (**CAs**) and certificates, users, and group management, locally and externally, and you can, for example, use it as a robust OpenVPN server solution for hundreds and maybe thousands of simultaneous users, using proper hardware. It will cover two-factor authentication and many features that can be enabled to work with a VPN, such as web filtering and DNS filtering.

- **Web proxy and filtering**: Using Squid as the web proxy server, OPNsense can act as a powerful web proxy and web filter. As a well-known and ubiquitous web proxy service, Squid allows you to do web proxying in transparent or explicit modes, HTTPS/SSL intercepting, web filtering with an external categories database, and so on.

There are other possible deployments, such as a web application firewall, a next-generation firewall, an advanced network router, a DNS filtering appliance, and **Software Defined WAN (SD-WAN)**. It's not possible to cover all the possibilities in one book, but we will explore the most common ones in the following chapters.

Where to get help?

Have you had trouble or don't know how to use a feature? Didn't understand what the official documentation says? Take it easy! We are a big and strong community, and someone in it will try to help you for sure!

Some facts

While selling OPNsense-based firewall solutions in comparison to commercial firewall closed-source solutions, customers commonly ask us a question such as, *"Okay! It seems that this open source solution you are offering in your service can do anything that other closed source commercials listed in the 'some magic geometrical guide'* (which I will not mention the name of here) *solution can do, but what about support? When we need support or an urgent security fix, who are we going to call?"* Our answer? *"Not the Ghostbusters!"*

Okay, just kidding – but let me tell you about some of the myths that the closed source firewall vendors teach customers about open source support and how to answer them!

- **Open source has no professional support**: So, let me tell my story – I've been paying my bills for almost the last 15 years by offering professional support! Deciso is the company that founded and maintained the OPNsense project, and it is a company, not just a couple of genius guys (although there are genius guys there too)! But they are doing an outstanding job as a company! They provide professional services, hardware, and so on. If we look at the open source world, we can see many companies providing support and, making money. Yes, you can make money with open source! Just search on Google for open source professional support, and you will see that it is not a rare service.

- **Security fixes/software improvements**: Some commercial vendors fail to perform quick security fixes – you can find some examples on Google; some took more than a year to fix a known vulnerability. However, if you repeat the search and use OPNsense as an example, you will see that security fixes are done quickly! Talking about the software improvements, let's suppose your customer asks you about custom features. Try to ask a big vendor to know if you even will be heard! Probably not! I know what I'm talking about, and I used to work with them a long time ago! With an excellent open source project such as the OPNsense project, you can ask (open an issue) or even write code and submit it on GitHub (a pull request). Most of the time, you will be heard, and sometimes after a lengthy discussion and code review, the community will probably accept it! Again, I know what I'm talking about!

So, after this introduction, let's see where else we can find help:

- **OPNsense docs**: The OPNsense documentation is incredible! You will probably find there a lot of answers to your questions already. So, before you start typing questions anywhere, read the docs!

- **Official forum**: Always search first in the forums for questions like the one you intend to ask. Maintaining a helping platform, such as a forum, demands a lot of work, so please respect that and avoid duplicating questions. I'm not against WhatsApp or Telegram groups, but they aren't the best medium to get help. Think about it – if you just arrived in one of those groups, you can't see the message history, so all of the effort done before answering questions like yours is lost. Some of my OPNsense course students often ask me about those groups, and I always say, *"No! And I wouldn't!"* and discourage them from using those groups as a trusted source of information. Please prefer using the official forum!

- **IRC**: There is an OPNsense channel on IRC Libera (`https://web.libera.chat/#opnsense`), and you can chat about it there!

- **Commercial support**: Suppose you are in a hurry or have some critical issue and can't wait for the community to answer. In that case, you can count on the commercial support provided by several reliable companies that support the OPNsense project. As we discussed in this section, there are many ways to get help with OPNsense, maybe more than some of its commercial competitors; this is the advantage of an open source project. You can always count on the community and the companies that support it and are not left with just one option!

This brings us to the end of the chapter, which has provided an overview of OPNsense.

Summary

In this chapter, you were introduced to OPNsense, its project history, versions, and the improvements made by pfSense and m0n0wall. I also shared a little bit of my history with you and why I decided to use it, with some examples. We learned about FreeBSD and what it is and is not, and about its hardened fork, HardenedBSD. We explored the OPNsense core features and typical deployments, with some examples of how you can deploy them in your network environment. Finally, we dived into support options and how to use each of them. Now that we know how OPNsense works, we can find help if needed, which versions are available, and the most common scenarios to deploy it. We will now move on to the following chapters, which will help you better understand some of the concepts discussed in this chapter more practically. In the next chapter, we will see how to install OPNsense and explore some of its features!

2
Installing OPNsense

This chapter will be more hands-on than the first one. We will see how to install OPNsense on VirtualBox, update it, install plugins, access it from **Secure Shell** (**SSH**) and a **command-line interface** (**CLI**), and explore FreeBSD packages.

In this chapter, we will cover the following topics:

- Versions and requirements
- Downloading and installing OPNsense
- Updates and plugins
- SSH and CLI access
- FreeBSD packages

Technical requirements

You will need VirtualBox, VMware, Parallels, Qemu, or another hypervisor installed and basic knowledge of a virtual machine installation to follow the instructions in this chapter. You also will need an SSH client installed on your host machine.

Versions and requirements

To start to talk about the installation process, we first need to know more about the OPNsense versions. First of all, it is important to know that in this book, we'll focus on the Community edition. The OPNsense Business edition has some advantages for those who want to be more *selective in the upgrade path*, as the official documentation defines it.

Versioning

As I briefly mentioned in the first chapter, the OPNsense versioning is very simple to understand. At the time of writing this chapter, the current version is *22.1*, but what does it mean?

The first part, before the dot, is the *Year* of the version release: 2022.

The second part, after the dot, is the *Month*: January (1).

In the Community edition, the versions are always released in January (1) and July (7) every year. So, for example, the next version will be in January 2022: 22.1.

But what about the Business edition? Well, the first version released this year was the *22.4* version. The next one, according to Deciso, will be in October: *22.10*.

In this way, it's easy and predictable to know and plan when you will need to upgrade your OPNsense firewall. While managing firewalls, it is important to have a plan to apply security fixes without compromising network availability.

As we discussed in the last chapter, this is the firmware flavor we have:

- **OpenSSL**: The default and most used and supported flavor. This flavor uses the OpenSSL TLS/SSL toolkit.

As I mentioned in the last chapter, if you don't have any special reason to choose LibreSSL, maybe it's better to pick the default one, OpenSSL, which we'll use in this book.

Hardware

OPNsense will install on most modern *x86*-based hardware, but it is important to understand some concepts before installing your future firewall. We will explore the minimum requirements to install it, how to choose a good network card, and how to choose the right install image depending on the kind of disk you want to use.

Architecture

Since version 20.1, *i386*, also known as Intel 32 - bit architecture, was dropped, the only currently supported architecture is *x86-64* or *amd64* while choosing on the Download page.

Disk

If you are installing on a disk or media with limited write cycles, such as SD or CF memory cards, then it's better to choose the nano image. OPNsense will write a lot of logs on disk and it isn't a good idea to use media with limited write cycles, as it will very quickly damage disk media that wasn't designed for that. So, the nano image will by default mount the /var and /tmp directories to a RAM disk. The same can be done with other images, such as DVD, VGA, or Serial. Let's talk about them here:

- **nano**: To install OPNsense using the nano image, you will need a USB stick and an SD or CF card with at least 4 GB capacity. The main console will output in the serial console port in 115.200 bps mode. If the hardware you are using doesn't have a serial port, you can still use the VGA console, but without the output of kernel messages. After being installed, it will expand the filesystem automatically to the disk capacity.

- **DVD**: This is the ISO burnable installer image that contains a live system and the installer; we must have hardware with a display card to install OPNsense using this image. We can use it to install in a virtual machine environment too – we'll use this one in this chapter to describe the installation process.

- **VGA**: This is almost the same as the DVD with the difference that it is intended to be used on USB media.

- **Serial**: This is a USB media installer intended to use the serial console port. It's a choice you make while installing in a network appliance that doesn't have a display port, for example, and the only option is a serial console port.

Except for nano, all the other images support **Unified Extensible Firmware Interface** (**UEFI**) hardware, usually available in most modern hardware, and are the more common choices for most hardware or virtualization environments.

Network

Since OPNsense made use of the *netmap* network framework for better performance while using IPS (Suricata) or the Sensei plugin, it is also important to choose the right network card for a good performance firewall.

According to netmap's FreeBSD manual page, the currently supported network devices are as follows:

- **cxgbe**: Chelsio T4 and T5, and T6-based 100Gb, 40Gb, 25Gb, 10Gb, and 1Gb

- **em**, **igb**, **ixgbe**, and **ixl**: Intel's network interfaces of 1Gb, 10Gb, 25Gb, and 40Gb

- **re**: RealTek 8139C+/8169/816xS/811xS/8168/810xE/8111

- **vtnet**: Virtio ethernet for virtual machines

- **iflib**: All the drivers utilizing the Network Interface Driver Framework (`iflib`)

In my personal experience, for an average OPNsense network appliance, a good option will be to choose an Intel network card, as it is a cost-effective choice.

Minimum hardware requirements

According to the OPNsense official documentation, the minimum requirements to run it are a 1 GHz dual-core CPU with 2 GB RAM for a full image type (serial, DVD, and VGA), and 4 GB RAM and disk for nano image installation. Will it work on something with fewer resources than that? Probably, yes! But, I'll not recommend you do that in a production environment!

On the official documentation page, you can find a sizing guide that helps to find the appropriate hardware for your needs:

`https://docs.opnsense.org/manual/hardware.html#throughput`.

Downloading and installing OPNsense

Now that we know the OPNsense versions and options available, we can proceed with our installation. In this book, I'll use Oracle's VirtualBox as the virtualization platform. I picked this one because it is easy to download and it runs on Windows, Mac, or Linux.

The first step is to download the OPNsense image:

1. Go to the `https://opnsense.org/download/` download page.
2. Select the options as shown in the following screenshot:

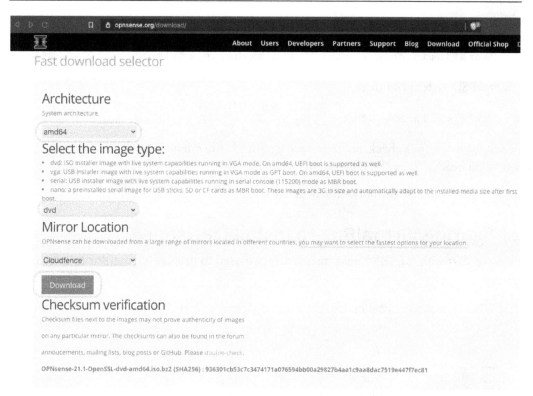

Figure 2.1 – The OPNsense download page

After downloading, you must uncompress the file that will be saved in a `bzip2` format. In Windows, you can use the 7 - zip tool for that from `https://www.7-zip.org/download.html`.

If you are on Linux or macOS, use the following command line:

```
$ bzip2 -d <filename>.bz2
```

It's recommended to run a checksum in the download file, just to check its integrity.

On Linux you can run this:

```
$ sha256sum <filename>.bz2
```

On a mac OS run this:

```
$ shasum -a 256 <filename>.bz2
```

On a Windows machine, open the command prompt:

```
C:\Users\julio.camargo>certutil -hashfile <filename>.bz2 sha256
```

On a FreeBSD system run this:

```
$ sha256 <filename>.bz2
```

After running the file's checksum, compare it with the download page at the Checksum verification header (after the **Download** button).

Now let's go ahead and check out the installation steps.

Configuring VirtualBox to install OPNsense

Before we start with the OPNsense installation, we need to prepare VirtualBox to install it properly:

1. With VirtualBox installed and running, click on the **New** button:

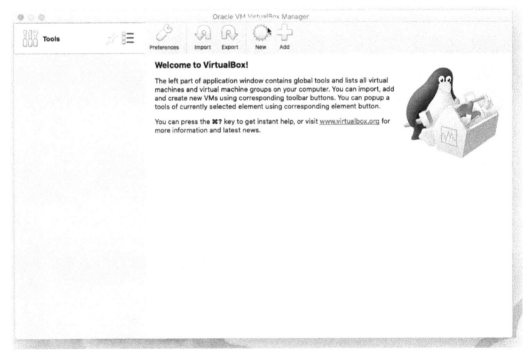

Figure 2.2 – Adding a new virtual machine

2. Pick a name for your virtual machine and select the type as **BSD** and the version as
    FreeBSD (64-bit):

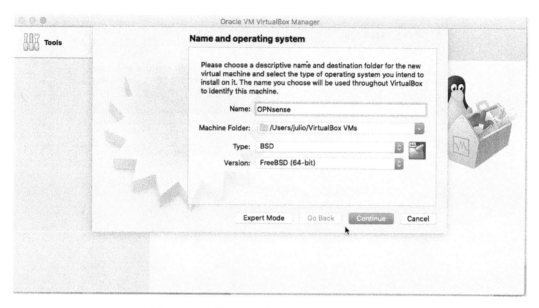

Figure 2.3 – Choosing the virtual machine name

3. Select the RAM amount. For our initial installation, 2 GB (2048 MB) should be
    enough:

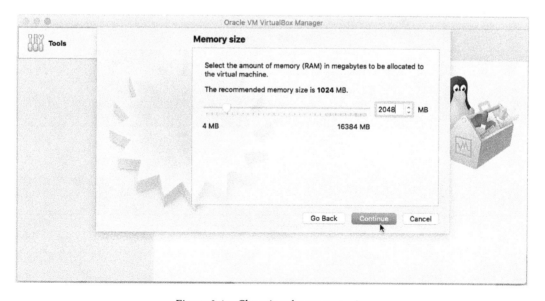

Figure 2.4 – Choosing the memory size

4. For **Hard disk**, you can keep the defaults.

5. Create a virtual hard disk now with the default size of 16 GB:

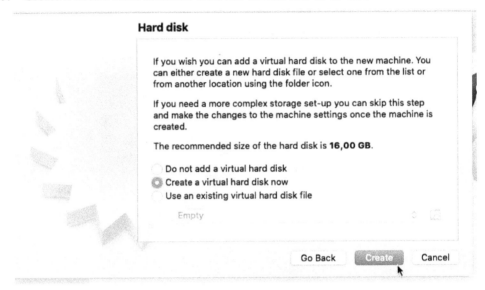

Figure 2.5 – Creating a hard disk

6. In the **Hard disk file type** step, keep the **VDI (VirtualBox Disk Image)** default option:

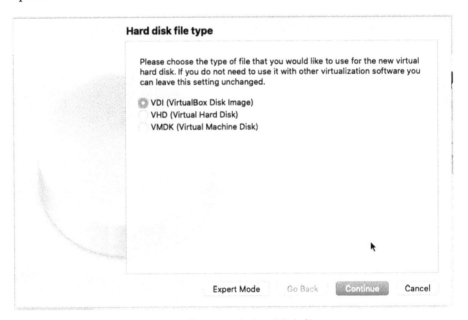

Figure 2.6 – Choosing the hard disk file type

7. In the **Storage on physical hard disk** step, I recommend that you keep the **Dynamically allocated** default option:

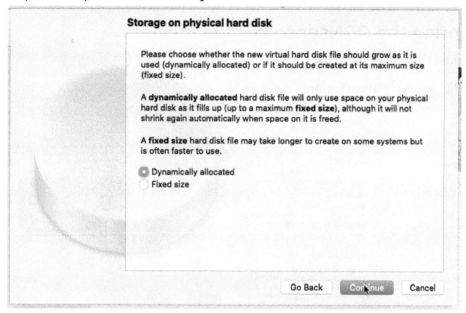

Figure 2.7 – Configuring storage on the physical hard disk

8. Next, select your VM hard disk file path and click **Create**:

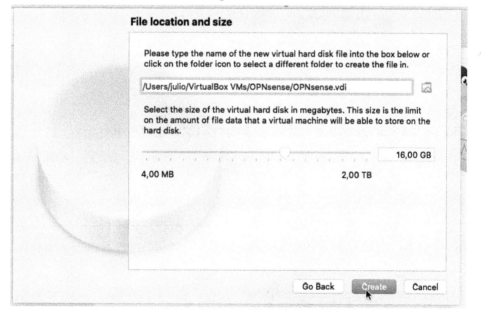

Figure 2.8 – Choosing the hard disk file location

Before we can start our new VM, we need to mount the OPNsense ISO file in the **Optical Drive** option.

Mounting the OPNsense ISO file

Let's mount the ISO file so the new virtual machine can boot using it:

1.   Click on the **Settings** button:

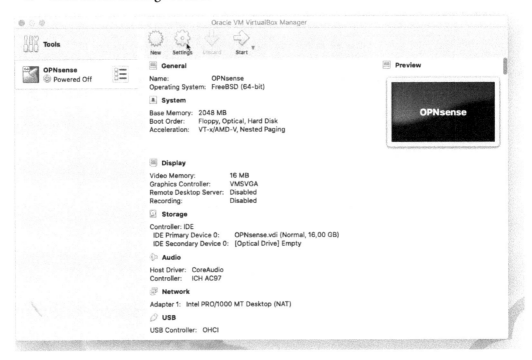

Figure 2.9 – Editing the VM settings

2.   Click on the **Storage** button and select **Empty** in **Storage Devices**:

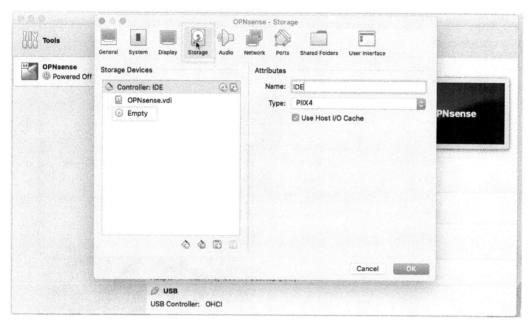

Figure 2.10 – Mounting the ISO image file

3. In the **Optical Drive** option, click on the CD disk icon and select **Choose a disk file…** and select the ISO file path we downloaded before, then just click **OK** to proceed with our VM boot:

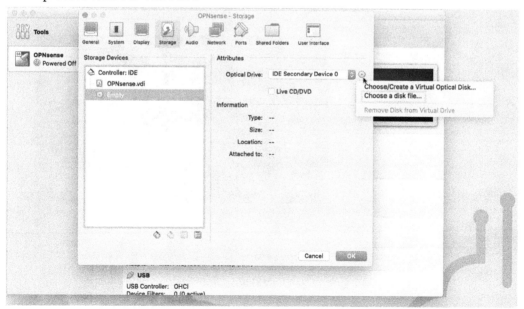

Figure 2.11 – Mounting the ISO image file

4.   Now, just click on the **Start** button and confirm the ISO file location to start the live system boot:

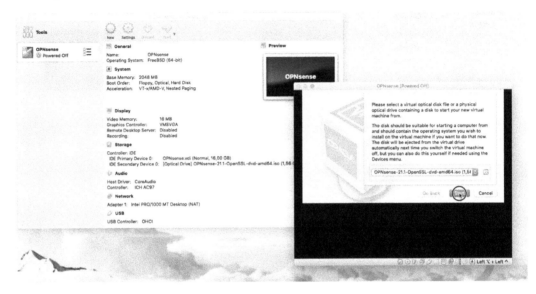

Figure 2.12 – Mounting the ISO image file

Our VM will start the *boot* process:

Figure 2.13 – The VM boot process

In this first boot, you need to just wait for the login prompt:

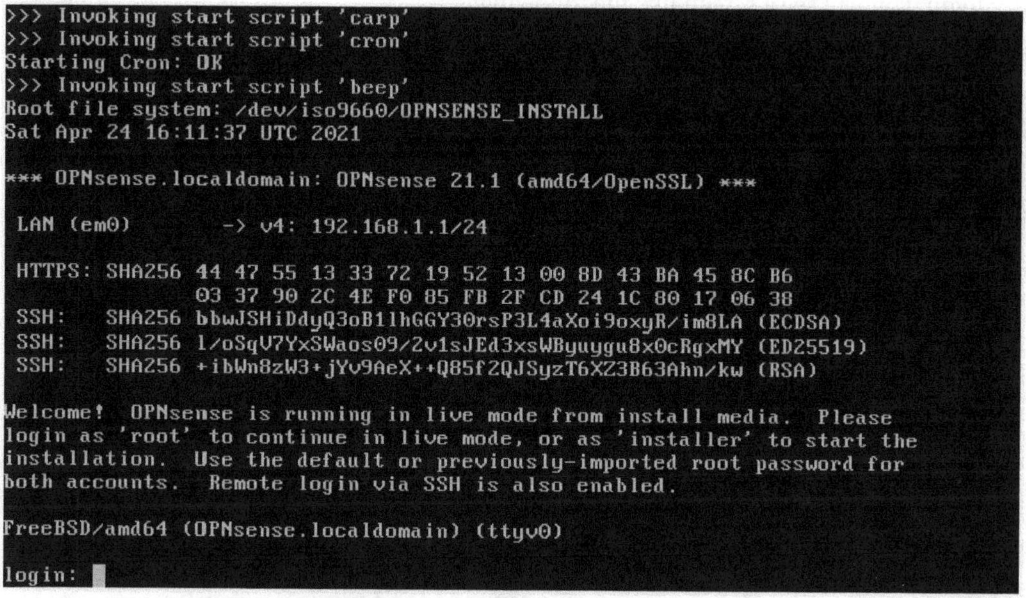

```
>>> Invoking start script 'carp'
>>> Invoking start script 'cron'
Starting Cron: OK
>>> Invoking start script 'beep'
Root file system: /dev/iso9660/OPNSENSE_INSTALL
Sat Apr 24 16:11:37 UTC 2021

*** OPNsense.localdomain: OPNsense 21.1 (amd64/OpenSSL) ***

 LAN (em0)          -> v4: 192.168.1.1/24

 HTTPS: SHA256 44 47 55 13 33 72 19 52 13 00 8D 43 BA 45 8C B6
               03 37 90 2C 4E F0 85 FB 2F CD 24 1C 80 17 06 38
 SSH:    SHA256 bbwJSHiDdyQ3oB11hGGY30rsP3L4aXoi9oxyR/im8LA (ECDSA)
 SSH:    SHA256 1/oSqV7YxSWaos09/2v1sJEd3xsWByuygu8x0cRgxMY (ED25519)
 SSH:    SHA256 +ibWn8zW3+jYv9AeX++Q85fZQJSyzT6XZ3B63Ahn/kw (RSA)

Welcome!  OPNsense is running in live mode from install media.  Please
login as 'root' to continue in live mode, or as 'installer' to start the
installation.  Use the default or previously-imported root password for
both accounts.  Remote login via SSH is also enabled.

FreeBSD/amd64 (OPNsense.localdomain) (ttyv0)

login:
```

Figure 2.14 – The OPNsense login prompt

Now, we have OPNsense running in *live* mode. To begin the installation process, we must log in using the following details:

- Username: `installer`
- Password: `opnsense`

> **Important Note**
> If you log in with root and start using it in live mode, all the modifications done in this mode will be lost after a reboot.

Installing OPNsense

After login, you will see a screen like the following:

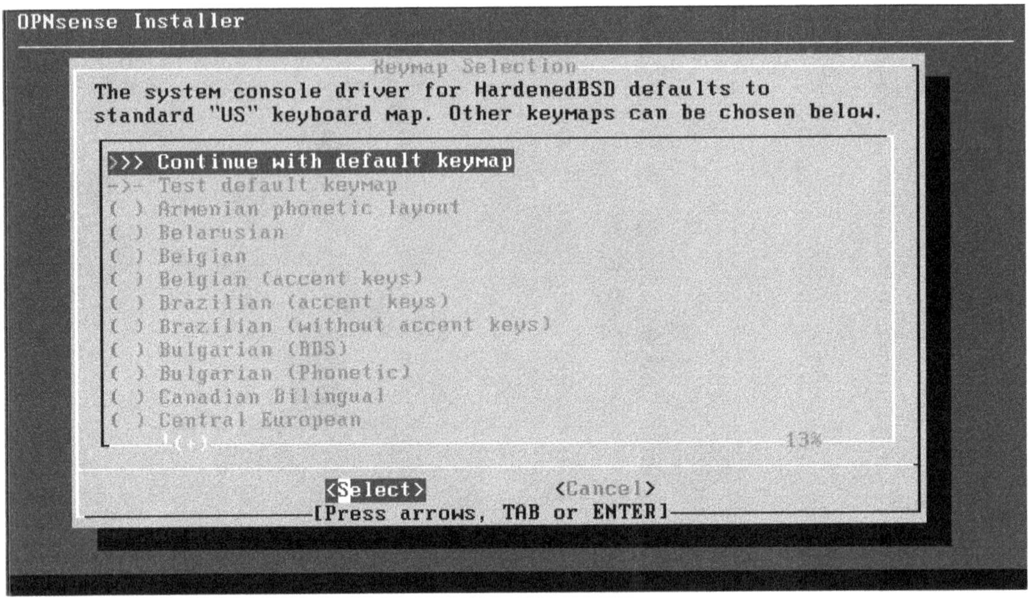

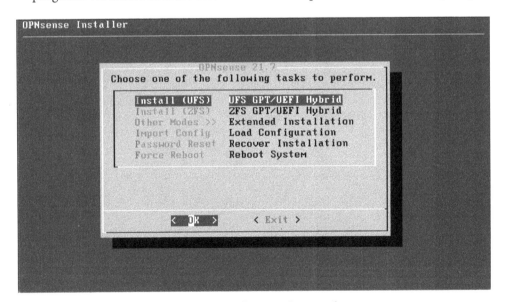

Figure 2.15 – Keymap Selection

1. This screen will display a few possible operations that can be performed by the setup program. We must select the installation tasks to proceed. Select **Install (UFS)**:

Figure 2.16 – Installation tasks to perform

2. Next, select the desired disk to install OPNsense:

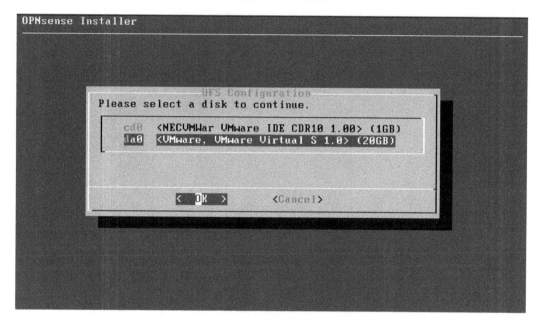

Figure 2.17 – Selecting a disk to install OPNSense

OK! Let's go! Just press *Enter* to begin the installation process.

3. On the next screen, confirm the selected disk to install by choosing **YES**:

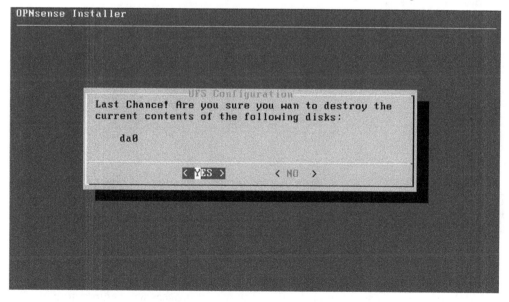

Figure 2.18 – Confirming the previously selected disk to install OPNsense

The **Select install mode** screen will display the options related to install mode on the disk:

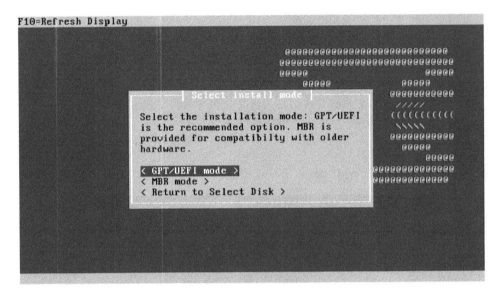

Figure 2.19 – Installation – Select install mode

After this step, the OPNsense installer will start to copy files to the hard disk.

4. On the **Final Configuration** screen, select the **Root Password** option to change the root user password:

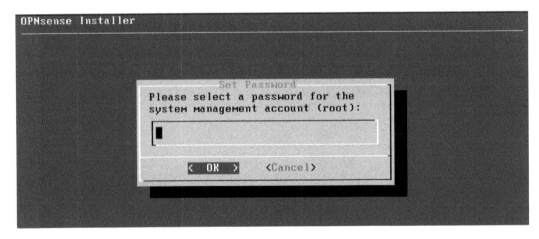

Figure 2.20 – Setting the root password

5. Back on the **Final Configuration** screen, select the **Exit** option to exit the installation process and reboot the VM. The reboot process will begin automatically:

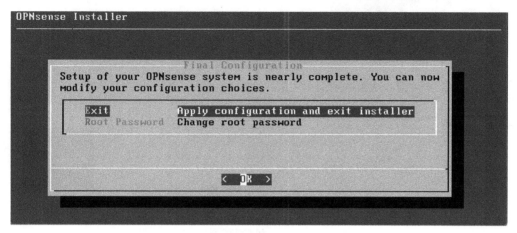

Figure 2.21 – Installation – Final Configuration step

Once you have rebooted the system, let's go ahead and unmount the ISO installation file.

Unmounting the ISO installation file

Before you start using your new OPNsense VM, you must unmount the ISO installation file image to not boot from it again:

1. Go to **Settings**.

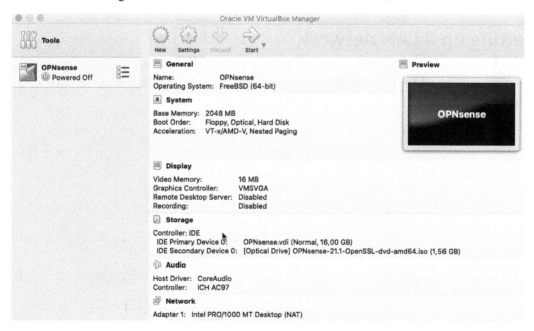

Figure 2.22 – Editing the VM settings

2. Click on **Storage**, select the ISO file, and click on the remove icon as shown in the following screenshot:

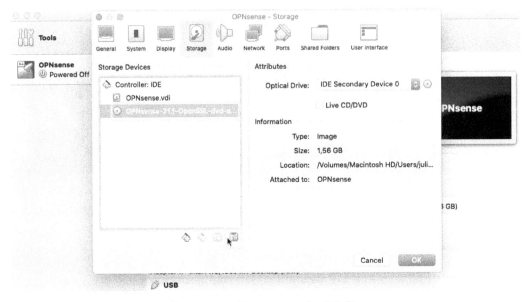

Figure 2.23 – Unmounting the ISO file

After this final step, the VM will now boot from the hard disk.

Setting up a LAN network

Before we can start to use our OPNsense VM, it is necessary to add one more network interface, which will be the LAN. We will access OPNsense through this local network interface.

Let's go through the steps:

1. Go to the **File** menu and click on **Host Network Manager….**

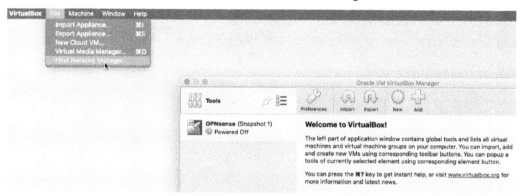

Figure 2.24 – Configuring the network

2. Click on the **Create** button:

Figure 2.25 – Creating a new network interface

3. A new network interface will be created. Now click on the **Properties** button:

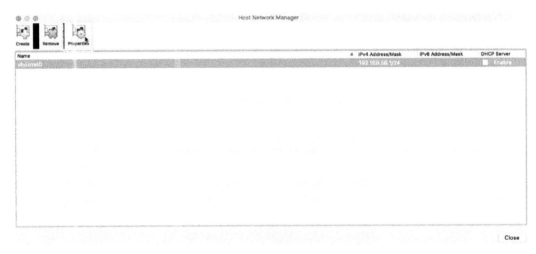

Figure 2.26 – Editing new interface properties

4. Click on the checkbox to enable **DHCP Server** and click on the **Close** button:

Figure 2.27 – Enabling DHCP server in the created interface

5. Go back to the VM settings:

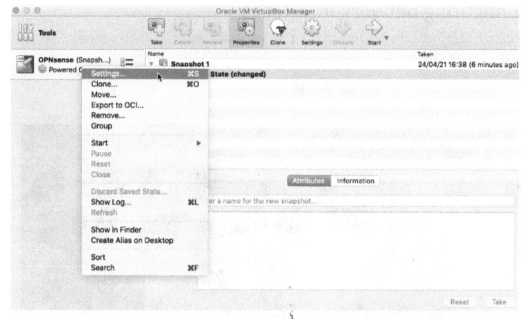

Figure 2.28 – The VM settings page

6. Click on **Network | Adapter 2 | Enable Network Adapter**.

7. In the **Attached to:** option, select **Host-only Adapter** and then select the network interface created previously, as shown in the following screenshot:

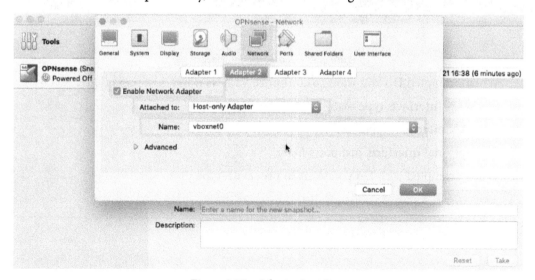

Figure 2.29 – Adapter 2 settings

Now you can start your OPNsense virtual machine!

Configuring network interfaces

With your OPNsense firewall up and running, we need to set up the new network interface we've just created:

1. Log in as root and type 1 for **Assign interfaces**:

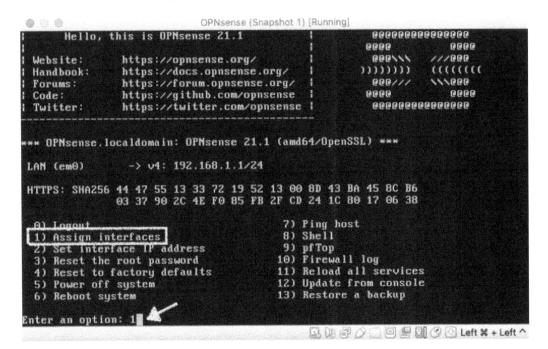

Figure 2.30 – Assigning interfaces

2. When prompted **Do you want to configure VLANs now?**, just press *Enter*.

3. For **WAN interface**, type em0.

4. For **LAN interface**, type em1.

5. For **Optional interface**, just press *Enter*.

6. When prompted with **Do you want to proceed?**, type y:

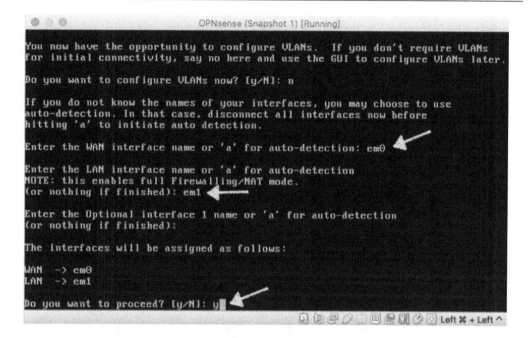

Figure 2.31 – OPNsense LAN network setup

The network configurations will be reloaded and then we're back to the CLI menu:

7. Now, select the **Set interface IP address** option by typing 2:

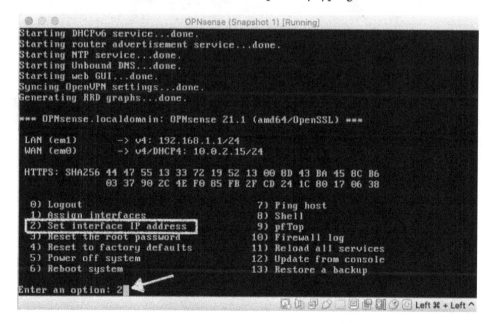

Figure 2.32 – Setting the LAN interface IP address

8. Select **LAN interface** by typing 1 and answer the questions as follows.

9. **Configure IPv4 address LAN interface via DHCP?** Type y.

10. **Configure IPv6 address LAN interface via WAN tracking?** Type n.

11. **Configure IPv6 address LAN interface via DHCP6?** Type n.

12. **Enter the new LAN IPv6 address.** Just press *Enter*.

13. **Do you want to revert to HTTP as the web GUI protocol?** Type n (or just press *Enter*).

14. **Do you want to generate a new self-signed web GUI certificate?** Type n.

15. **Restore web GUI access defaults?** Type n:

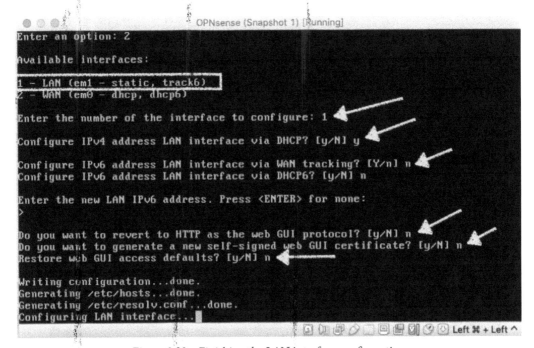

Figure 2.33 – Finishing the LAN interface configuration

Finally, we have our LAN interface configured and running! Let's test it:

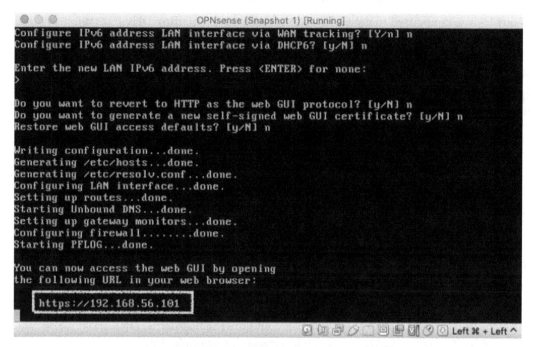

Figure 2.34 – The LAN interface configuration is done

Now check which IP address your OPNsense LAN took and open it in your favorite web browser.

Voilá! You have access to WebGUI through the LAN interface (`https://<ip_address_of_your_VM>`).

Access the OPNsense Web User Interface through the web browser:

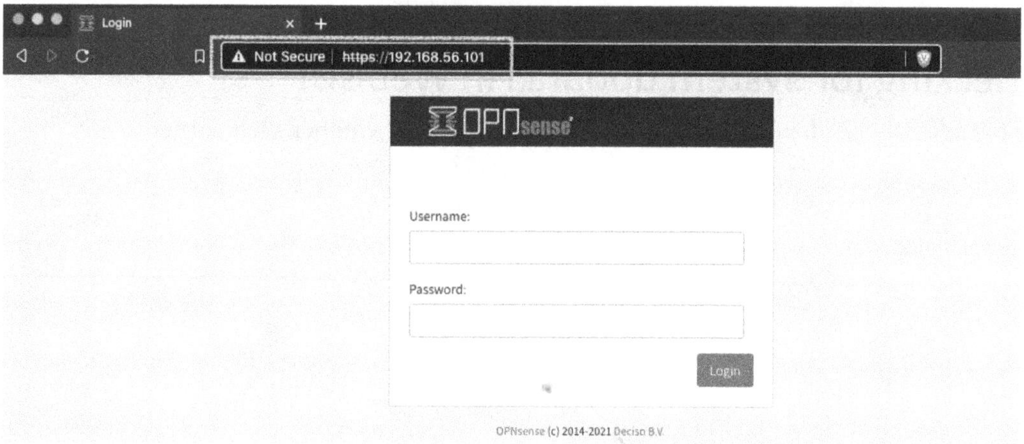

Figure 2.35 – Accessing the OPNsense WebGUI by using the LAN IP address

As our LAN is configured to get an IP address from the VirtualBox DHCP server, how do we know which IP is configured to our network interfaces? Easy! Just check in your VM console; the login screen shows all you need to know! Maybe you are wondering why the heck we are using DHCP to set a LAN address. Patience, my good reader; at the right moment we will make things more normal, but for now, it's good to experience new unusual settings and see that they can work well too!

Now that we have installed OPNsense and configured the basics to access it, we can explore how to update it and extend its features by adding plugins.

Updating firmware

Once we have successfully installed OPNsense, we can explore it using either the CLI or WebGUI. Let's start with WebGUI to see the update process and how to install some plugins using it.

When we log in for the first time in WebGUI, the first screen that appears is the configuration wizard, which will help you to do the first configurations in your OPNsense. Through a next-next-finish process, you can configure the hostname, DNS servers, and the Resolver, Timezone, WAN, and LAN interfaces, for example. I will not explore the wizard in this book, because the goal here is to explore each configuration in detail and the wizard will not help us with that. But, if you are curious about the configuration wizard, you can explore it by accessing the **System | Wizard** menu.

> **Important Note**
> If you want to test the Configuration Wizard, I recommend you take a VM snapshot beforehand, so that if something gets broken you can easily fix it.

Checking for system updates in WebGUI

To bypass the wizard, just click on the OPNsense logo in the upper - left corner.

To check for system updates in WebGUI, you can follow two paths:

- The long one is by going to the **System | Firmware | Status Updates (tab)** menu and then clicking on **Check for updates**:

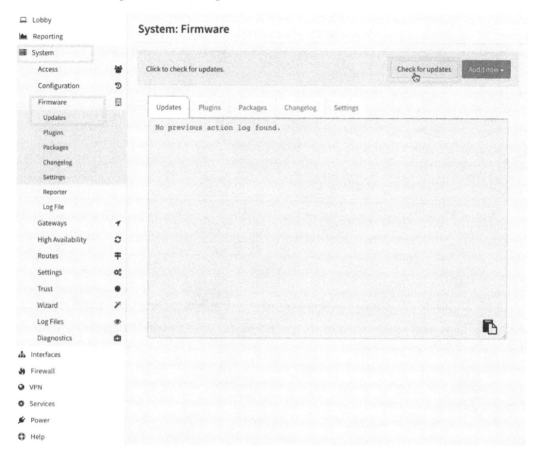

Figure 2.36 – Updating system firmware

- The shortest one can be done by clicking the **Click to check for updates** link in the **System Information** widget:

Figure 2.37 – Checking for updates in the System Information widget

Notice that the second option already starts checking for updates without clicking on the **Click to check for updates** button.

After the system has checked for updates in the configured repository, if there are any packages to update, you must click on the **Update now** button:

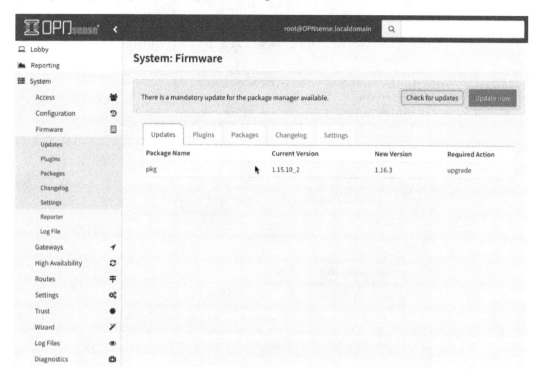

Figure 2.38 – OPNsense firmware updating

Otherwise, if there is a new release available, a dialog will pop up with the release notes:

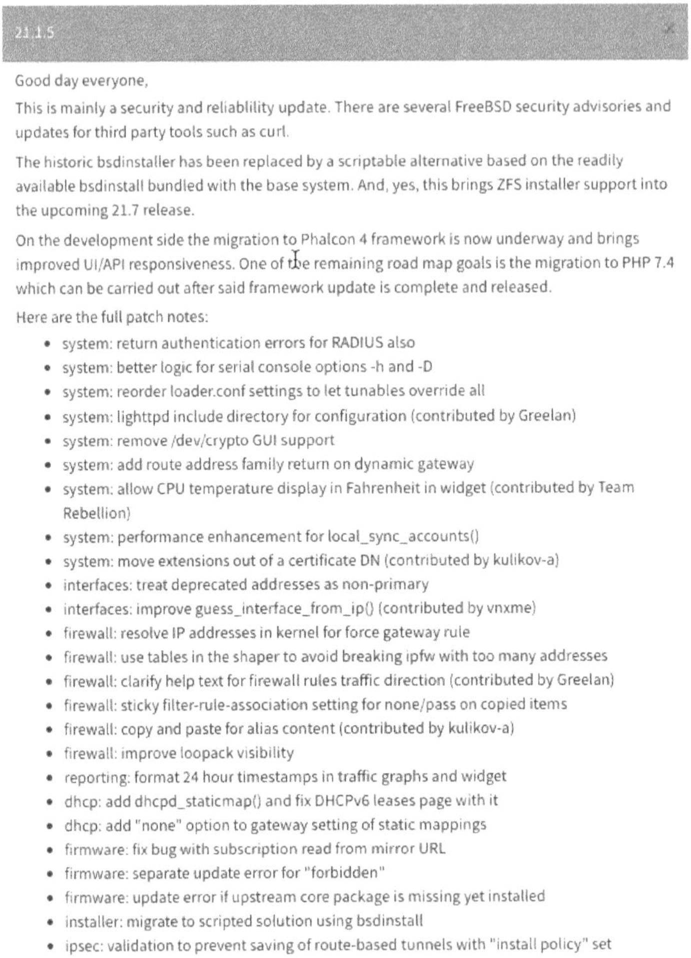

21.1.5

Good day everyone,

This is mainly a security and reliablility update. There are several FreeBSD security advisories and updates for third party tools such as curl.

The historic bsdinstaller has been replaced by a scriptable alternative based on the readily available bsdinstall bundled with the base system. And, yes, this brings ZFS installer support into the upcoming 21.7 release.

On the development side the migration to Phalcon 4 framework is now underway and brings improved UI/API responsiveness. One of the remaining road map goals is the migration to PHP 7.4 which can be carried out after said framework update is complete and released.

Here are the full patch notes:

- system: return authentication errors for RADIUS also
- system: better logic for serial console options -h and -D
- system: reorder loader.conf settings to let tunables override all
- system: lighttpd include directory for configuration (contributed by Greelan)
- system: remove /dev/crypto GUI support
- system: add route address family return on dynamic gateway
- system: allow CPU temperature display in Fahrenheit in widget (contributed by Team Rebellion)
- system: performance enhancement for local_sync_accounts()
- system: move extensions out of a certificate DN (contributed by kulikov-a)
- interfaces: treat deprecated addresses as non-primary
- interfaces: improve guess_interface_from_ip() (contributed by vnxme)
- firewall: resolve IP addresses in kernel for force gateway rule
- firewall: use tables in the shaper to avoid breaking ipfw with too many addresses
- firewall: clarify help text for firewall rules traffic direction (contributed by Greelan)
- firewall: sticky filter-rule-association setting for none/pass on copied items
- firewall: copy and paste for alias content (contributed by kulikov-a)
- firewall: improve loopack visibility
- reporting: format 24 hour timestamps in traffic graphs and widget
- dhcp: add dhcpd_staticmap() and fix DHCPv6 leases page with it
- dhcp: add "none" option to gateway setting of static mappings
- firmware: fix bug with subscription read from mirror URL
- firmware: separate update error for "forbidden"
- firmware: update error if upstream core package is missing yet installed
- installer: migrate to scripted solution using bsdinstall
- ipsec: validation to prevent saving of route-based tunnels with "install policy" set

Figure 2.39 – New firmware release notes

> **Important Note**
> It's recommended that you read the release notes before updating.

After reading the release notes, you can proceed, if you want to, by clicking on the **Update now** button.

Checking system updates using the CLI

If you prefer to update via the CLI, just log in to the VM console and select the **Update from console** option by typing 12:

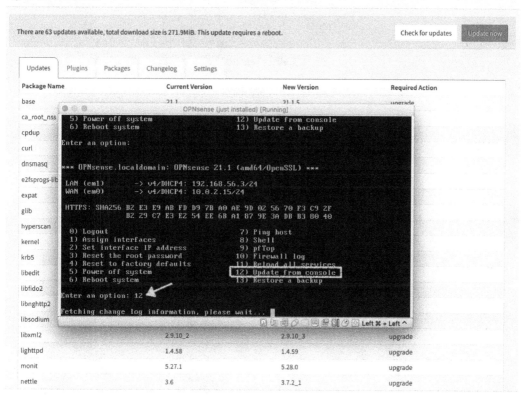

Figure 2.40 – Updating from the CLI

> **Important Note**
> Depending on the update, it might require a reboot, so pay attention to that before updating!

For major upgrades, an extra unlock update step will be required. You'll need to type the next major release to unlock the update (22.1 to update for the 22.1 version, for example). The OPNsense developer team did this to avoid accidental updates that sometimes could have implications.

The time taken to update will basically depend on your hardware and bandwidth, so just sit back, have a coffee, and relax, watching the beauty of open source doing its job!

After the update is done, you can check the updated version in the WebGUI dashboard as shown in the following screenshot:

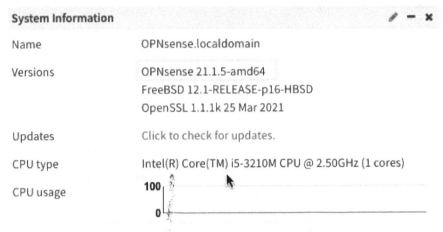

Figure 2.41 – The WebGUI System Information widget

We can also check the latest firmware version installed on the CLI as follows:

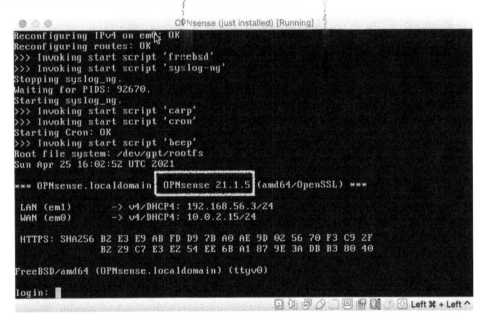

Figure 2.42 – Checking the newly installed firmware version on the CLI menu

Now that we have an updated version of OPNsense, let's dive into the plugins installation process.

Installing plugins

Plugins can be very helpful to extend OPNsense features or even to just customize them. In this section, we will install a theme plugin that will change the WebGUI look and feel.

You can find the available plugins in the **System | Firmware | Plugins** menu:

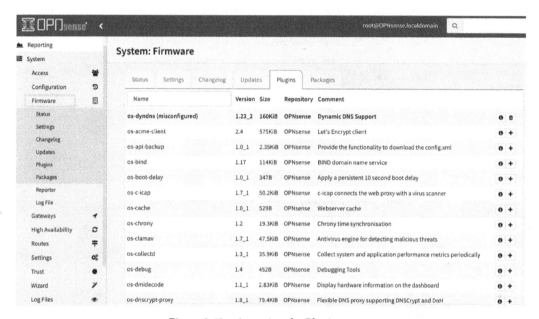

Figure 2.43 – Accessing the Plugins menu

There are dozens of plugins listed and the list is growing. To test a plugin installation, we will get a new OPNsense WebGUI theme, a dark one!

To do that, select a plugin theme and just click on the + button:

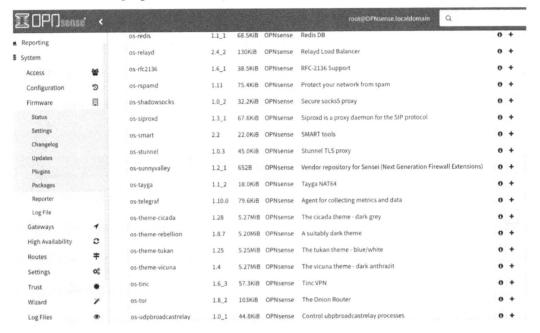

Figure 2.44 – Installing a new WebGUI theme plugin

After that the plugin should be installed:

Figure 2.45 – Plugin installation process output

When a plugin is installed, it will appear at the top of the list in bold as shown in the following screenshot:

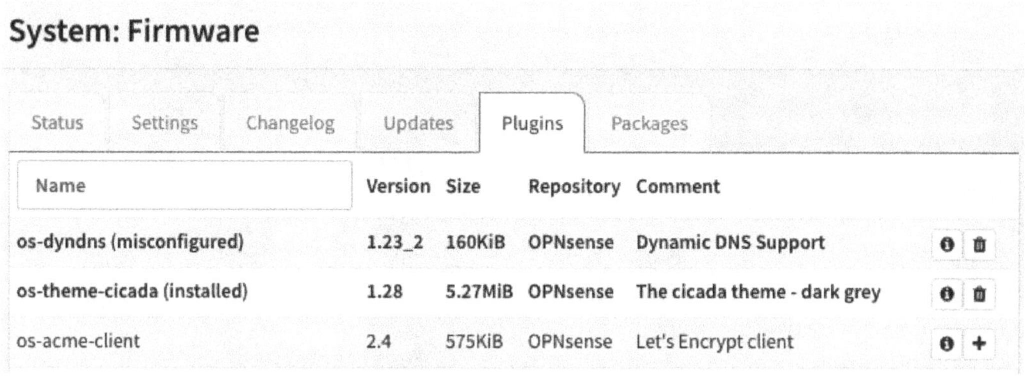

System: Firmware

Status	Settings	Changelog	Updates	**Plugins**	Packages				

Name	Version	Size	Repository	Comment		
os-dyndns (misconfigured)	**1.23_2**	**160KiB**	**OPNsense**	**Dynamic DNS Support**	ⓘ	🗑
os-theme-cicada (installed)	**1.28**	**5.27MiB**	**OPNsense**	**The cicada theme - dark grey**	ⓘ	🗑
os-acme-client	2.4	575KiB	OPNsense	Let's Encrypt client	ⓘ	+

Figure 2.46 – Installed plugins

Congratulations, you just installed your first plugin!

Want to see your newly installed theme? Just go to **System | Settings | General** and change it on the **Theme** option and then click on the **Save** button:

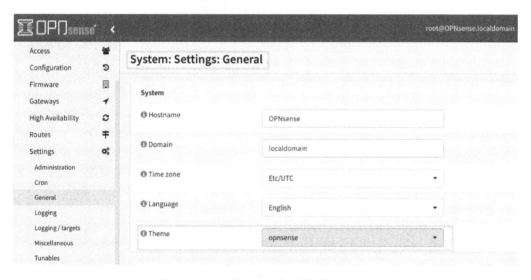

Figure 2.47 – Changing the WebGUI theme

Your new dark theme is installed and working!

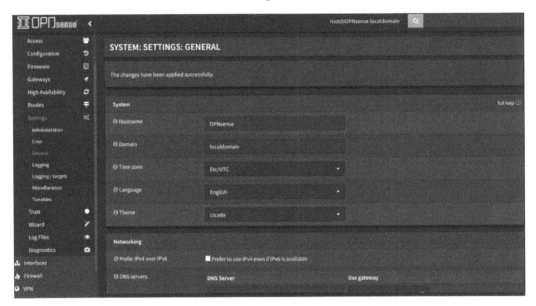

Figure 2.48 – A newly installed theme

In the next chapters, we'll take a look at the settings of the **Firmware** menu.

As we can see, it's very easy to install new plugins on OPNsense. With them, you can add new features fairly quickly. Now that you have been introduced to WebGUI, it's time to meet the command-line interface, also known as the CLI.

Advanced – Accessing the CLI through SSH

We already saw some CLI options through the VM console, which emulates a local display or serial console. Now, it's time to explore them using a more flexible option: SSH remote access! With CLI remote access, we can execute commands, filter logs, and do some other things that are not possible via WebGUI. Sometimes you can even solve a WebGUI access problem through the CLI, so, having the CLI, also known as shell access, will empower you to do a lot of advanced things. But remember, *with great power comes great responsibility*, to quote a well-known superhero's uncle; using the CLI can break things! So, if this is your first time reading this book, read it twice before starting to use the CLI like a superhero firewall administrator, OK?

Before we start to configure our SSH access, let me introduce you to the Quick Navigation search box. This search box is located at the top-right corner of the screen. You can either click on it or if the mouse focus isn't on some menu, just press *Tab*.

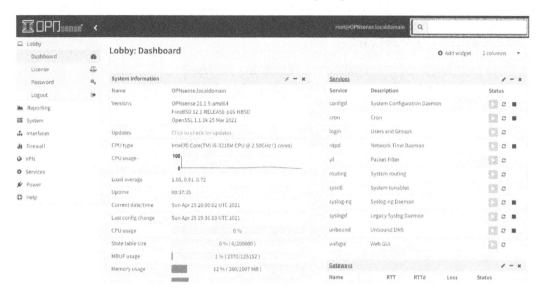

Figure 2.49 – The OPNsense WebGUI dashboard

Try typing the options presented on the menu here. It will list the available menu shortcuts, as shown in the following screenshot:

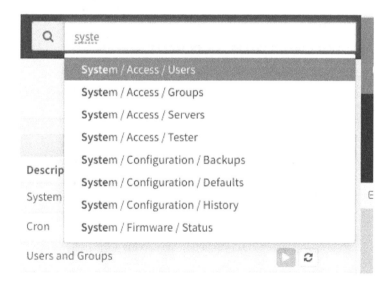

Figure 2.50 – The Quick Navigation search box

Faster than clicking on menus, right? If you don't like it, that's not a problem; you can always navigate through the menu by clicking on it. They are leveled by slashes /; for example, **System / Access / Users** will correspond to the **System** menu, the second level **Access** menu, and the third level **Users** menu. From this point on, I'll refer to the menus using the Quick Navigation search box. I spent a lot of money on a mechanical keyboard, OK? Let me use it as fast as I can!

Now, let's go through the SSH enabling configuration process:

1. In Quick Navigation, type `administration` and choose the **System | Settings | Administration** menu.

2. From here, go to the **Secure Shell** options, enable as follows, and click on the **Save** button:

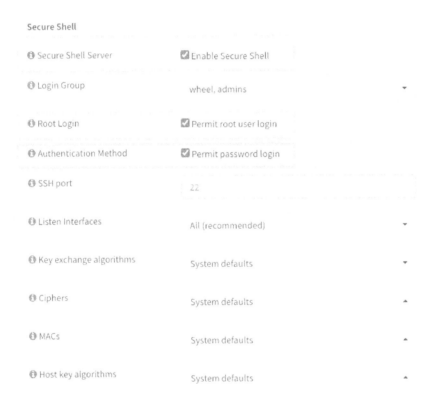

Figure 2.51 – Enabling SSH access

If you need to change any SSH server parameter, you'll be able to do it on this screen. Configurations such as changing the listen port or interface, ciphers, MACs, or algorithms can be done here. A common issue that will demand a different algorithm setting is when you are trying to connect from an old SSH client or SFTP software, some deprecated algorithms will not be allowed by default. It's always a good practice to keep only updated software that makes use of recommended algorithms.

> **Note**
>
> These settings aren't advised for production usage. We will use them just to simplify our lab access to SSH. For OPNsense with internet access, you should use PKI authentication (certificates), firewall rules restricting the source IPs that have access to SSH and webGUI, or even only allow access through a VPN.

3. Click on the OPNsense logo to go back to the dashboard and check if the SSH service is running:

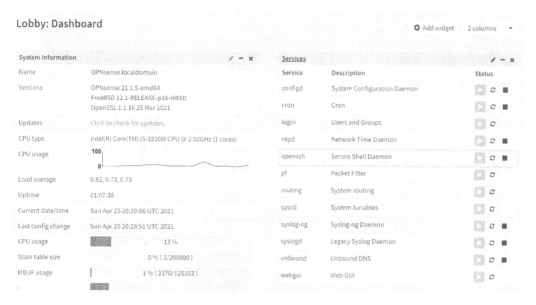

Figure 2.52 – Checking the OpenSSH service is running

4. Now we can log in using an SSH client. If your computer is running Linux or macOS, just go to the terminal and type the following:

```
$ ssh root@<OPNsense_LAN_IP_address>
```

If you are running Windows, you will need an SSH client such as Putty or Termius, or any other of your choice. Just fill your client with *username*: root and your OPNsense LAN IP address to connect.

The following screenshot shows my machine example (running macOS):

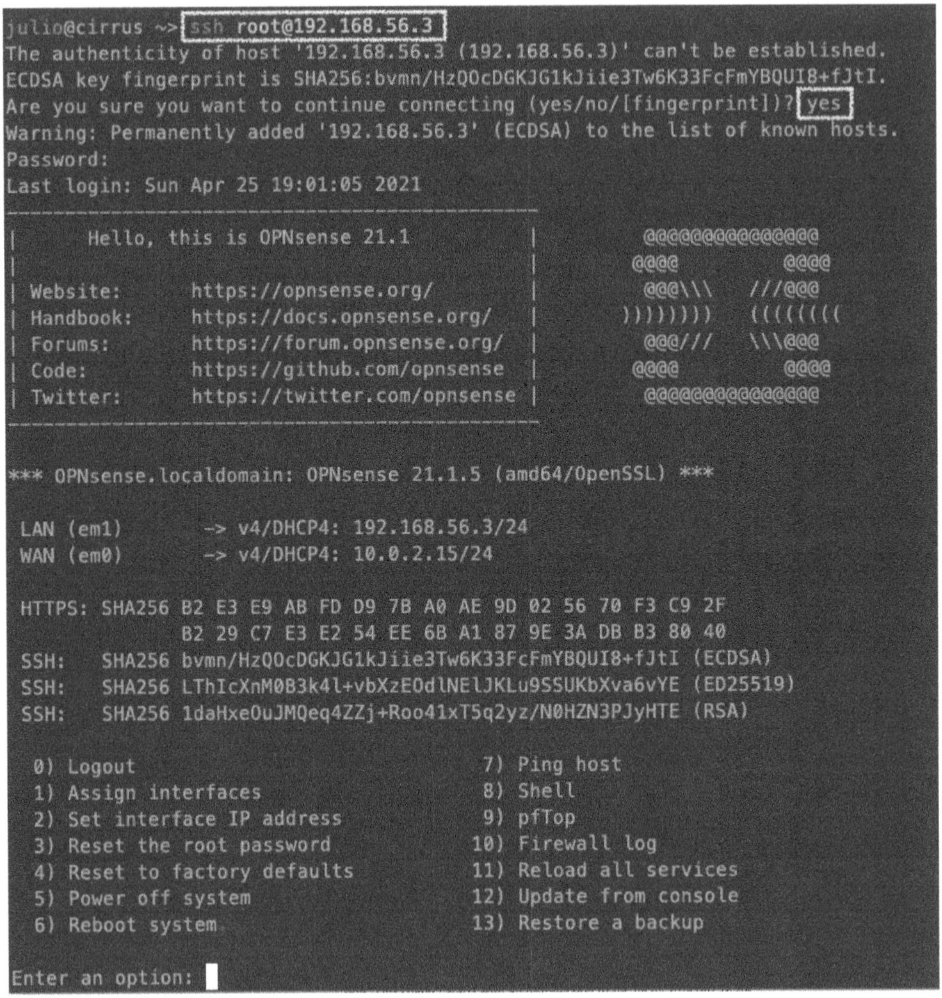

```
julio@cirrus ~> ssh root@192.168.56.3
The authenticity of host '192.168.56.3 (192.168.56.3)' can't be established.
ECDSA key fingerprint is SHA256:bvmn/HzQOcDGKJG1kJiie3Tw6K33FcFmYBQUI8+fJtI.
Are you sure you want to continue connecting (yes/no/[fingerprint])? yes
Warning: Permanently added '192.168.56.3' (ECDSA) to the list of known hosts.
Password:
Last login: Sun Apr 25 19:01:05 2021

|      Hello, this is OPNsense 21.1      |    @@@@@@@@@@@@@@@@
|                                        |   @@@@          @@@@
| Website:     https://opnsense.org/     |   @@@\\\    ///@@@
| Handbook:    https://docs.opnsense.org/|   ))))))))   (((((((((
| Forums:      https://forum.opnsense.org/|  @@@///    \\\@@@
| Code:        https://github.com/opnsense|  @@@@          @@@@
| Twitter:     https://twitter.com/opnsense| @@@@@@@@@@@@@@@@

*** OPNsense.localdomain: OPNsense 21.1.5 (amd64/OpenSSL) ***

 LAN (em1)       -> v4/DHCP4: 192.168.56.3/24
 WAN (em0)       -> v4/DHCP4: 10.0.2.15/24

 HTTPS: SHA256 B2 E3 E9 AB FD D9 7B A0 AE 9D 02 56 70 F3 C9 2F
               B2 29 C7 E3 E2 54 EE 6B A1 87 9E 3A DB B3 80 40
 SSH:    SHA256 bvmn/HzQOcDGKJG1kJiie3Tw6K33FcFmYBQUI8+fJtI (ECDSA)
 SSH:    SHA256 LThIcXnM0B3k4l+vbXzEOdlNElJKLu9SSUKbXva6vYE (ED25519)
 SSH:    SHA256 1daHxeOuJMQeq4ZZj+Roo41xT5q2yz/N0HZN3PJyHTE (RSA)

 0) Logout                          7) Ping host
 1) Assign interfaces               8) Shell
 2) Set interface IP address        9) pfTop
 3) Reset the root password        10) Firewall log
 4) Reset to factory defaults      11) Reload all services
 5) Power off system               12) Update from console
 6) Reboot system                  13) Restore a backup

Enter an option: █
```

Figure 2.53 – Connecting in OPNsense with an SSH client

We are in!

> **Important Note**
>
> The default configuration has a full allow firewall rule on the LAN interface; in this case, we did not create a rule for that and the access was permitted. The same doesn't apply to other network interfaces, such as WAN, for example.

Now, let's take a look at the FreeBSD packages and how to manipulate them.

FreeBSD packages

The FreeBSD operating system has two basic ways to install new software on it: using ports or packages.

Ports contain the source code and need to be compiled to be used by the system, which means it can take a long time to have new software running on a system, but we can customize it in a way a binary will not permit. When we think about a firewall system, every resource must be saved to process packet filtering and other almost real-time tasks it has to do. So, it is not suitable to compile software, taking precious processing resources, in a firewall system, right? We need something that is ready to use, fast, and can be installed quietly on the operating system. Are FreeBSD packages the right choice? Let's take a look!

A FreeBSD package is an archive file that contains binaries that are ready to use. It resolves dependencies automatically (ports do too!). It's smaller than a port and can be installed using the pkg tool. Seems to be a good option to use on a firewall system, right? Yes! OPNsense uses it to install and manage packages and plugins. We'll see it in action with some pkg command examples. Actually, we already used it before on our update and plugin installation practices, but now we will explore it the shell:

1. Log in through SSH or the console and type 8 to have the shell prompt.

2. Type pkg info to list the installed packages:

```
$ pkg info
```

Now, go to WebGUI, go to **System | Firmware | Packages**, and check that this page lists the packages such as the `pkg info` command:

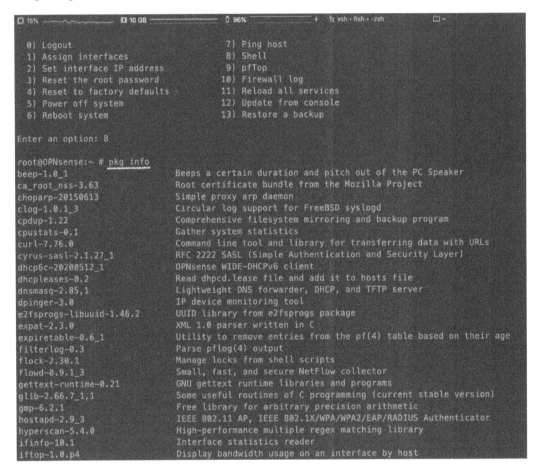

Figure 2.54 – pkg info output

Looks familiar? These are two different ways to list installed packages.

Let's take a look at the installed plugins:

1. Try typing `pkg info | grep "os-"`:

    ```
    $ pkg info | grep "os-"
    ```

2. Now, go to WebGUI and go to **System | Firmware | Plugins** (remember you can just type `plugin` in the Quick Navigation search box).

It's the same list of installer plugins, right? So, plugins are packages too, and can be manipulated using the pkg tool.

PKG basic operations

Let's take a look at the basic FreeBSD pkg operations:

- To install a package locally:

```
$ pkg add <local_package_file>
```

- To install a package from a repository:

```
$ pkg install <package_name>
```

- To list installed packages:

```
$ pkg info
```

- To remove a package:

```
$ pkg delete <package_name>
```

  You can also use the following for this:

```
$ pkg remove <package_name>
```

You can check out more about pkg on FreeBSD's pkg main page: https://www.freebsd.org/cgi/man.cgi?query=pkg&sektion=&n=1.

Summary

In this chapter, we reviewed the OPNsense versioning concepts, explored some basic hardware considerations for better performance of your OPNsense installation, and how to download and install it on VirtualBox. We also configured VirtualBox to add a LAN on an OPNsense VM, and after that, we looked at how to update and install plugins on it. We learned how to enable SSH access and even explored a little bit of the CLI world, practicing with the pkg command while we learned about FreeBSD packages.

In the next chapter, we will dive into networking configurations in OPNsense.

3
Configuring an OPNsense Network

In this chapter, we will learn more about some hardware considerations concerning networking and how to configure a network interface using the OPNsense WebGUI, and we will explore the network interface types available to configure in it. We will also learn about the virtual IPs and the differences between them, and finally we will look at some common troubleshooting scenarios and how to solve issues using the diagnostics tools available in OPNsense.

In this chapter, we will cover the following topics:

- Hardware considerations
- Configuring a basic network
- Types of network interfaces
- Exploring virtual IPs
- Networking diagnostics and troubleshooting

Technical requirements

This chapter requires that you have a clear understanding of the TCP/IP network stack, OSI model concepts, and some experience of using a command shell, or CLI. You will need a running OPNsense to practice some of the exercises proposed in this chapter. If you have followed all the steps from the previous chapters of this book, you can use your newly configured OPNsense virtual machine.

Hardware considerations

It is crucial to choose the right hardware for your OPNsense, so that it functions well as a firewall, and the most important components are the CPU and the **network interface cards** (**NICs**). For example, choosing the right NIC for the network environment can make the difference between a successful deployment and a complete disaster. Of course, the high-quality code of OPNsense can do an outstanding job while securing networks, but without a good hardware bundle, it won't make any magic.

As we discussed in the last chapter, the kernel OPNsense runs on has the netmap framework implemented in it, which can be used by network IPS and Sunny Valley's Sensei plugin, for example, to process network packets with better performance and low CPU usage. If one of your OPNsense installation tasks uses one or maybe both of these (network IPS and Sensei plugin in different NICs), then you should choose a network card with a driver that is supported by netmap to get the job done; otherwise, the whole system will suffer from high CPU load when processing network packets while using these components.

Before starting your OPNsense firewall implementation, let's take into account the following considerations:

- **Network size and overall throughput**: Is it expected to have a lot of concurrent connections being processed by OPNsense? Consider that every new connection state will consume 1 KB of RAM; for example, if you need 1,000,000 concurrent connections, 1 GB of the firewall's RAM will be consumed.

- **Features that may demand more hardware power**: As an example, we can talk about a web proxy with HTTPS inspection or an **Intrusion Detection System/Intrusion Prevention System (IDS/IPS)** enabled that consumes a lot of CPU and RAM resources. Another example is if you have a lot of VPN tunnels; maybe you could consider a CPU with **Advanced Encryption Standards-New Instructions (AES-NI)** support, which is an instruction set that some CPUs have to improve performance while using some algorithms using the **Advanced Encryption Standard (AES)**. Using this algorithm will mean better performance.

- **The number of services that will be running in the firewall**: If you are considering building a firewall with a lot of services running on it, for example, a proxy server, a VPN, an IDS/IPS, and a firewall, and you install a bunch of plugins to do other tasks, then aside from a good CPU and network card, consider using a lot of RAM too.

These are some examples of possibilities you may consider in your OPNsense deployment. It is not within the book's scope to give you basic recipes on how to deploy each one; however, I will show you the right path to know what to look for before building your OPNsense firewall.

FreeBSD NIC names

FreeBSD uses the network interface driver to name each installed card followed by a number, starting at 0. For example, if you install two Intel PRO/1000 network cards, they will be identified as **igb0** and **igb1**; the **igb** part is from the driver used by FreeBSD to manage this type of Intel Gigabit network card, followed by the number to identify each one: in this example, **0** for the first network card and **1** for the second one.

Here is a reference: `https://docs.freebsd.org/doc/7.3-RELEASE/usr/share/doc/en/articles/linux-users/network.html`.

The ifconfig command

The `ifconfig` command is a native FreeBSD network tool that can help you in managing network interfaces in the CLI. This tool can be considered as the Swiss army knife of network managing tools in CLI.

> **Important Note**
>
> Changes made using the `ifconfig` command will not survive a reboot!

In the CLI, an easy way to list the installed network interfaces is using the `ifconfig` command:

```
root@bluebox:~ # ifconfig
igb0:
flags=8943<UP,BROADCAST,RUNNING,PROMISC,SIMPLEX,MULTICAST>
metric 0 mtu 1500
        options=8500b8<VLAN_MTU,VLAN_HWTAGGING,JUMBO_MTU,VLAN_
HWCSUM,VLAN_HWFILTER,VLAN_HWTSO>
        ether 00:90:0b:4d:25:b0
```

```
        inet6 fe80::290:bff:fe4d:25b0%igb0 prefixlen 64 scopeid
0x1

        groups: LAN
        media: Ethernet autoselect (1000baseT <full-duplex>)
        status: active
        nd6 options=21<PERFORMNUD,AUTO_LINKLOCAL>
igb1: flags=8843<UP,BROADCAST,RUNNING,SIMPLEX,MULTICAST> metric
0 mtu 1500
        options=8500b8<VLAN_MTU,VLAN_HWTAGGING,JUMBO_MTU,VLAN_
HWCSUM,VLAN_HWFILTER,VLAN_HWTSO>
        ether 00:90:0b:4d:25:b1
        inet6 fe80::290:bff:fe4d:25b1%igb1 prefixlen 64 scopeid
0x2
        inet 192.168.15.9 netmask 0xffffff00 broadcast
192.168.15.255
        media: Ethernet autoselect (100baseTX <full-duplex>)
        status: active
        nd6 options=21<PERFORMNUD,AUTO_LINKLOCAL>
```

To list a specific network interface, just type `ifconfig` followed by the interface name:

```
root@bluebox:~ # ifconfig igb1
igb1: flags=8843<UP,BROADCAST,RUNNING,SIMPLEX,MULTICAST> metric
0 mtu 1500
        options=8500b8<VLAN_MTU,VLAN_HWTAGGING,JUMBO_MTU,VLAN_
HWCSUM,VLAN_HWFILTER,VLAN_HWTSO>
        ether 00:90:0b:4d:25:b1
        inet6 fe80::290:bff:fe4d:25b1%igb1 prefixlen 64 scopeid
0x2
        inet 192.168.15.9 netmask 0xffffff00 broadcast
192.168.15.255
        media: Ethernet autoselect (100baseTX <full-duplex>)
        status: active
        nd6 options=21<PERFORMNUD,AUTO_LINKLOCAL>
```

In the preceding example, you selected the `igb1` network interface, and as we can see, the IPv4-configured IP address is `192.168.15.9` with a 24-bit netmask (`0xffffff00`) and a network broadcast address of `192.168.15.255`. But wait a moment! A 24-bit netmask? Where?! Don't worry! Let's understand this part. If you came from the Linux world, like me, you are probably wondering why on earth this netmask is in hexadecimal format.

Well, the default option in the `ifconfig` command in FreeBSD is hex notation, but we can change that with some parameters, but before we change it in `ifconfig`, let's remember some concepts.

If you convert the value `0xffffff00` to binary notation, you will get the following:

```
11111111 11111111 11111111 00000000
```

Counting the 1 bit, you have 24 bits, right? Remembering the network basic classes, the IPv4 address has 4 octets. Converting that to decimal, we get the following:

- 1st octet: 8 bits in 1 state = `255` decimal

- 2nd octet: 8 bits in 1 state = `255` decimal

- 3rd octet: 8 bits in 1 state = `255` decimal

- 4th octet: 8 bits in 0 state = `0` decimal

This equals `255.255.255.0` decimal.

> **Important Note**
>
> This chapter requires that you have a basic IPv4 understanding: how to convert hex numbers to binary and decimal will not be detailed here. If you don't know how exactly to do that, I strongly recommend you read some Packt books about network basics before moving on with firewalling. It's just advice, but you can always make use of calculators to convert these numbers too.

Now that we have remembered some of the basics of IPv4 concepts, we can try to convert a hex to a decimal using `ifconfig`:

```
root@bluebox:~ # ifconfig -f inet:dotted igb1
igb1: flags=8843<UP,BROADCAST,RUNNING,SIMPLEX,MULTICAST> metric
0 mtu 1500
        options=8500b8<VLAN_MTU,VLAN_HWTAGGING,JUMBO_MTU,VLAN_
HWCSUM,VLAN_HWFILTER,VLAN_HWTSO>
        ether 00:90:0b:4d:25:b1
```

```
        inet6 fe80::290:bff:fe4d:25b1%igb1 prefixlen 64 scopeid
0x2
        inet 192.168.15.9 netmask 255.255.255.0 broadcast
192.168.15.255
        media: Ethernet autoselect (100baseTX <full-duplex>)
        status: active
        nd6 options=21<PERFORMNUD,AUTO_LINKLOCAL>
```

Adding the `-f inet:dotted` parameter to the `ifconfig` command, now we can see the netmask in decimal notation: `255.255.255.0`.

Better? But you want more? OK! Let's try now with *CIDR* notation. For that, just change the `ifconfig` parameter to `-f inet:cidr`:

```
root@bluebox:~ # ifconfig -f inet:cidr igb1
igb1: flags=8843<UP,BROADCAST,RUNNING,SIMPLEX,MULTICAST> metric
0 mtu 1500
        options=8500b8<VLAN_MTU,VLAN_HWTAGGING,JUMBO_MTU,VLAN_
HWCSUM,VLAN_HWFILTER,VLAN_HWTSO>
        ether 00:90:0b:4d:25:b1
        inet6 fe80::290:bff:fe4d:25b1%igb1 prefixlen 64 scopeid
0x2
        inet 192.168.15.9/24 broadcast 192.168.15.255
        media: Ethernet autoselect (100baseTX <full-duplex>)
        status: active
        nd6 options=21<PERFORMNUD,AUTO_LINKLOCAL>
```

Now we can see the IP address followed by CIDR notation:

```
inet 192.168.15.9/24 broadcast 192.168.15.255
```

Here are some other important parts of the output of the `ifconfig` command as regards hardware:

- `ether 00:90:0b:4d:25:b1`: Here, `ether` shows the physical address of the NIC, also known as the MAC address.

- `options=8500b8<VLAN_MTU,VLAN_HWTAGGING,JUMBO_MTU,VLAN_HWCSUM,VLAN_HWFILTER,VLAN_HWTSO>`: Here, we can check which options are supported by the network interface driver. In this example, we have options such as VLAN hardware tagging support (`VLAN_HWTAGGING`) and jumbo frames (`JUMBO_MTU`), which is the support required for a **maximum transmission unit** (**MTU**) with more than 1,500 bytes, and some other options available to this NIC. The available options may vary depending on the selected network card.

- `media: Ethernet autoselect (100baseTX <full-duplex>)`: Here, `media` shows the active media type used to connect the NIC physically to the network; for example, `100baseTX` is a twisted-pair cable that supports 100 Mbps.

- `status: active`: Here, `status` shows that this NIC is active, in this example, with the cable connected.

Want to learn more about the `ifconfig` command? Take a look at FreeBSD's `ifconfig` man page: `https://www.freebsd.org/cgi/man.cgi?ifconfig(8)`.

Now that we have explored the NIC considerations and learned how to get some information about the NIC in the CLI, we can move on to configuring a basic network in OPNsense.

Basic network configuration

Local network configuration begins with a good IP addressing plan! Always try to follow the RFC1918 reserved address space (`10.0.0.0/8`, `172.16.0.0/12`, and `192.168.0.0/16`) for local networks, using private IP addresses. In this way, you avoid future issues with local addresses overlapping with public addresses on the internet. I have seen many times network administrators not paying attention to this rule of thumb and creating problems for themselves.

Another good practice is not using huge broadcast domains. If you are projecting a small network, then why use a `10.0.0.0/8` network? Avoid doing that! This can save you time in the future; for example, while connecting two or more networks using a VPN tunnel, there will be a smaller chance of network addresses overlapping with other connected networks. If you choose `10.10.10.0/24`, which means 254 usable IP addresses, instead of choosing `10.0.0.0/8`, which has more than 16 million IP addresses, which one do you think has more chances of conflict with another `10.x.x.x` address networks? Smaller broadcast domains means smaller chances of conflicts happening.

Some good tools to start your firewall and network project are a piece of paper and a pencil. Draw your network plan and discuss it with your work buddies before starting to set IP addresses for network devices.

After these brief considerations, let's go to the network configuration.

WebGUI – network interface configuration

Everything that we need to configure a network interface is in the **Interfaces** menu. OPNsense will set the LAN and WAN interfaces by default, as we explored in *Chapter 2, Installing OPNsense*, where we already configured both, so let's explore the available network configuration options in WebGUI.

The following options are available both in LAN and WAN interface configuration or in any new network interface you may add later in your OPNsense.

To explore the network interface configuration options, head to the **Interfaces | LAN** menu (you can also try **Interfaces | WAN**; the options are the same):

Figure 3.1 – Network interface configuration options

The available basic configuration options are as follows:

- **Enable Interface**: When checked, this will enable the network interface.
- **Lock**: This will prevent the network interface being removed. This option can protect against accidental removal.
- **Device**: Just a label showing the name of the FreeBSD network interface.
- **Description**: Here, you can set an interface name that will be used in WebGUI.

Here are the generic configuration options:

- **Block private networks**: When checked, this will block all traffic from the RFC1918 private network space as well as the loopback reserved address (127.0.0.0/8) and carrier-grade NAT addresses (100.64.0.0/10). *Don't use it in local network interfaces!* You may check this option in WAN interfaces.
- **Block bogon networks**: When checked, this will block reserved addresses (not yet assigned by IANA), which does not include RFC1918. You may check this option in WAN interfaces.
- **IPv4 Configuration Type**:
  - **None**: This will not assign any IP address to the interface.
  - **Static IPv4**: This will set a static IP address to the interface.
  - **DHCP**: This will use the DHCP protocol to set a network interface IP address.
  - **PPP**: This will set this network interface using the PPP protocol; common usage is connecting using a mobile networking modem.
  - **PPPoE**: This will configure this network interface using the PPP over the Ethernet protocol; common usage is connecting through DSL networks.
  - **PPTP / L2TP**: This will use the PPTP or L2TP protocol to set up a connection in the interface; both will require the same configuration options, such as username and password, for example.
- **IPv6 Configuration Type**:
  - **Static IPv6**: This will set a static IPv6 address to the interface.
  - **DHCPv6**: This will use the DHCPv6 protocol to set a network interface IPv6 address.
  - **SLAAC**: This will configure an interface using stateless address autoconfiguration.

- **6rd Tunnel**: This will set the interface using IPv6 rapid deployment. This configuration will probably be served by your ISP.

  - **6to4 Tunnel**: This will set the interface using tunneling IPv6 inside an IPv4 mechanism.

  - **Track interface**: This will track configuration from another interface configured using IPv6.

- **MAC address**: Use this to spoof the interface physical address; it will be set in the network interface. A common usage of it is when you have to change a network card or the entire OPNsense hardware without changing the MAC address from the old hardware. Take care when using this option!

- **MTU**: This will set the MTU in the interface; the default value for Ethernet is 1,500 bytes.

- **MSS**: This option changes the maximum segment size for TCP connections to be used by the interface.

- **Speed and duplex**: Keeping this as the default option, **Default (no preference, typically autoselect)**, will mean auto-negotiating between the network card and the switch or other connected device.

- **Dynamic gateway policy**: If there is no gateway interface and this option is set to false, no "route-to" settings are generated. This means that you can load balance in multi-WAN setups if the connections are to public networks, but it may break in many cases.

After configuring the network interfaces, let's proceed to assign network interfaces.

Assigning network interfaces

If you want to add more NICs to your OPNsense, you can do this by heading over to the **Interfaces | Assignments** menu.

Proposed exercise – adding a new network interface to OPNsense

To practice adding more network interfaces using WebGUI, you can follow the steps in this exercise:

1. To start, we need to repeat *steps 23 to 29* covered in *Chapter 2, Installing OPNsense*, under the *Setting up LAN network* section to create as many new network interfaces as you wish using **VirtualBox Host Network Manager**.

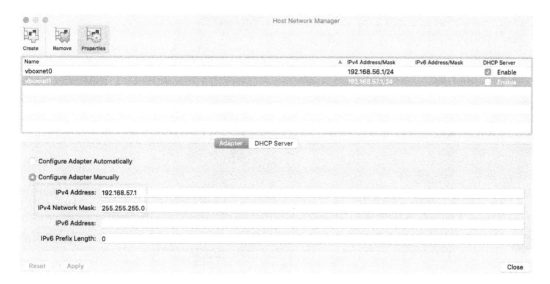

Figure 3.2 – VirtualBox Host Network Manager: creating a new network interface

> **Applying a New Network Configuration to Your OPNsense Virtual Machine**
>
> You must first shut down your **virtual machine** (**VM**) to apply the new network settings. Remember to note down the new IP address assigned in your new network interface on VirtualBox *to use a different one in your new OPNsense interface configuration*.

2. After you have applied the new network interface in your VirtualBox, start your OPNsense VM again and log in to WebGUI to set it up. To do that, point to the **Interfaces | Assignments** menu.

3. In **New interface**, select the desired NIC and then click the + icon to add the new interface, and after that, just click the **Save** button.

Figure 3.3 – Adding a new network interface in the OPNsense WebGUI

4. After adding it, you can just click on the interface name or go to **Interfaces** | [**New Interface Description**] (**OPT1** in the following example) to configure it.

Figure 3.4 – New network interface added

Interface added! Let's practice a little bit now.

Proposed exercise – setting a static Ipv4 in added NIC

Now you have added a new network interface, let's add a static IPv4 to it:

1. Go to **Interfaces** | [*Name given by you*] and set things as follows:

    Enable: Checked.

    Description: Name your interface as you wish.

    Block private networks and Block bogon networks: Unchecked.

    IPv4 Configuration Type: Select **Static IPv4**.

    IPv6 Configuration Type: Select **None**.

2. Jump to **Static IPv4 configuration**:

    IPv4 address: Fill with an address that is in the network range you noted down when you created the network interface on VirtualBox Network Manager. Pay attention to the network CIDR.

    IPv4 Upstream Gateway: Leave as the default: **Auto-detect**.

3. Click on the **Save** button.

Now that you have your new network interface configured in your OPNsense, let's take a closer look at it in our next section.

Overview of the network interface

Go to **Interfaces | Overview** and click on the interface you just configured. You should see something like this:

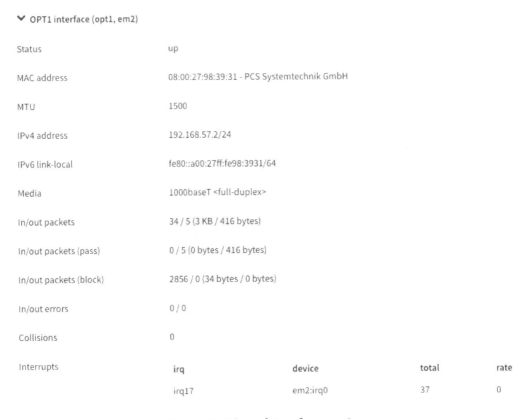

Figure 3.5 – Network interface overview

As we can see, there is some information in **Interfaces | Overview** that is quite similar to the `ifconfig` output, but with a much better look! We can check here, for example, whether the network interface is up, which IP address is configured in it, the MTU size, the MAC address, and a lot more useful information.

What about your interface overview configuration? Is it the same as the one you just configured in the previous exercises? If yes, congratulations! If no, then no problem, repeat the exercise, paying attention to the details, and I'm sure you will find the mistake and fix it!

Now that we have concluded the basic network configuration steps, we can continue our networking saga and head to the next section to explore the different types of network interfaces you can configure in OPNsense.

Types of interfaces

As a complete firewall solution, OPNsense supports many different types of network interfaces. We will explore each one available for configuration in WebGUI next.

Bridge

A **bridge** can connect two different network interfaces in the same network segment. For example, you can connect a LAN interface connected by cable to a Wi-Fi interface using a bridge. In this way, the devices connected in both interfaces will rely on the same broadcast domain or the same network segment.

GIF

The **Generic Tunnel Interface**, or just **GIF**, is a type of interface configuration that can be used to tunnel IPv6 via IPv4. An example of its usage is configuring it with the IPv6 tunnel broker from Hurricane Electric, with which you can reach IPv6 internet using an existing IPv4 connection. If you want to learn more about this service, you can access it at `https://tunnelbroker.net/`.

GRE

The **Generic Routing Encapsulation (GRE)** is a network configuration that allows two hosts to tunnel traffic without encryption. An example of its usage is when you need to traverse some protocols that aren't supported by the intermediate systems.

LAGG

Link aggregation, also known as port-channel sometimes, can provide, as the name suggests, aggregation using multiple network interfaces. It can be used to increase bandwidth in local networks with the prerequisite of a compatible device, such as a switch, for example. It can support failover since at least one interface is up. To perform these functions, LAGG will need to be configured on both sides of the connection, with the same protocol. The current protocols supported by OPNsense are as follows:

- LACP
- Failover
- FEC
- LoadBalance
- RoundRobin

Before starting LAGG configuration, check whether your switch supports one of these protocols. This feature is awesome when combined with VLANs, so that you can have a reliable trunk with a lot of networks passing through it.

Loopback

The **loopback** is a virtual interface that is commonly used to test local communications in a host. You can easily test it by trying to ping the `127.0.0.1` address (or `::1` for IPv6); this is supported in most modern operating systems, and FreeBSD is no different. In OPNsense, we can create additional loopback interfaces for different applications; for example, configuring a service only locally with the port listening in a loopback interface address.

VLAN

With VLANs, you can make better network segmentation, dividing one big network or broadcast domain into many smaller ones. By doing so, you can save your network from issues such as traffic flowing between all the hosts without any packet filtering, making it difficult for malware to spread to the whole network and preventing lateral movement, for example. Another example is to use a single network port to configure multiple network segments. OPNsense supports VLAN 802.1Q standard permitting VLAN tagging and priority code point, which can be used for **Quality of Service** (**QoS**). For extra protection, you may add network ports authentication using 802.1x. To learn more about 802.1X, please refer to `https://en.wikipedia.org/wiki/IEEE_802.1X`.

To learn more about network lateral movement, refer to `https://en.wikipedia.org/wiki/Network_Lateral_Movement`.

VXLAN

The **Virtual Extensible Lan** (**VXLAN**) was created to overcome VLAN limitations in this era of the cloud. It can address up to 16 million logical networks, while VLANs can do only 4,096. Most modern virtualization technologies support it, and its usage is more common in cloud scenarios. Now that we have seen the different types of interfaces, let's go ahead and create one of them, for example, the loopback interface.

Proposed exercise – creating another type of network interface

To practice how to configure another type of network interface, we will configure a new **loopback** interface using WebGUI:

1. Go to **Interfaces | Other Types | Loopback**.
2. Click on the + icon.

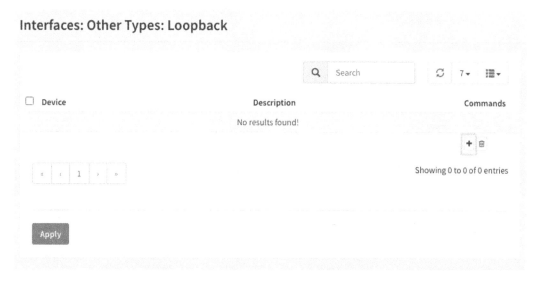

Figure 3.6 – Adding a new loopback interface

3. In **Description**, enter something that will make sense to you, for example, loopback2. Then, click on the **Save** button and click **Apply**.

Figure 3.7 – Creating a new loopback interface

4. Go to **Interfaces | Assignments** and add the new interface.

5. Name your interface, adding a description such as `LB2`.

6. Click on the new interface name.

7. Enable it and set the static IPv4 address to `127.0.0.10/8` and leave the other options as default. Then, click on the **Save** button and click on the **Apply changes** button, which will appear in the top right.

Figure 3.8 – Assigning the new loopback network interface

8.   Go to **Interfaces | Overview** and check your newly created interface.

✔ LB2 interface (opt2, lo1)

Status	up
MAC address	00:00:00:00:00:00 - XEROX CORPORATION
MTU	16384
IPv4 address	127.0.0.10/8
IPv6 link-local	fe80::1/64
In/out packets	1 / 1 (133 bytes / 133 bytes)
In/out packets (pass)	1 / 1 (133 bytes / 133 bytes)
In/out packets (block)	0 / 0 (0 bytes / 0 bytes)
In/out errors	0 / 0
Collisions	0

Figure 3.9 – Newly created loopback interface. Notice the In/out packets section

9. Test it by sending a ping to the `127.0.0.10` address in **Interfaces | Diagnostics | Ping**.

Figure 3.10 – Pinging the created loopback interface

10. Repeat *step 8* and check whether the **In/out packets** value was incremented.

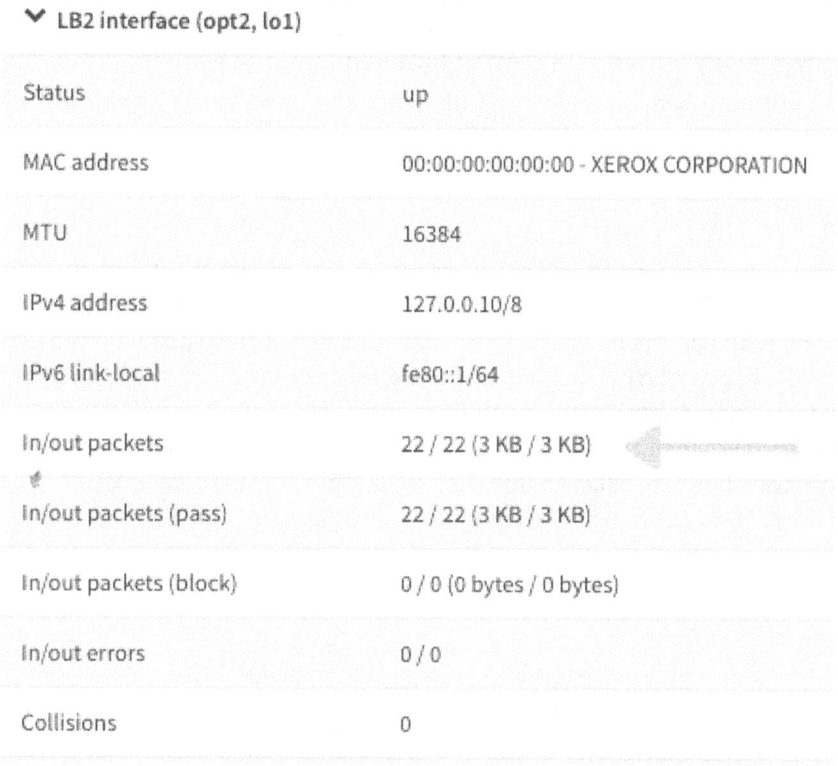

❯ LB2 interface (opt2, lo1)

Status	up
MAC address	00:00:00:00:00:00 - XEROX CORPORATION
MTU	16384
IPv4 address	127.0.0.10/8
IPv6 link-local	fe80::1/64
In/out packets	22 / 22 (3 KB / 3 KB)
In/out packets (pass)	22 / 22 (3 KB / 3 KB)
In/out packets (block)	0 / 0 (0 bytes / 0 bytes)
In/out errors	0 / 0
Collisions	0

Figure 3.11 – Loopback interface overview: packets incremented after ping test

---

Adding a New Interface to OPNsense

You can review the steps on how to add a new network interface by looking at the *Proposed exercise – adding a new network interface to OPNsense* section in *Chapter 2, Installing OPNsense*.

Congratulations, you have created your first loopback network interface! You can repeat this exercise with other types of interfaces if you wish; the only difference is that each type will demand its specific configurations. I proposed using the loopback type to save your time to repeat all the required steps for creating a new network interface on VirtualBox, but it's up to you! You can repeat this exercise, creating as many types of network interfaces as you want!

As we explored in this topic, OPNsense can support a variety of types of network interface configurations. Now that you know about each one, let's dive into IP addressing, starting with how to add extra IPs in network interfaces.

Exploring virtual IPs

A virtual IP address can be used for a high-availability configuration, such as creating a **Network Address Translation** (**NAT**) in many different services on the same network port, or just for adding more than one IP address in the same network interface. In this section, we'll dive into each type of virtual IP that is supported by OPNsense and when to choose which virtual IP configuration type.

IP alias

An **IP alias** can be used as an additional IP address in a configured network interface. It will behave like the address configured in the interface, replying to ICMP requests (ping) and generating ARP packets on the network. The netmask must match with the network interface the IP alias will be created on; otherwise, you can set it as a single address (/32 CIDR for IPv4). You can even set an IP address from another network. There are some special cases, such as when the ISP delivers a small network (/30 CIDR) and you need to set up a high-availability installation. Instead, you should consider using VLANs, which will isolate the packets from the different networks configured within the same network interface.

> **Loopback Interfaces**
>
> An IP alias is the only type of address accepted to be used as a virtual IP address on loopback interfaces.

CARP

The **Common Address Redundancy Protocol** (**CARP**), as its name suggests, is a protocol used for network redundancy, and in OPNsense, we use it for high-availability deployments. We will explore this protocol and the high-availability configuration in detail later in *Chapter 17, Firewall High Availability*.

Proxy ARP

Wikipedia defines ARP as follows: "*The Address Resolution Protocol (ARP) is a communication protocol used for discovering the link layer address, such as a MAC address, associated with a given internet layer address, typically an IPv4 address.*" A proxy ARP uses this protocol (ARP) to reply to queries for an IP address set in the network interface using this configuration. The IP address will not be added to the interface. OPNsense will use the cheap and omitted proxy ARP daemon (**choparp** – https://www.freebsd.org/cgi/man.cgi?query=choparp) to reply to requests to the configured proxy ARP IP address.

> **Proxy ARP – Important Notes**
>
> To configure a single IP address, you must use /32 CIDR. Only network addresses will be accepted with smaller CIDR numbers.
>
> When a proxy ARP IP address is configured, it will not be shown in **Interfaces | Overview** or in the ifconfig command output.
>
> It does not respond to ICMP requests.

Other

The exclusive usage for this type of virtual IP is in NAT, since it will respond neither to ICMP nor ARP requests.

The **Interfaces | Virtual IPs | Status** page is exclusively used to show the status of CARP virtual IPs; hence, we will explore it later in *Chapter 17, Firewall High Availability*.

Proposed exercise – creating a virtual IP address

Let's practice how to create a virtual IP address by adding an extra IP to our previously created network interface, using the following steps:

1. Go to **Interfaces | Virtual IPs | Settings** and click on the **Add** button.

2. For **Mode**, leave the default option: **IP Alias**.

3. In **Interface**, select the previously created loopback interface (see the *Proposed exercise – creating another type of network interface* section).

4. In **Address**, enter 127.0.0.20/8.

5. Leave the other options with their default values and click on the **Save** button and then on **Apply changes**.

6. Go to **Interfaces | Overview** and see the newly created virtual IP address configured in the created loopback interface.

∨ LB2 interface (opt2, lo1)	
Status	up
MAC address	00:00:00:00:00:00 - XEROX CORPORATION
MTU	16384
IPv4 address	127.0.0.10/8 ⊕ 127.0.0.20/8
IPv6 link-local	fe80::1/64
In/out packets	2330 / 2330 (191 KB / 191 KB)
In/out packets (pass)	2330 / 2330 (191 KB / 191 KB)
In/out packets (block)	0 / 0 (0 bytes / 0 bytes)
In/out errors	0 / 0
Collisions	0

Figure 3.12 – Interfaces overview: virtual IP alias added

7. Go to **Interfaces | Diagnostics | Ping** and try to ping the newly created IP alias.

If the ping replies, you just created a new IP alias successfully! Otherwise, repeat the exercise by reviewing each step carefully to get it working.

Now that we have learned about the OPNsense networking configuration, it's important to see how to troubleshoot some common issues involving network and connectivity. Let's explore some common issues in the next section.

Network diagnostics and troubleshooting

It's important to know how to configure network interfaces in OPNsense, but it is more important to know how to solve problems related to it. In this section, we'll explore the common issues and see how to solve each one.

First of all, let's see what diagnostic tools are available in WebGUI.

All of the following options can be accessed from **Interfaces | [tool name]**:

- **ARP Table**: This page lists the operating system ARP table, in which you can find the IP and MAC addresses listed in the following fields:

  - **Device Manufacturer**, based on MAC OUI (organizationally unique identifier)

  - **Interface**, using the FreeBSD NIC name

  - **Interface Name**, using the configured one in the OPNsense interface description

  - **Hostname**, if it resolves hostnames

- **NDP Table**: Like the **ARP Table** page, this page will show the same information fields, but for IPv6 addresses based on the **Neighbor Discovery Protocol** (**NDP**). Due to the similarity with ARP, you can use the previous ARP troubleshooting examples to solve issues but using the **NDP Table** page for IPv6 networks.

Let's now explore troubleshooting examples.

True story – how to use ARP Table diagnostics

Someone calls you and complains that the firewall stopped working. You ask about the symptoms that made this person believe that. Immediately the person tells you that all the local network hosts lost connectivity with the firewall, but they can access the firewall from another network interface, a WAN interface, for example, and it appears to be working as it was meant to be.

Well, based on a true story, I can tell you that it is a good prompt to start looking at the ARP table – why? Because if you look at it, you will probably find the possible reasons and see whether they are related to OPNsense, or you can call the network infrastructure team to check the switch connected to this firewall.

If the ARP table isn't showing the amount of entries that you expected, the next step will be to check whether the LAN interface is up; if it's not, it is probably a problem with the network cable or switch. Otherwise, the LAN interface port may be damaged, and you will need a new one; only a physical test will help in this case. Even if the LAN interface is up, it's a good idea to ask for help from the network infrastructure team, if there is a high chance that it's a problem with network connectivity.

These are some possible solutions:

- Check whether the network port is up. Where do you find it? As we learned earlier in this chapter, go to **Interfaces | Overview** and check its status.
- Check cable connections, switches, and test the LAN physical network port in your OPNsense hardware.

The **DNS Lookup** page will resolve DNS names to IP addresses; you can use it to find issues with DNS resolution.

Common issue – local network hosts can't open websites

After the initial tests, such as pinging the firewall and other internet IP addresses, the users are still complaining that they can't access websites or receive emails on their Outlook. Is this another true story? Most of the stories in this book are!

As you are not using a web proxy (because we haven't got there in this book yet), you can start trying to resolve the internet DNS names in the **DNS Lookup** page. If the DNS Lookup did not succeed, then ask a friendly user to try to do the same on their computer, you're probably dealing with a DNS issue. This page can be more useful than a command-line ping because it shows the query time for each DNS server configured in OPNsense; sometimes, the issue is related to slow responses from one of the configured DNS servers.

Here are some possible solutions:

- Check whether the hosts with DNS issues are using the same DNS servers configured in OPNsense or using one of OPNsense's local network addresses as the DNS server.
- Test connectivity between OPNsense and the configured DNS servers.
- Try to change the DNS servers configured in OPNsense.
- See whether there are any DNS names resolving to IPv6 addresses (while the host isn't configured to use IPv6); if your network is IPv4-based, this can cause some issues.

Let's go back to the diagnostics tools:

1. The **Netstat** page shows the following:

   - network interface
   - bpf

- protocol, sockets
- Netisr
- memory statistics
- status

Since this topic is focused on troubleshooting, let's use the `netstat` command in the CLI instead of using WebGUI page.

Some examples of the `netstat` command usage are as follows:

- List the TCP ports state:

```
root@OPNsense:~ # netstat -ap tcp
Active Internet connections (including servers)
Proto Recv-Q Send-Q Local Address          Foreign
Address          (state)
tcp4        0        0 OPNsense.ssh
192.168.56.1.58376      ESTABLISHED
tcp4        0        0 localhost.rndc        *.*
LISTEN
tcp4        0        0 *.domain              *.*
LISTEN
tcp6        0        0 *.domain              *.*
LISTEN
tcp6        0        0 *.http                *.*
LISTEN
tcp4        0        0 *.http                *.*
LISTEN
tcp6        0        0 *.https               *.*
LISTEN
tcp4        0        0 *.https               *.*
LISTEN
tcp4        0        0 *.ssh                 *.*
LISTEN
tcp6        0        0 *.ssh                 *.*
LISTEN
```

- List the UDP ports state:

```
root@OPNsense:~ # netstat -ap udp
Active Internet connections (including servers)
Proto Recv-Q Send-Q Local Address          Foreign
Address          (state)
udp4      0        0 127.0.0.20.ntp         *.*
udp6      0        0 fe80::1%lo1.ntp        *.*
udp4      0        0 OPNsense.ntp           *.*
udp4      0        0 localhost.ntp          *.*
udp6      0        0 fe80::1%lo0.ntp        *.*
udp6      0        0 localhost.ntp          *.*
udp6      0        0 fe80::a00:27ff:f.ntp   *.*
udp4      0        0 OPNsense.ntp           *.*
udp4      0        0 OPNsense.ntp           *.*
udp6      0        0 fe80::a00:27ff:f.ntp   *.*
udp4      0        0 OPNsense.ntp           *.*
udp6      0        0 fe80::a00:27ff:f.ntp   *.*
udp4      0        0 *.ntp                  *.*
udp6      0        0 *.ntp                  *.*
udp4      0        0 *.domain               *.*
udp6      0        0 *.domain               *.*
udp6      0        0 *.dhcpv6-client        *.*
```

When you are not sure of whether a configured service is listening at the correct port, you can use the preceding commands.

- List active network routes on OPNsense:

```
root@OPNsense:~ # netstat -nr
Routing tables

Internet:
Destination      Gateway          Flags        Netif
Expire
10.0.2.0/24      link#1           U            em0
10.0.2.15        link#1           UHS          lo0
127.0.0.1        link#5           UH           lo0
127.0.0.10       link#8           UH           lo1
127.0.0.20       link#8           UH           lo1
```

```
Internet6:
Destination                               Gateway
Flags        Netif Expire
::1                                       link#5
UH           lo0
fe80::%em0/64                             link#1
U            em0
fe80::a00:27ff:feeb:a6ff%em0     link#1
UHS          lo0
fe80::%em1/64                             link#2
U            em1
```

If you want to know whether an added route or the correct default gateway is active, this command can help you!

2. **Packet Capture**: This page will use the `tcpdump` command to capture packets, using which you can do some network packet sniffing to look for possible problems with protocols or the network. It's a lot easier than using the `tcpdump` CLI syntax and can help you in a lot of situations.

    If you want to know whether a packet is flowing in some specific interface, with **Packet Capture**, you can find out.

3. **Ping**: The classic network test tool; this uses the ICMP protocol to test whether a host is responding or not.

    If you need to test whether a remote host is online or not, a good starting point is trying to ping it.

    > **Note**
    >
    > Windows' default configuration blocks ICMP packets. If you are testing against Windows machines, be sure that they are allowing ICMP packets.

4. **Port Probe**: This page will try to connect to a TCP port to test whether it is connecting or not. It uses `netcat` (`nc`) to do that and is very efficient. If some TCP port isn't responding to a NAT rule, then you can use it to test the destination host port from OPNsense.

5. **Trace Route**: This page will use `traceroute`(6) to test a network packet's path trying to reach a host. Windows uses ICMP in the `tracert` utility; be sure that ICMP packets are allowed before using it.

Say your users are complaining about a single website that isn't accessible. One good starting point is trying to reach it using `traceroute`. We already saw cases where changing the WAN, or the network route, solved the problem. Sometimes the ISP's cloud can be very tricky, and when you have another ISP available, changing the route to a second one is worth trying! Another good example is when the `traceroute` output shows the same IP multiple times; in this case, we have a routing loop.

We'll explore some of these diagnostic tools in depth in later chapters, which will present you with some examples of how to use them practically, so don't worry – we have a long journey ahead of us!

Summary

In this chapter, we learned how to design an OPNsense firewall for a production environment, considering which hardware and network card to use depending on the features that will be enabled. We learned about the FreeBSD network interfaces and practiced some CLI network commands. We also saw how to add new IP addresses to OPNsense using virtual IPs, choosing the right type depending on its usage. In the last section, on diagnostics and troubleshooting, I tried to share with you a little bit of my experience, telling you some true stories about network issues with some examples of how to solve them.

In the next chapter, we'll learn how to manage our OPNsense system configuration using WebGUI.

4
System Configuration

In this chapter, we will learn more about system configuration and administration, adding users and groups, authenticating on LDAP servers, managing certificates, changing some advanced operating system settings, and how to back up all system configurations.

As we already learned, OPNsense is a complete security platform with a great framework to manage the operating system it runs on – FreeBSD. We will continue exploring WebGUI and learn how to perform some system administration tasks on it. We are going to create users, groups, and certificates, see the steps to add an external authentication server, change some system settings for testing, and perform backup and restore tests. In this chapter, we will look at the following topics:

- Managing users and groups
- External authentication
- Certificates – a brief introduction
- General settings
- Advanced settings
- Configuration backup

Technical requirements

To follow this chapter, you will need to have a basic knowledge of digital certificates and how they work, an understanding of user and group privileges, LDAP and SSH usage and configuration, and logging concepts, and a running OPNsense to practice the proposed exercises.

Managing users and groups

Before we can start managing users and groups, it's important to understand the least privilege principle. It defines that a user must have only the privileges necessary to complete a task, so it isn't a good idea to have all firewall users as admins with full privileges or even to share the root password with a lot of users. This will break the least privilege concept.

A better approach is to define profiles and apply these profiles to users so that unnecessary privileges for some users can be avoided. A good way to do this is by creating groups and assigning the required privileges to each one. After that, you can add new users or assign existing ones to each group based on the least privilege principle.

Let's see how to create users and groups, and assign privileges to them.

Creating users and groups

Users created in WebGUI can be used for authentication in services such as Captive Portal, the proxy, IPsec, and OpenVPN and will be authenticated to log into OPNsense using the **Command-Line Interface (CLI)** or WebGUI.

The users can be managed using the **System | Access | Users** menu:

Figure 4.1 – System | Access | Users page

Some important properties are shown in the screenshot:

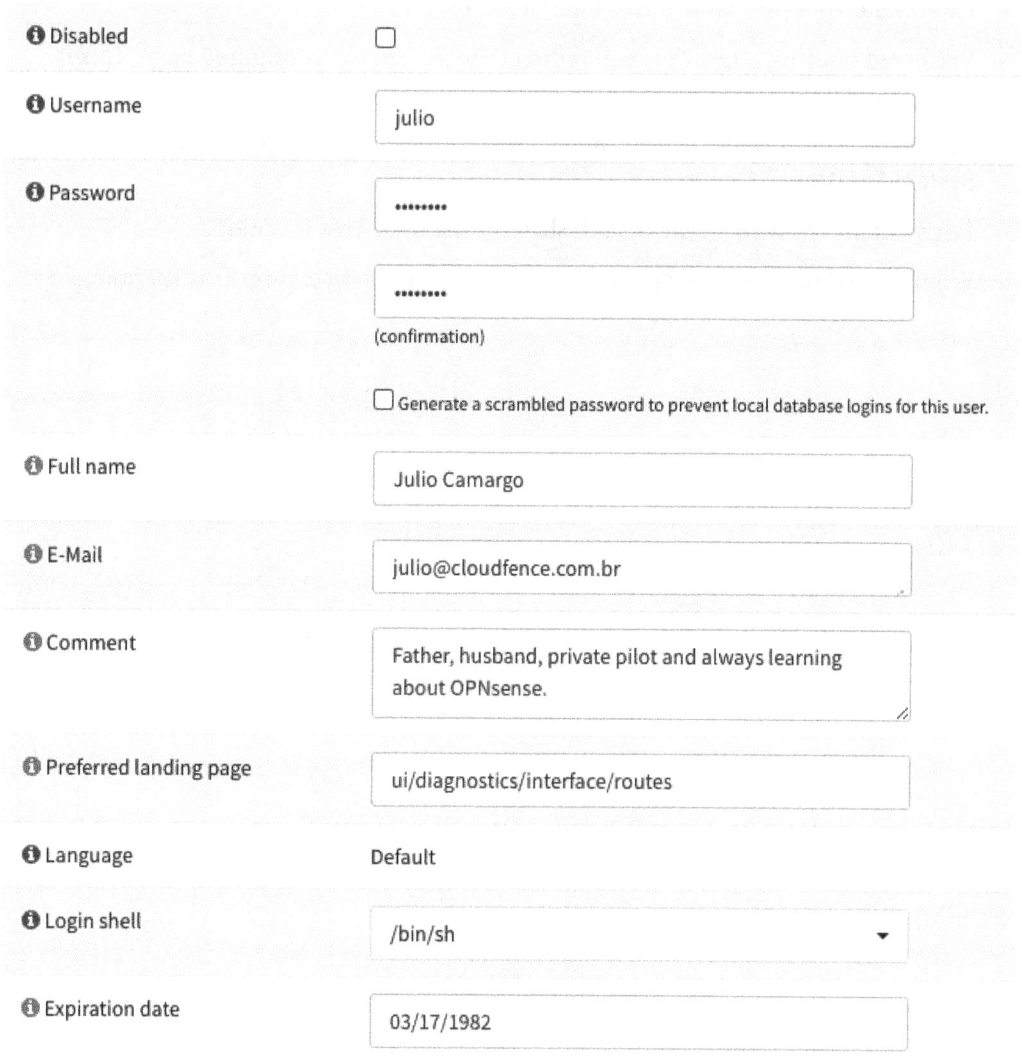

ⓘ Disabled   ☐

ⓘ Username
> julio

ⓘ Password
> ••••••••

> ••••••••

(confirmation)

☐ Generate a scrambled password to prevent local database logins for this user.

ⓘ Full name
> Julio Camargo

ⓘ E-Mail
> julio@cloudfence.com.br

ⓘ Comment
> Father, husband, private pilot and always learning about OPNsense.

ⓘ Preferred landing page
> ui/diagnostics/interface/routes

ⓘ Language   Default

ⓘ Login shell
> /bin/sh   ▾

ⓘ Expiration date
> 03/17/1982

Figure 4.2 – User account properties

The following properties are depicted in the preceding screenshot:

- **Disabled**: This option will disable the user account.
- **Username**: The name that will identify the user.
- **Password**: The password that will be set for the user in the local database.
- **Full name**: The user's full name. This is optional.

- **E-Mail**: The user's email address. This is optional.

- **Comment**: An optional comment about the user.

- **Preferred landing page**: This will redirect the user to the configured page. This is optional, and privileges are required to access the page.

- **Language**: This is optional and can't be set to users individually yet. You can set it in **System | Settings | General**.

- **Login shell**: This will be the default shell for the user. This is optional.

- **Expiration date**: The date on which the user account will expire. This is optional.

Figure 4.3 – User account properties

- **Group Memberships**: Groups that the user can be associated with. This is optional.

- **Effective Privileges**: The user's WebGUI privileges, which are mandatory. We will explore these later in this chapter.

- **User Certificates**: User certificates that can be used in some OpenVPN setups, for example.

- **API keys**: The API keys and secrets are used to access the web API. We will explore it later in *Chapter 20, API – Application Programming Interface*. This is optional.

- **OTP seed**: Generate a **One-Time Password (OTP)** seed to be used by **Two-Factor Authentication (2FA)** such as Google Authenticator or the FreeOTP app, for example. This is optional.

- **OTP QR code**: If you selected the **Generate new secret** (160 bit) option, after clicking on the **Save** button, this option will be shown. To see the generated QR code, required by the OTP app, just click on the **Click to unhide** button and point your mobile camera to it to add it to your preferred OTP app.

- **Authorized keys**: This option is used for SSH access using a public key. This is optional.

- **IPsec Pre-Shared Key**: An optional IPsec pre-shared key used in an IPsec Road Warrior setup. This is optional.

Let's look at the user privileges.

User privileges

The user privileges are privileges related to WebGUI and define what a specific user can do or see while using it.

Figure 4.4 – Editing user privileges

After clicking on the pencil icon, a new page will show, as in the following screenshot:

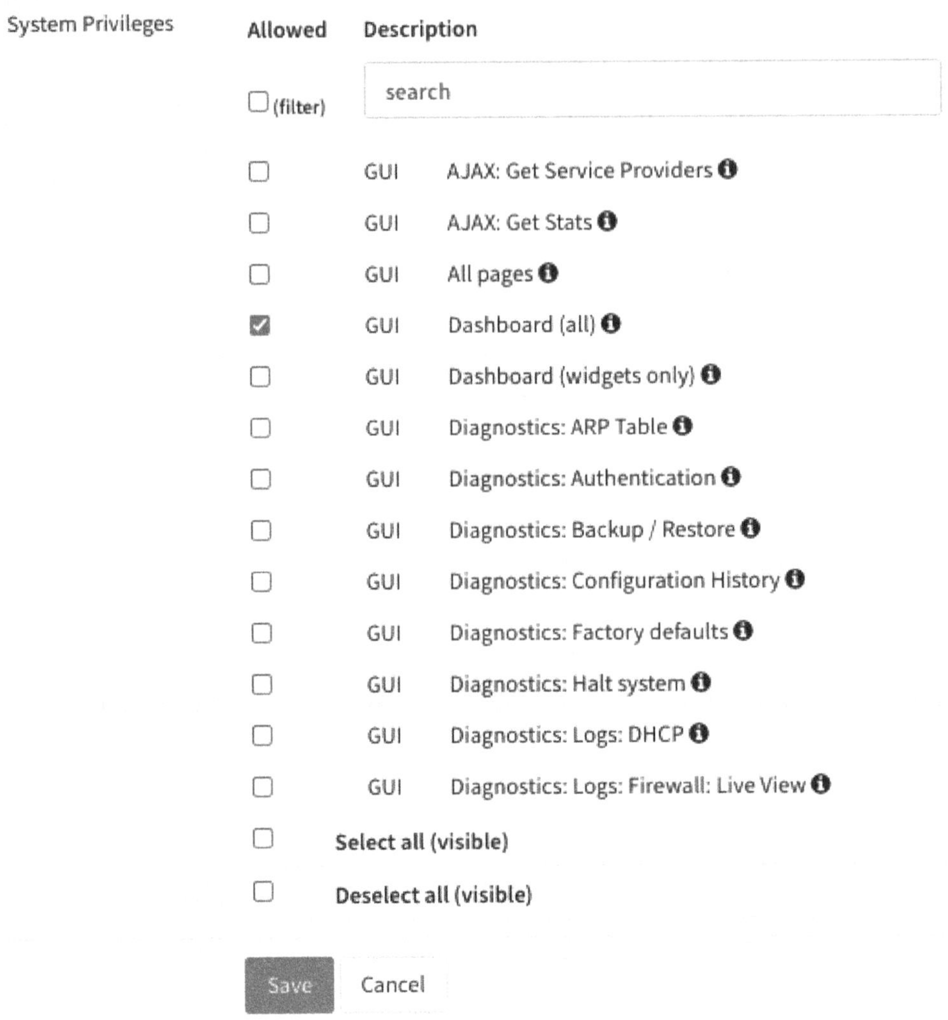

System Privileges

Allowed   **Description**

☐ (filter)   [search   ]

☐   GUI   AJAX: Get Service Providers ⓘ

☐   GUI   AJAX: Get Stats ⓘ

☐   GUI   All pages ⓘ

☑   GUI   Dashboard (all) ⓘ

☐   GUI   Dashboard (widgets only) ⓘ

☐   GUI   Diagnostics: ARP Table ⓘ

☐   GUI   Diagnostics: Authentication ⓘ

☐   GUI   Diagnostics: Backup / Restore ⓘ

☐   GUI   Diagnostics: Configuration History ⓘ

☐   GUI   Diagnostics: Factory defaults ⓘ

☐   GUI   Diagnostics: Halt system ⓘ

☐   GUI   Diagnostics: Logs: DHCP ⓘ

☐   GUI   Diagnostics: Logs: Firewall: Live View ⓘ

☐   **Select all (visible)**

☐   **Deselect all (visible)**

[Save]   Cancel

Figure 4.5 – The System Privileges page

Now that we have learned how to create and manage users, it's time to see how to group users.

Proposed exercise – creating a new user and assigning privileges to them

To practice how to create a new user and assign WebGUI privileges, follow these steps:

1. Go to **System | Access | Users** and click on the **+ Add** button.
2. Fill in the username and password, you can leave the other options with the default values, click on **Save**, and then click the back button.
3. In the new user row, click on the pencil icon to edit the user.

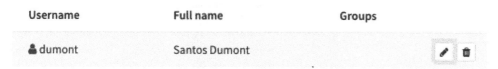

Figure 4.6 – Editing the user

4. In **System Privileges**, select only **Dashboard (all)**.

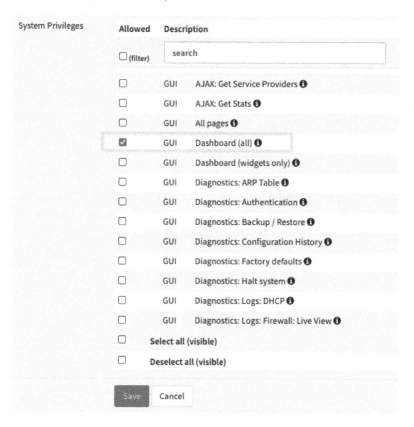

Figure 4.7 – Assigning user privileges

5.   Now try to log out and log in again with the new user we just created.

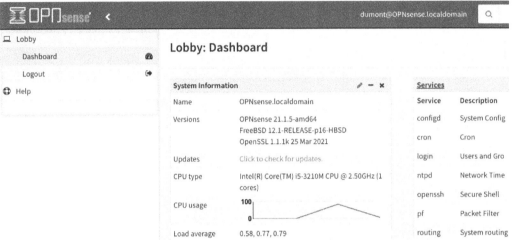

Figure 4.8 – New user dashboard access

Note that the new user only has access to the **Dashboard** screen. Now you can practice selecting other privileges.

> **Important Note**
> If you don't select any privileges, the user won't be able to log into WebGUI.

Adding groups

Groups, as the name suggests, will help us to group users and apply privileges to them, instead of defining each one individually. As with user privileges, we can only assign privileges to a group its users can access on WebGUI.

To add a new group, go to **System | Access | Groups** as shown in the following screenshot and click on the **+ Add** button:

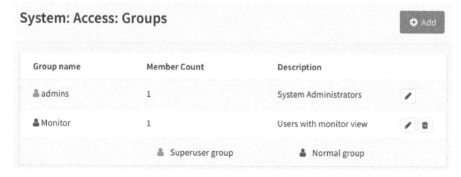

System: Access: Groups

ⓘ Defined by

ⓘ Group name
> Monitor

ⓘ Description
> Users with monitor view

ⓘ Group Memberships

Not Member Of
> root
> dumont

Member Of
> julio

Save   Cancel

Figure 4.9 – Adding a new group page

The following are the options depicted in the preceding screenshot:

- **Group name**: This is the name of the group.

- **Description**: Enter a description of the new group.

- **Group Memberships**: Select which users will be part of this group.

As usual, when you finish creating the group, just click on the **Save** button. As you can see in the following screenshot, the new group is listed as **Normal group**:

System: Access: Groups   ● Add

Group name	Member Count	Description
🔒 admins	1	System Administrators	
🔒 Monitor	1	Users with monitor view	

🔒 Superuser group   🔒 Normal group

Figure 4.10 – List of existing groups

This means that it doesn't have privileges to all WebGUI pages like the admins, for example. We need to give some privileges to this new group since it was created with no privileges assigned.

Assigning group privileges

To assign privileges to a group, you must click on the pencil icon on the **Groups** page. On the group editor page, now the **Assigned Privileges** option will be visible. Just click on the pencil icon to assign group privileges:

Figure 4.11 – Editing group privileges

In the system privileges, try selecting only **All pages** and click on the **Save** button and this will take you back to the group editor page. Click on **Save** again. As you can see in the following screenshot, the new group is shown as **Superuser group**, because we selected the **All pages** option:

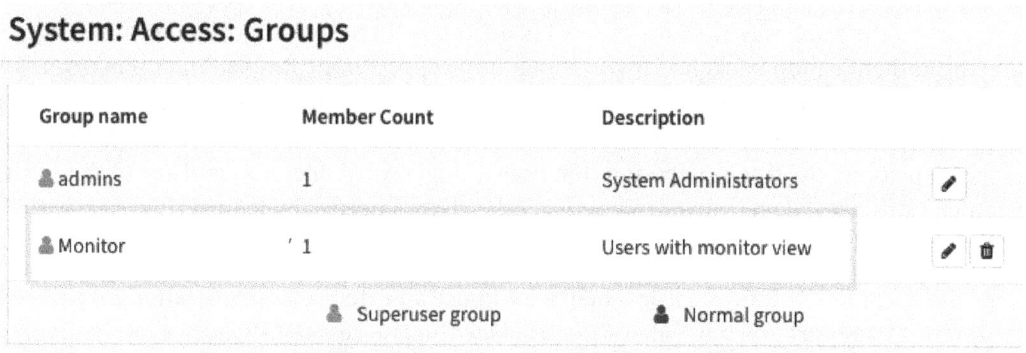

System: Access: Groups

Group name	Member Count	Description		
👤 admins	1	System Administrators	✏️	
👤 Monitor	' 1	Users with monitor view	✏️	🗑️
	👤 Superuser group	👤 Normal group		

Figure 4.12 – Newly created group

We just did this as an example to show the difference between a normal group and a superuser group. Now you can go back and edit your new group with the privileges you want.

We have learned how to manage users from the local database. Now, let's see how to use external authentication services.

External authentication

Besides local database authentication, OPNsense also supports an external authentication backend, such as **Microsoft Active Directory** or **OpenLDAP**, for example. The currently supported protocols are **RADIUS** and **LDAP**. There is a special authentication backend that is only used for the Captive Portal service: *Voucher Server*, which we will explore in detail in *Chapter 14, Captive Portal*. You can combine these backends with a **Time-Based One-Time Password** (**TOTP**) using Google Authenticator, for example, to enable **2FA**.

If you aren't familiar with any of these protocols, you might be asking, *When do I need to use an external authentication backend?* Let's start with one common example.

VPNs

Try to imagine the following scenario: You need to set up a new VPN tunnel that will be used by one of your customers. The IT team told you that they need to provide secure access to employees – a few hundred. Most of them will work from home, but they already access all network resources using their Microsoft Active Directory credentials and this must work in the same way with the new VPN access. Is there another option other than using the authentication backend? If you thought to suggest importing all the users from the Microsoft Active Directory server, this may appear to help at first glance, but the credentials would not be synced with the privilege groups and the password changes will be forgotten about! So, think twice and don't say a single word. Just nod and say to the customer, *Consider it done!* How can you meet the client's needs? By using OPNsense with external authentication, of course!

So, for the client to site tunnels, it will help a lot to use an external authentication service. In this way, IT support and administration teams without access to OPNsense can manage the users just by accessing the authentication server. Cool, huh? That means peace for you and you'll save time because you don't have to deal with support tickets such as *please change the password of so-and-so user*.

Currently, the VPN services that can use external authentication on OPNsense are OpenVPN and IPsec. Other services that can use the external authentication backend are Captive Portal, the proxy, and some plugins such as NGINX as well.

You can also use external authentication to manage users that will be allowed to log into WebGUI and SSH with limited access.

Enabling external authentication

To enable an external authentication server, go to **System | Access | Servers** and click on the **+ Add** button.

The following fields will be available:

- **Descriptive name**: The name of the authentication server.
- **Type**: The options are **LDAP, LDAP + Time-based One Time Password, RADIUS, Local + Time-based One Time Password**, and **Voucher**. They are described in the following section.
- **LDAP**: Its server will use the LDAP protocol to authenticate the users. Common LDAP servers are Microsoft Active Directory, OpenLDAP, and Red Hat's 389 Directory Server.

LDAP options

The following are the available LDAP options:

- **Hostname or IP address**: Fill this with the IP address or hostname of the LDAP server.

- **Port value**: The LDAP port configured in the LDAP server. The default is 389. For security reasons, it's recommended to use **LDAPS** (**LDAP over SSL**), which uses TCP/ port 636.

- **Transport**: This is the protocol mode. It can be defined as **TCP - Standard**, which is used for unencrypted communication, or **StartTLS** and **SSL – encrypted** communications.

- **Protocol version**: This can be configured using version 2 or 3 of LDAP.

- **Bind credentials**: The user credentials used to authenticate queries on the LDAP server.

- **Search scope**: This can be configured as **One Level** or **Entire Subtree** in the LDAP tree.

- **Base DN**: This is the LDAP distinguished name. Usually, you will fill it with your internal domain configured in the LDAP server, for example, FQDN cloudfence.intranet becomes dc=cloudfence,dc=intranet in the base DN format.

- **Authentication containers**: This click-and-select option will list the existing **Organizational Units (OUs)** or the **Common Name (CN)** components from the configured server.

- **Extended Query**: This is a special syntax query that will limit the results for the LDAP server.

- **Initial Template**: Select which type of template your server is compatible with: OpenLDAP, Microsoft AD, or Novell eDirectory.

- **User naming attribute**: This may vary depending on the selected option in **Initial Template**. It will use **cn** if the selected options are OpenLDAP or Novell eDirectory and **sAMAccountName** for Microsoft Active Directory. You can also type a custom option in this field, which is only recommended if you know what you are doing.

- **Read properties**: This option will enable fetching the account details after a successful login.

- **Synchronize groups**: This option depends on the preceding option and will synchronize groups that are specified in the `memberOf` attribute on the server. Consider using it if you've already created a group with the same name in OPNsense to synchronize which users will be part of the synchronized group.

- **Limit groups**: This multi-selection option will define which groups will be used by the LDAP protocol to sync with local ones.

- **Match case insensitive**: This option will turn the case sensitivity off and will allow mixed case input while getting local user settings.

- **LDAP + Time-Based One Time Password**: This option will enable external LDAP authentication plus the OTP mechanism using the Google Authenticator app, for example. You must enable the **OTP seed** option for the users to be able to use it.

The extra options when using LDAP and time-based OTP are as follows:

- **Token length**: This is the length of the token in digits. The options are **6** or **8**.

- **Time window**: The time that the token will be valid for before changing to a new one. The default time is 30 seconds.

- **Grace period**: The time difference that will be allowed between the token in the user app and the server.

- **Reverse token order**: The default will require the user to type the token before the password. Checking this option will reverse this order, which can make things easy for the user. The token will be valid for only 30 seconds. In this way, the user can type the password without being concerned about the token changing.

- **Local + Time-based One Time Password**: This will enable the OTP feature of the local database authentication.

- **RADIUS**: This will use RADIUS as the authentication protocol.

The following are the available options for the RADIUS protocol:

- **Hostname or IP address**: Fill this with the IP address or hostname of the RADIUS server.

- **Shared Secret**: This is the shared secret configured in the RADIUS server.

- **Services offered**: The available options for this are **Authentication** to just authenticate users in the RADIUS server, and **Authentication and Accounting**, which will send a request packet to the RADIUS server to start the accounting session. This is useful while using Captive Portal with RADIUS authentication.

- **Authentication port value**: The configured port in the RADIUS server.

- **Accounting port value**: The configured port in the RADIUS server.

- **Authentication Timeout**: The timeout for the RADIUS server to respond to requests.

- **Voucher**: This option is used in combination with Captive Portal. We'll explore it in *Chapter 14, Captive Portal*.

Now we have explored how to manage users, groups, and authentication, it's time to look at digital certificates and how to use them in OPNsense.

Certificates – a brief introduction

OPNsense uses certificates to ensure secure communication between nodes in services such as OpenVPN and IPsec, and HTTP services such as Captive Portal, web proxies, and WebGUI.

The available types of certificates in OPNsense are the following:

- **Certificate Authority (CA)**: A CA is a trusted entity that issues trusted certificates. On OPNsense, this will be used to issue self-signed certificates.

- **Server type certificate**: This certificate must be used on the server side of encrypted communication. On OPNsense, you will enable it in services such as OpenVPN or WebGUI, for example.

- **Client type certificate**: This type of certificate will be used on the other side of client-server encrypted communication. On OPNsense, we use it, for example, on OpenVPN clients' certificates and on some plugins such as NGINX or HAProxy.

- **Combined Client/Server**: This type of certificate can be used on both the server side and client side.

We will dive into the entire process of creating and using certificates in *Chapter 8, Virtual Private Networking*.

Meanwhile, let's move on to learn how to configure the OPNsense general settings.

General settings

In the **System | Settings** menu, you will find several submenus that will allow you to configure some settings in OPNsense. Let's go through them one by one. To access each one, access **System | Settings | <Submenu's name>** as follows.

The administration page

On this page, we will find options related to accessing OPNsense and administration such as WebGUI and SSH configuration and authentication options.

WebGUI options

The following options are available in WebGUI:

- **Protocol**: Which protocol will be used to access WebGUI: HTTP or HTTPS? It isn't recommended that you use HTTP as it will pass all data in cleartext, which means no security; prefer HTTPS instead.

- **SSL certificate**: The certificate that will be used to enable HTTPS to have access to WebGUI.

- **SSL ciphers**: The ciphers that will be used to access WebGUI. This can limit access from old web browsers, for example. Caution: If you select ciphers incompatible with your available browsers, you could lose access to WebGUI.

- **HTTP Strict Transport Security**: Known as **HSTS**, this policy mechanism can help to protect against **man-in-the-middle** attacks such as protocol downgrade and cookie hijacking. It is disabled by default.

> **Note**
>
> To learn more about HSTS, please refer to `https://developer.mozilla.org/en-US/docs/Web/HTTP/Headers/Strict-Transport-Security`.

- **TCP port**: The port used by WebGUI. It is desirable to change this when you plan to use the 443 port for another service running in the same IP used by OPNsense, such as a NAT rule pointing to a web server, or an NGINX or HAProxy plugin installed and in use. I would recommend you always change it to another port number that is unpredictable.

- **HTTP redirect**: When checked, this option will disable automatic redirection while trying to access WebGUI with the default HTTP protocol in TCP port 80. This is another option to leave checked if you plan to use the HTTP port for another service.

- **Login messages**: When checked, this will disable logging after successful logins, which is a bad idea. When you want to trace and audit access to WebGUI, it is recommended to leave it as the default option: unchecked.

- **Session timeout**: The period that a WebGUI session will last before it expires; the default value is 4 hours.

- **DNS rebind check**: WebGUI has protection against DNS rebinding attacks as default. It is recommended to leave this unchecked. For more information about DNS rebinding attacks, please refer to `https://en.wikipedia.org/wiki/DNS_rebinding`.

- **Alternate hostnames**: In this option, you can fill in the hostnames that will bypass the DNS rebinding check.

- **HTTP compression**: The level of compression can be low, medium, or high. When compression is enabled, it will save bandwidth while transferring data from WebGUI to the client web browser but will demand more CPU power from the OPNsense hardware. If you are accessing the web-browser from a high-speed WAN or by LAN, it won't be necessary.

- **Access log**: This option will enable the access log file for WebGUI. The process that serves WebGUI access is `lighttpd`, thus if you enable it, the access log will be available in `/var/log/lighttpd.log` and can be accessed in real time using the `clog -f /var/log/lighttpd.log` command. If you want to audit access to WebGUI, I recommend you enable this option.

> **Note**
>
> In the old OPNsense versions (until 21.7 version), the default log format was the circular log (`clog`), which changes the manner to access the logs using the CLI:
>
> Ex.: `clog -f /var/log/lighttpd.log`

- **Listen interfaces**: This option will configure the WebGUI to listen for requests only in the selected interfaces. The default is to listen to all interfaces. Be careful changing that. If you are accessing from the LAN interface, for example, and unmark it here, you could lose access to WebGUI.

- **HTTP_REFERRER enforcement**: By default, WebGUI will check the **HTTP_REFERRER** header to protect itself from redirection attempts. If you use an application that doesn't work well with that, for example, password managers that access WebGUI from it, maybe it would be desirable to disable this option.

WebGUI caveats

Nowadays, it is common to read in cybersecurity news pages about flaws in WebGUI/ managers that allow attackers to gain privileged access to them, so I would advise you – even with the excellent security history of the OPNsense project – to avoid allowing WebGUI to be accessed from the internet (WAN interfaces). This could expose your firewall unnecessarily. Always prefer to use a VPN or limit the source IP in a firewall rule.

Another relevant topic is the default ports used by WebGUI, HTTP, and HTTPS – TCP/80 and TCP/443 (HTTP/3 should use the UDP protocol) respectively are common ports used by web servers that might be without the protection of OPNsense. So, to avoid ports conflicting, change the WebGUI default port and disable the HTTP redirect. Let's now learn how to enable CLI remote access using the Secure Shell protocol.

Secure Shell (SSH)

The SSH protocol is very popular, useful, and secure and it is the way to access the OPNsense CLI remotely. The available options to configure are the following:

- **Secure shell server**: Check this option to enable SSH access to OPNsense. This is disabled by default.

- **Login group**: The allowed groups that will have access through the SSH protocol. It is a good idea to limit these only to groups that really need access via SSH. Once allowed, these groups will have full SSH access, which means accessing files using SFTP or SCP methods that belong to the SSH protocol.

- **Root login**: This option will enable root access to SSH, which sometimes can be dangerous. You can leave it disabled and enable sudo (explained in the *Authentication* section) to log in with a normal user and change to the root user afterward or just disable password authentication.

- **Authentication method**: When checked, this will permit logging in using passwords, otherwise, you will need to configure authorized keys for each user that has SSH access.

- **SSH port**: The port that will listen to SSH requests.

- **Listen interfaces**: This option will configure the SSH service to listen for requests only for the selected interfaces. The default is to listen in on all interfaces.

- **Key exchange algorithms**: Select which algorithms will be used to generate keys used in each SSH connection.

- **Ciphers**: Select the ciphers that will be used to encrypt the SSH connections.

- **MACs**: The message authentication codes that the SSH service will use.

- **Host key algorithms**: Select which host key algorithms will be served by the SSH service.

Console access

The following are the available options on the **Console access** page:

- **Console driver**: When this is checked, OPNsense will prefer the *vt* terminal driver instead of the older *sc* driver.

- **Primary console**: The available options include **VGA console**, for displaying boot messages in a display output such as VGA, HDMI, DVI, and others. The **Serial console** option will display boot messages to an available serial port. This will require hardware support. The EFI console is available on modern hardware and will output messages to the display port configured in the hardware setup utility, also known as BIOS setup on older hardware. Only select this option if your hardware is UEFI compatible and if it is enabled on the hardware setup utility. The **Mute console** option, as the name suggests, will mute boot messages.

> **Important Note**
> If you select an option that isn't supported by the hardware on which OPNsense is installed, you could lose console access.

- **Secondary console**: If the hardware supports multiple console outputs, you can select a second one in this option. The same options from the primary console apply here.

- **USB-based serial**: Instead of using `/dev/ttyu0`, USB-based serial will use `/dev/ttyU(X)` for the respective connected USB serial console device, where X is the number of the connected device. For example, `/dev/ttyU0` for the first, `/dev/ttyU1` for the second, and so on.

- **Console menu**: When checked, this will protect the console access with a login prompt, requiring you to log in with an existing user.

Authentication

The authentication options are as follows:

- **Server**: Using centralized authentication makes sense to configure an external authentication backend here. This way, administrators can authenticate on OPNsense to manage it using an Active Directory authentication service, for example. If the authentication server or servers become unavailable, it's a good idea to let the local database always be selected as a fallback. The default option when none is selected is the local database.

- **Disable integrated authentication checkbox**: Check this option if you want to use only local accounts for system services such as SSH, console login, and others (based on Unix authentication).

- **Sudo**: This will enable sudo usage in the CLI and SSH access. You can select **disallow** (default) or the **Ask password** option to prompt the user to type the password every time sudo is invoked or **No password** to allow the user to call sudo without prompting for the password.

- **sudo group combo box**: The groups that will be allowed to use the sudo command. The special *wheel* group is always allowed. This is the root group.

On the **System | Settings | Administration** page, we can define how to connect to OPNsense to manage it, whether via the CLI, console, or WebGUI. Next, we will explore the **General** page options.

The General page

To access it, go to **System | Settings | General**. On this page, we can configure some basic system and network settings. Let's explore each of these.

System settings

The system options are the following:

- **Hostname**: The OPNsense hostname. Just type here the hostname without the domain or suffix part.

- **Domain**: Type here the network's domain.

- **Time zone**: Select the time zone that OPNsense will be configured in.

- **Language**: Select the preferred language that WebGUI will use.

- **Theme**: You can select here your preferred theme for WebGUI. Additional themes can be installed using plugin additions.

Network settings

The following are the networking options:

- **Prefer IPv4 over IPv6**: If OPNsense is installed on an IPv4 network, check this option to avoid that name resolution using IPv6 as the preferred address.

- **DNS server**: Here, you can set the DNS that OPNsense will use to name the resolution service.

> **Important Note**
>
> If you have multiple WANs configured and it will set the ISP's DNS servers, you should select the respective ISP gateway to avoid resolution requests flowing to another ISP that has been blocked at the destination DNS server.

- **DNS server options**: Check this option if you want WAN connections configured via DHCP or the PPP protocol to override the configured DNS servers. In the **Exclude interfaces** option, you can select which interfaces will be excluded from receiving DNS servers' configurations automatically. To not use local DNS services such as *Dnsmasq* and *Unbound* as the DNS server, check the **Do not use the local DNS service as nameserver for this system** option. This will only apply to OPNsense, not the entire network, for example, when OPNsense has the DHCP server enabled.

- **Gateway switching**: This is a very important option to check if you are using multiple WANs. If a WAN goes offline, then the next one available will be used as the default gateway.

Let's now discuss the logging process in OPNsense.

About OPNsense logging

OPNsense has inheritance from **pfSense**, which used the circular logging format. This means that the logs will have a predefined size, never exceeding it. This can be good when you need to keep small logs but isn't suitable when you want a longer event history in each log file. This applies to the core log files, which means that log files from plugins don't use the circular logging approach. Later in this chapter, we will see how to read circular logs in the CLI. Once in WebGUI, the visualization doesn't differ from regular log files.

Since version 20.7 of OPNsense, it is possible to disable the circular logs. By doing that, the one-file-per-day approach will be used for log file rotation. Let's see how to configure this and other logging options in OPNsense.

Go to **System | Settings | Logging** to open the log settings page:

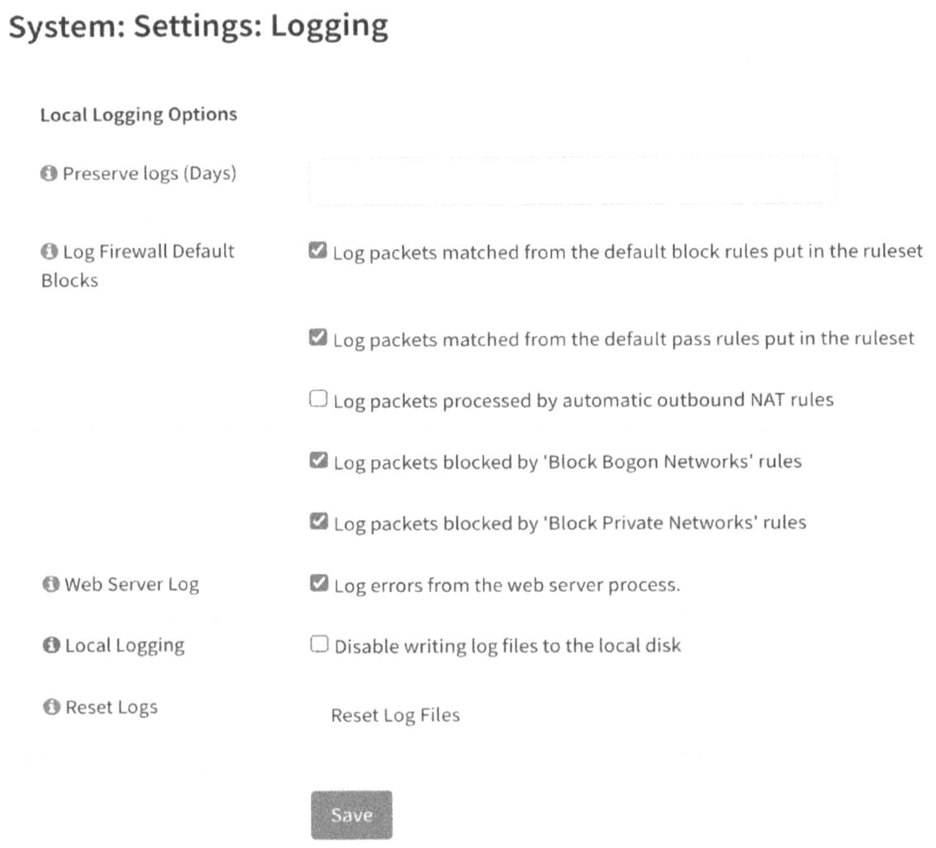

System: Settings: Logging

Local Logging Options

🛈 Preserve logs (Days)

🛈 Log Firewall Default Blocks

☑ Log packets matched from the default block rules put in the ruleset

☑ Log packets matched from the default pass rules put in the ruleset

☐ Log packets processed by automatic outbound NAT rules

☑ Log packets blocked by 'Block Bogon Networks' rules

☑ Log packets blocked by 'Block Private Networks' rules

🛈 Web Server Log   ☑ Log errors from the web server process.

🛈 Local Logging   ☐ Disable writing log files to the local disk

🛈 Reset Logs   Reset Log Files

Save

Figure 4.13 – Log settings page

The options in the log settings page are as follows:

- **Preserve logs**: This will set how many log files will be preserved, one per day. The default number of logs preserved is 31.

Note – Disabling the Circular Log

If you disable the circular log, pay attention to the size of your log files. They can become very large and then occupy the entire OPNsense disk. The size of log files can vary depending on the volume of traffic OPNsense is processing.

- **Log Firewall Default Blocks**: The options present here are as follows:

  - **Log packets matched from the default block rules put in the ruleset**: Checking this option will log every packet blocked by the default block firewall rule. You can always decide which firewall rule will be logged or not, adjusting the logging option present in each rule.

  - **Log packets matched from the default pass rules put in the ruleset**: Checking this option will log the implicit pass rules created by OPNsense automatically. *Wait... default pass rule? There is a hole in our firewall?!* Stay calm! There are some implicit rules such as permitting traffic, for example, when OPNsense needs to obtain a dynamic IP address in an interface configured as a DHCP client. So it is desirable for some pass rules to be created to avoid things from breaking. As Franco Fichtner says, *Stay safe. Don't worry, with OPNsense your network is safe and sound!*

  - **Log packets processed by automatic outbound NAT rules**: checking this option will log the automatic outbound NAT rules created by default on OPNsense installation.

  - **Log packets blocked by 'Block Bogon Networks' rules**: As the option's label says, it will block the packets dropped by the rules generated by the **Block Bogon Networks** option present in the network interface configuration, as we learned in *Chapter 3, Configuring an OPNsense Network.*

  - **Log packets blocked by 'Block Private Networks' rules**: As the option's label says, it will block the packets dropped by the rules generated by the **Block Private Networks** option, when the **Block Private Networks** option is checked on the network interface configuration.

- **Web Server Log**: OPNsense has an embedded web server used by WebGUI and by Captive Portal, which is the lighttpd web server. When this option has been checked, errors related to it will be presented in /var/log/system.log. You can also access this log by going to **System | Log Files | General**.

- **Local Logging**: Disable writing log files to the local disk (checkbox). Checking this option will disable all local disk logging. I don't recommend you do that, even with remote logging enabled. Without logs, all visibility of errors or other system events is lost.

- **Reset Logs**: Clicking on this button's option will empty all the local disk log files. It will also restart the DHCP service.

Let's now look at the remote logging options.

Remote logging

Remote logging is very useful when we want to preserve log entries, whether from losing them in a disk crash or from a system compromise due to a cyberattack. It is a good practice to enable remote logging in all managed hosts and send the log entries to a system that will process each log entry. For example, at *Cloudfence*, our SOC analyzes all log entries using a **Security Information and Event Management** (**SIEM**) system that helps us to detect and respond to threats. We can do that thanks to the remote logging feature present in OPNsense, protecting dozens of networks. If you are curious which system we are using to do that, look at the Wazuh project: `https://www.wazuh.com`. Another good option is to send logs to the ELK Stack, which you can create nice dashboards with. You can learn more about it by visiting `https://www.elastic.co`. If you need help in sending logs from OPNsense to the ELK Stack, look at Fabian Franz's repository on GitHub: `https://github.com/fabianfrz/opnsense-logstash-config`.

Now that I think that you are convinced that it's a good idea to enable remote logging, let's move on and see how to do that. Go to **System | Settings | Logging / targets** and click on the + icon to add a new syslog server, also called **target** here:

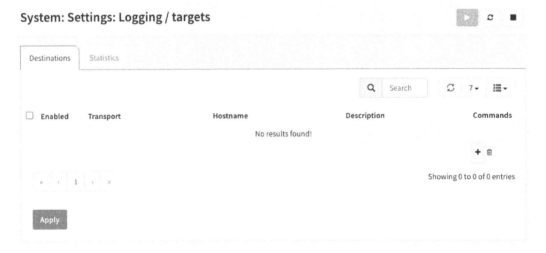

Figure 4.14 – Adding a new logging target

A new page will appear with details that must be filled in to configure the remote logging target as shown in the following screenshot:

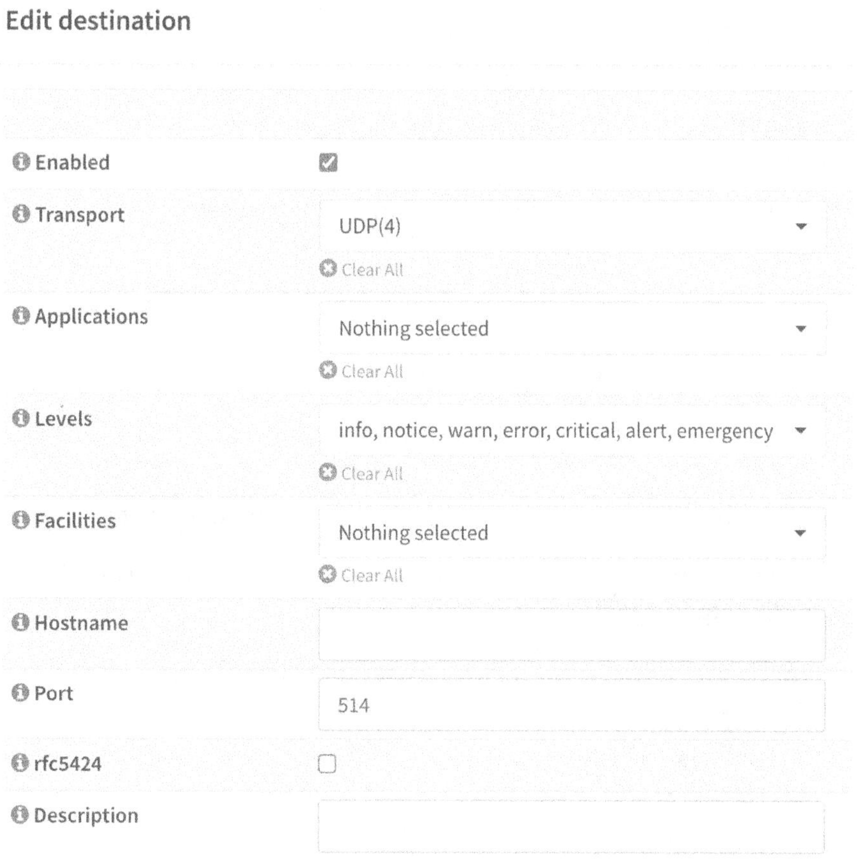

Edit destination

❶ Enabled	☑
❶ Transport	UDP(4) ▾
	⊗ Clear All
❶ Applications	Nothing selected ▾
	⊗ Clear All
❶ Levels	info, notice, warn, error, critical, alert, emergency ▾
	⊗ Clear All
❶ Facilities	Nothing selected ▾
	⊗ Clear All
❶ Hostname	
❶ Port	514
❶ rfc5424	☐
❶ Description	

Figure 4.15 – Configuring a new logging target

Before detailing each field, it is important to say that we will not cover here how to set up and configure a remote logging server, but if you want, you can do that using a syslog server such as **syslog-ng**, **Graylog**, or any other syslog compatible server. Googling it, for syslog, you will find a lot of options to use as a syslog-server. OPNsense uses the **syslog-ng** server to send remote events using the syslog standard. At `https://en.wikipedia.org/wiki/Syslog-ng`, you will find explanations about terms used in syslog message logging, such as facility, severity, or level. If you don't know anything about syslog messages, then I recommend you spend some time reading about this before continuing.

The following are the fields present in the preceding screenshot:

- **Enabled**: Check this to enable this remote logging target.

- **Transport**: Select between the **UDP** and **TCP** options. They are available in the IPv4 and IPv6 versions.

- **Applications**: Select which OPNsense internal applications will send events to this remote logging target.

- **Levels**: Select which levels will be included in the logging events that will be sent to the remote target.

- **Facilities**: Select the facilities that will be configured for the select remote logging services.

- **Hostname**: Fill this with the logging server IP or hostname.

- **Port**: Fill this with the port used by the remote logging server. The default for syslog is UDP 514.

- **Description**: An optional description to help you identify this server.

Lastly, before we move on from the logging topic, let's look at the log visualization options available in the **System | Log Files** menu:

- **Backend**: Here, you can visualize `/var/log/configd.log`, which is the log of OPNsense's backend daemon.

- **General**: This page will show you the `/var/log/system.log` events. This log will show general system events.

- **Web GUI**: This page will output `/var/log/lighttpd.log`, which is the WebGUI web server.

You can always check the files presented in this chapter in the CLI using the `clog` command while circular logs are enabled. If they're disabled, then change the command to `tail`.

An example of reading the `system.log` CLI is shown here.

Circular logs can be enabled, using the following command:

```
tail -f /var/log/openvpn/openvpn_`date *%Y%m%d`
```

Circular logs can be disabled, using the following command:

```
tail -f /var/log/system/system_20210605.log
```

Now that we have explored the OPNsense general settings and logging, we are ready to learn about advanced settings available in WebGUI.

Advanced settings

OPNsense's WebGUI is a powerful management interface that allows us to configure even the most advanced features available in the FreeBSD operating system. Let's explore them and learn how to optimize and customize our firewall with available advanced options.

In this topic, we will explore two menu options: **System | Settings | Miscellaneous** and **System | Settings | Tunables**. Let's start with the first one.

On the **Miscellaneous** page, we will find options related to hardware, system file backups, disk and memory settings, and the hardware buzzer control when it is present on the hardware.

Cryptography settings

There are the following cryptography options:

- **Diffie-Hellman parameters**: These cryptographic keys exchange methods used in services such as VPN tunnels. In the default configuration, it will be updated at least twice a year by software updates. If you need to change it from the default option, you will be able to set it as RFC7919, which defines static recommendations. Otherwise, it can be set to a renewal period such as **Weekly**, **Monthly**, or **Custom**, which will require that you create a custom CRON entry at **System | Settings | Cron**. Only change it if you know what you are doing.

- **Hardware acceleration**: Here, you can set a cryptographic accelerator if present in OPNsense hardware. As we discussed back in *Chapter 3*, *Configuring an OPNsense Network*, in the *Hardware considerations* topic, you can, for example, select the **AES-NI CPU-based Acceleration** option if the CPU supports it. If your OPNsense hardware doesn't support any of the listed options, choose **None**.

Thermal sensors

Hardware requirements: it will load the driver to read the CPU's temperature. For Intel CPUs, select **Intel Core CPU on-die thermal sensor (coretemp)**. If the CPU is an AMD, then select **AMD K8, K10 and K11 CPU on-die thermal sensor (amdtemp)**, otherwise, leave it as the default **None/ACPI** option, then the operating system will try reading the temperature from the motherboard sensor if it has **Advanced Configuration and Power Interface (ACPI)** support.

Periodic backups

In the following options, you can select how many times you want OPNsense to run a backup of the data to be restored in the next boot. These options are especially useful when OPNsense is running based on a nano image that saves the volatile data only in RAM or when the power is cut off and data not saved in the disk will be lost:

- **Periodic RRD Backup**: This option is related to the graphs presented in the **Reporting** menu.

- **Periodic DHCP Leases Backup**: This option is related to the DHCP server leases data when it is enabled.

- **Periodic NetFlow Backup**: This option is related to the NetFlow data, which will only save data if this service is enabled. You can configure NetFlow at **Reporting | NetFlow**.

- **Periodic Captive Portal Backup**: This option is related to Captive Portal. When it is enabled, it will save the session data.

Power saving

The power saving options are related to power control. There are modes available and three options, each one representing a power mode: **on AC**, **on Battery**, and **on Normal**. This last one is used when the power control can't determine whether the power source is connected.

The power modes are as follows:

- **Adaptive**: This mode will try to balance power saving and performance. The performance will decrease while the system is idle and increase when it is under load.

- **Hiadaptive**: This works similarly to **Adaptive**, but in this mode will prioritize the system performance over power saving.

- **Minimum**: This will save power but with degraded hardware performance.

- **Maximum**: This will set the power control to the maximum hardware performance.

Disk/memory settings (reboot to apply changes)

The following are the options available in the disk/memory settings:

- **Swap file** – the **Add a 2GB swap file to the system** checkbox: This option will do as the name suggests, adding a swap file to the system. Select it only if your OPNsense installation doesn't create a swap partition.

- **/var RAM disk - Use memory file system for /var**: If this is checked, this option will move the `/var` directory to a disk mounted on RAM, which means that in the event of a reboot, all data in this directory will be lost. My advice is that it isn't a good idea to use it.

- **/tmp RAM disk**: The same as the previous option but related to the `/tmp` directory.

System sounds

Startup/Shutdown Sound - Disable the startup/shutdown beep: Check this if you no longer want to hear the sound played when OPNsense finishes the boot or when it starts the shutdown process.

Let's move on to the **System | Settings | Tunables** page to have a brief introduction to what we can set for the FreeBSD low-level settings.

Tunables

When FreeBSD boots into multi-user mode, which is the default mode for OPNsense, it reads a configuration file, `/etc/sysctl.conf`, and there we can set a lot of low-level settings that allow us to tweak the entire system behavior.

The OPNsense default configuration brings a lot of sysctl settings, or tunables as we will call them here. These tunables can configure settings such as network card driver options, TCP/IP protocol stack behavior, and so on. To get a complete list, you can run the `sysctl -a` command in the CLI.

> **Tunables Important Note**
>
> Only change settings on the **Tunables** page if you really know what you are doing. Common examples of changing tunable settings are recommended settings for specific hardware drivers, such as network interfaces or a network performance tweak. It is recommended that an experienced FreeBSD professional handles these settings.

If you want to learn more about FreeBSD sysctl and other advanced configurations, I recommend you look at Packt's available FreeBSD books.

Configuration backup

OPNsense saves all the configuration settings in an XML file, `/conf/config.xml`, and it is extremely important to save it in a safe place that will allow you to restore this configuration if it is necessary.

OPNsense offers some embedded options to save this configuration in external cloud drives such as Google Drive and Nextcloud (using an additional plugin). This book will not cover the configuration of how to use backup configuration, but you can find information at `https://docs.opnsense.org/manual/how-tos/cloud_backup.html`. Some plugins can help with this task also; check out the OPNsense plugins list at **System | Firmware | Plugins**.

To back up your system configuration files, you can go to **System | Configuration | Backups**.

System: Configuration: Backups

Download

☑ Do not backup RRD data.
☑ Encrypt this configuration file.

Password

Confirmation

Download configuration

Click this button to download the system configuration in XML format.

Restore

Restore area:

ALL

Choose File   No file chosen

☑ Reboot after a successful restore.
☑ Configuration file is encrypted.

Password

Restore configuration

Open a configuration XML file and click the button below to restore the configuration.

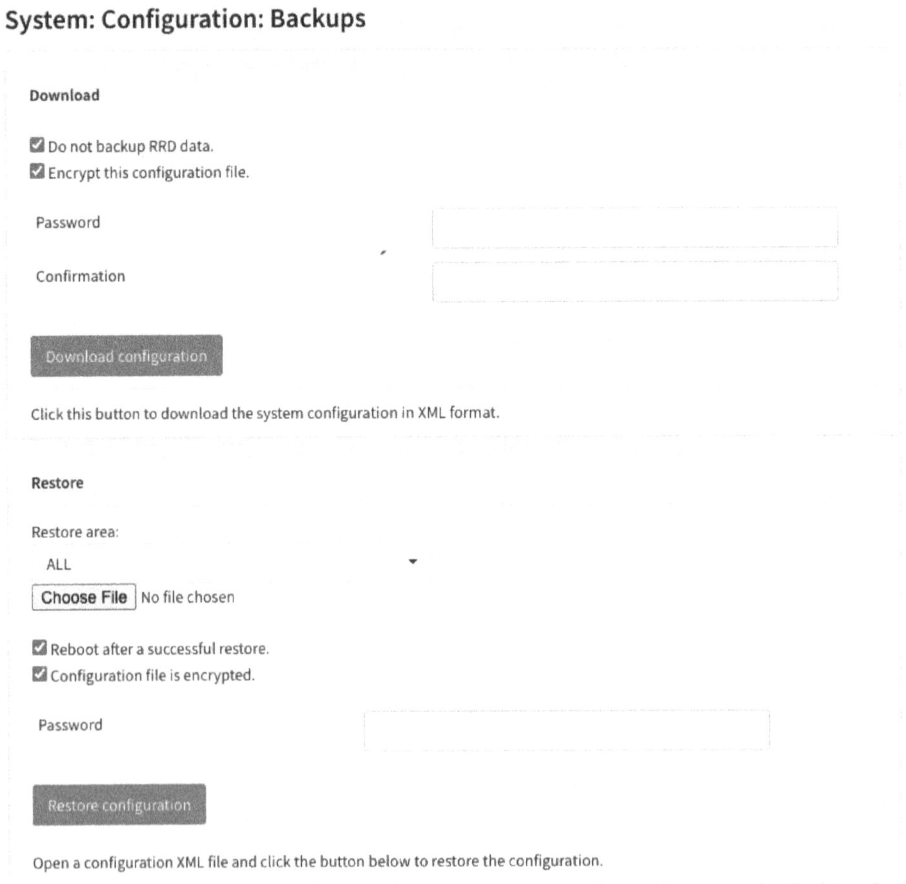

Figure 4.16 – Configuration backup and restore page

Click on the **Download configuration** button to download the configuration file to your computer's local disk. Check **Encrypt this configuration file.**, if you want to encrypt it. Doing that requires that you type a password twice. Fill in the **Password** and **Confirmation** textboxes.

To restore the file is quite simple. On the same page, you will find the **Restore** options. Click on the **Choose File** button, choose the backup `config.xml` file on your computer, and then click on the **Restore configuration** button. A system reboot is recommended, so you can leave the **Reboot after a successful restore.** option checked. If you saved this configuration file using encryption, then you need to check the **Configuration file is encrypted.** option and then fill it with the password used while backing up this configuration.

In the **Restore area** option, you can select which configurations you want to restore. If you need a full restore, just leave it as the default option: **ALL**.

Another very useful backup and restore feature is the configuration history. To use it, go to **System | Configuration | History**:

History

View differences | To view the differences between an older configuration and a newer configuration, select the older configuration using the left column of radio options and select the newer configuration in the right column, then press the button.

Diff	Date	Size	Configuration Change	
●	6/5/21 20:28:24	46 KB	root@192.168.56.1: /diag_logs_settings.php made changes	Current
● ○	6/5/21 17:16:47	46 KB	root@192.168.56.1: /system_general.php made changes	2nd 1st [icons] 3rd
○ ○	6/5/21 10:09:04	46 KB	(system): /usr/local/opnsense/mvc/script/run_migrations.php made changes	[icons]
○ ○	5/30/21 20:41:36	46 KB	root@192.168.56.1: /system_general.php made changes	[icons]
○ ○	5/30/21 20:39:57	46 KB	root@192.168.56.1: /system_general.php made changes	[icons]

Figure 4.17 – Configuration history page

This page will contain the last 100 configuration file versions. It can save you when you made a change by mistake and need to restore the configuration fairly quickly. To restore, just go to the line before your modification and click on the restore icon button (the first icon). It will restore the configuration without a reboot. Very nice, huh? I'm sure that this feature has saved a lot of jobs!

The other two icons are to exclude the configuration version – the trashcan icon (the second icon), or to download it (the third icon).

Figure 4.18 – Comparing configuration file versions

You can also see the differences between two configuration versions, in a different format, on this page. To do that, just select the two versions you want to compare and then click on the **View differences** button.

Summary

In this chapter, we've learned how to configure the OPNsense system settings, create users and groups, and how to back up the system configuration. You are now able to change OPNsense settings, add users and groups with different levels of privileges, enable external authentication for remote users' authentication in OPNsense, and change low-level settings using tunables. And you can do all these system modifications with confidence by doing a system backup before applying changes.

In the next chapter, we will dive into firewalling concepts and features available on OPNsense. We will learn how to manage rules, change firewall settings when necessary, and troubleshoot common issues using diagnostic tools and logs.

Section 2: Securing the Network

This part will dive into firewalling concepts and practices, how to filter packets, how to use network address translation, traffic shaping, setting VPNs, load balance and failover multi-WANs, and will also explore some reposting tools available on OPNsense.

This part of the book comprises the following chapters:

- *Chapter 5, Firewall*
- *Chapter 6, Network Address Translation (NAT)*
- *Chapter 7, Traffic Shaping*
- *Chapter 8, Virtual Private Networking*
- *Chapter 9, Multi-WAN – Failover and Load Balancing*
- *Chapter 10, Reporting*

5
Firewall

If you got this far, congratulations! We'll explore one of the main features of **OPNsense** – the **firewall**! In this chapter, we will learn which packet filter system OPNsense uses for firewalling, what type is it, how it works, what aliases or **Packet Filter** (**pf**) tables are, create our first rule, and explore diagnostics and troubleshooting. Through this chapter, you'll learn the firewalling basics that will be required to move on in this book, so pay attention reading it, practice the suggested exercises, and if you are unsure about some of the concepts presented here, please try to read it twice – it will be very important for you in the next chapters.

There will also be practical elements, so start your OPNsense virtual machine to follow all the topics explored here:

- Understanding firewalling concepts
- Firewall aliases
- Firewall rules
- Firewall settings
- Diagnostics and troubleshooting

Technical requirements

This chapter requires basic knowledge of networks, such as networking (for example, IP and IPv6) and transport protocols (for example, **Transmission Control Protocol (TCP)**, **User Datagram Protocol (UDP)**, **Internet Control Message Protocol (ICMP)**, and ICMPv6), how they work, how to execute commands in the CLI, and a running OPNsense machine.

Understanding firewalling concepts

The word *firewall* is one of the most used ones to define OPNsense; even with a lot of other features, it is very common to hear from someone curious about your network topology asking, *which firewall are you using in the network?* The firewall feature is so important that it defines a whole network security platform. Let's find out why, beginning with the basics.

A stateful firewall

Every connection that a stateful firewall permits to pass will create a **connection state**, which means that the firewall will monitor all the connection information, such as the source, the destination, the protocol, the port number, and the protocol state. The protocols that a stateful firewall can handle are the ones that run on layers 3 and 4, using the **OSI model** as a reference. OPNsense running only with core features is considered a stateful firewall.

For example, monitoring the connection states will prevent common attacks that use the packet spoofing technique. If a packet is sent to a stateful firewall with an established state but doesn't exist in the firewall state table, then it will be blocked by the firewall.

> **Note**
> By default, OPNsense reserves 10% of the system RAM to the firewall state table; this setting can be changed, as we will see in the *Firewall settings* section.

As OPNsense contains a stateful firewall, you're able to check connections states in **WebGUI** at **Firewall | Diagnostics | States Dump**.

You will be able to check every state that OPNsense is monitoring. It will show the following columns:

- **Int**: The network interface.
- **Proto**: The protocol.
- **Source | Router | Destination**: `Source address:port` | `(Network router:port)` | `Destination:port` (respectively).
- **State**: The connection state.
- The **X** icon: Clicking on this button will drop the state.

OPNsense has another page on WebGUI that summarizes the state table connections: **Firewall | Diagnostics | States Summary**.

On this page, you will be able to find connections summarized by **By Source IP**, **By Destination IP**, and **By IP Pair** (source and destination).

> **Note**
>
> Depending on the number of connections handled by your state table size, this page can take a long time to load.

Sometimes, it is necessary to reset all connections on the state table. You can do it without rebooting OPNsense; to do so, go to **Firewall | Diagnostics | States Reset**.

On this page, you have two checkboxes:

- **Firewall state table**: Checking this option (checked by default) will remove all connection entries from the state table, forcing connections to be recreated.

> **Note**
>
> It will reset your WebGUI connection too, so it will be necessary to refresh the page after resetting the connections.

- **Firewall source tracking**: This will remove all the connection associations (source/destination), which means that all hosts will need to reconnect to establish connections again.

To reset the connections state table, just click on the **Reset** button.

> **Note**
>
> Be cautious using this – it will interrupt connections between the hosts that are using OPNsense as a network gateway. This can cause a lot of complaints from your boss, so think twice before pressing this button!
>
> Now that we have learned what a stateful firewall is and how to check connection states, let's move on and talk about the packet filtering system used by OPNsense – the powerful pf!

The Packet Filter

pf's history starts from the OpenBSD project in 2001; it was designed to replace the **IPFilter**, which was removed from OpenBSD due to licensing concerns. The first FreeBSD version to have a ported version of pf was *5.3* in 2004, so, as you can see, this is a packet filtering platform with a long history, which brings reassurance of great reliability and trustworthiness. Despite FreeBSD's pf version originating from OpenBSD, the actual versions differ a lot on both platforms.

pf is what OPNsense uses to filter packets and to do network address translation, and you can find options related to it in WebGUI in the **Firewall** menu. It's not the purpose of this book to dive into pf's CLI utilities; one of the best features in OPNsense is its WebGUI, which makes managing firewalls very easy. But if you are interested in how to use pf command utilities, don't worry – during this chapter, we will learn a little bit about it.

Incoming and outgoing packets

This is an important thing to understand before we start creating firewall rules – what is an incoming packet? And what about the outgoing packets? Which ones are filtered by default? Let's understand from the firewall perspective first:

- **Incoming packets**: All traffic that enters the firewall through a network interface is considered an incoming packet. In regards to the firewall ruleset, we will use the direction of the traffic as a reference, so in this case, the incoming traffic will be considered as follows: `source | (in) firewall (out) | destination`.

> **Note**
>
> By default, all incoming traffic, if not matched with some `pass action` rule, will be blocked by the default *deny* rule.

- **Outgoing packets**: From a firewall perspective, traffic leaving will be considered as going in the outward direction by the pf system.

> **Note**
> By default, all outgoing traffic is allowed, if not matched with some block rule.

After finishing with these basic firewall concepts, let's learn about firewall aliases and pf tables.

Firewall aliases

To introduce you to **aliases**, let's start with a little story.

Let's imagine that you oversee an OPNsense firewall in a network with thousands of hosts, dozens of network segments, and a lot of VPN tunnels. This scenario will probably (considering good security practices) demand a lot of different rules to control host traffic, right? So, this will easily produce more than a thousand rules (yes – this part is based on a real example!), and managing each host individually using the *one rule per host* approach can increase this number to tens of thousands of rules incredibly quickly, becoming a nightmare to anyone in charge of managing the firewall!

Now, imagine it being possible to group hosts by access profile, selecting both source and destination hosts and adding them to lists, and then creating firewall rules based only on these lists.

This will drastically reduce the number of rules, turning the nightmare into the *firewall ruleset of dreams!* Now, you are probably feeling relieved thinking about the possibility of doing that. Well, we can do it!

Let me tell you about OPNsense aliases.

Aliases (also known as *tables* in the pf world) can group not only hosts but also networks, ports, fetch IPs from URLs, regions (using GeoIP), and much more! Let's see how to use them.

From the web browser, head to **Firewall** | **Aliases**:

Figure 5.1 – The Firewall: Aliases page

Click on the + icon to open the new aliases dialog; in it, you will be able to create a new firewall alias using the following options:

- **Enabled** (checked by default): Check this option to enable this alias.

- **Name**: A descriptive name to the new alias (without spaces).

- **Type**: These are explained as follows:

    - **Hosts**: A single IP address or a hostname using the **Fully Qualified Domain Name** (**FQDN**). To exclude a host in the alias, you need to put a ! sign before the name or IP address. The ! sign represents a NOT (negate) expression.

    - **Networks**: A network address using the CIDR notation (such as `192.168.0.0/24`).

    - **Ports**: Port numbers or a port range using the `first_port:last_port` format (such as `10000:20000`).

    - **URL (IPs)**: An IP addresses list that is fetched just once.

    - **URL Tables (IPs)**: An IP addresses list that is fetched periodically.

    - **Refresh Frequency**: When you select URL tables, this option will be shown with two textboxes – **Days**, which will define the interval in days to refresh the IP list, and **Hours**, which will define the interval of refreshing the list in hours.

- **GeoIP**: In this option, you can select what geographical regions will be matched to the rules using this alias. Each region will list the respective countries. You can also select IPv4, IPv6, or both to match this type of alias. Note that to use this type of alias, you will need to follow these instructions first: `https://docs.opnsense.org/manual/how-tos/maxmind_geo_ip.html`.

- **MAC address**: Fill with complete MAC addresses or with the **Organizationally Unique Identifier** (**OUI**) part of the MAC address (such as `F4:90:EA`, which will match all *Deciso*-registered MAC addresses).

> **Note**
>
> You can consult the MAC OUI on the internet using websites such as `https://www.wireshark.org/tools/oui-lookup.html`.

Users can easily spoof MAC addresses to bypass firewall rules based only on these criteria.

- **Network group**: This type of alias can nest other ones. For example, you can combine several types of aliases (host, GeoIP, and network) into just one.

- **External (advanced)**: This type of alias can't be managed using WebGUI and it is used by plugins, API calls, and so on. Practically speaking, this tells OPNsense to not touch this alias, which means that it will be managed by an external program using the `pfctl` command directly to manage the created table (or, as we are calling it here, the alias). An example is the NGINX plugin, which manages an alias to block bots and so on.

> **Note**
> Custom pf tables are not managed by the WebGUI and will not be saved in the configuration file (`config.xml`).

You can check the alias content in **Firewall | Diagnostics: Aliases** – we will see how to do this later in the chapter.

- **Statistics**: This option enables a set of counters of each table entry. You can check the alias content in **Firewall | Diagnostics**. Don't worry – as already mentioned, we'll explore it later in the chapter.

- **Description**: Type a description of your new alias here.

Now that we have learned how to create an alias, let's learn how to import and export it.

Importing and exporting aliases

Let's think about the following scenario: you need to migrate from another firewall platform – let's say a *brand name x*, for example – and the configuration of it is sent to you with a lot of existing objects and with tons of IP addresses inside each one. If you decide to add each object IP list using just the *WebGUI*, then you'll spend a lot of time converting these objects into aliases. So, if you have programming skills (or hire professional support from a company that does), you can convert it quickly to JSON file format, which is the one OPNsense uses, and import all these objects, saving a lot of time.

Let's take an example to practice this concept:

1. Create a new alias choosing the **Port(s)** type, naming it SMTP_ports.

2. Add in its content the following ports – 25 and 587.

3. Save it by clicking on the **Save** button.

> **Tip**
> While adding alias content, type a comma (,) or just hit the *Enter* key to add new values in the **Content** textbox.

Showing 1 to 5 of 5 entries

Figure 5.2 – The firewall alias download and upload icon buttons

Done! But wait – oh no! We forgot to add a port – the 465 SMTP port! Let's say we want to add the forgotten port in ascending order. So, we must add it between ports 25 and 587 – how can we do it? Let's see:

1. In **Firewall | Aliases**, click on the download icon.

2. A JSON file named aliases.json will be saved on your computer.

3. Open it with your favorite text editor and search for the alias:

```
"aliases": {
  "alias": {
    "c0f60f78-c62b-4768-b9cf-b5c34df1f3bb": {
      "enabled": "1",
      "name": "SMTP_ports",
      "type": "port",
      "proto": "",
      "counters": "1",
      "updatefreq": "",
      "content": "25\n587",
      "description": ""
    },
```

4. Find the following line:

```
"content": "25\n587",
```

5. With your favourite text editor, change the line to the following:

```
"content": "25\n465\n587",
```

> **Note**
>
> The \n special character represents a new line.

6. Save your changes and go back to *WebGUI*.

7. Click on the **upload** icon; a dialog will be shown. Click on the **Choose File** button, select your changed file, and click on the **upload** button:

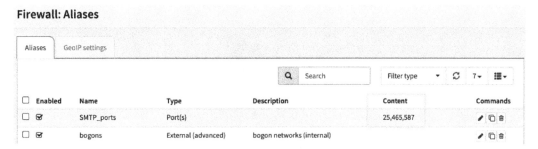

Figure 5.3 – A successfully imported alias JSON file

8. OPNsense will check the file syntax, and if everything passes, after refreshing the page, you will be able to see the updated aliases in WebGUI.

9. Congratulations! You did your first alias export and import successfully!

10. To remove or copy an existing alias is very simple:

Enabled	Name	Type	Description	Content	Commands
☑	SMTP_ports	Port(s)		25,465,587	✏ ⧉ 🗑

Figure 5.4 – Cloning an existing alias

11. If you need to copy an existing alias, just click on the **clone** icon.

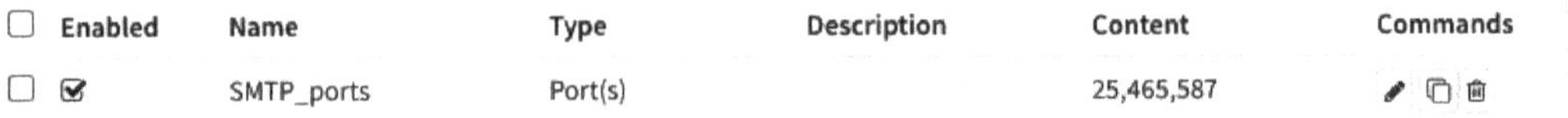

Enabled	Name	Type	Description	Content	Commands
☑	SMTP_ports	Port(s)		25,465,587	✏ ⧉ 🗑

Figure 5.5 – Removing an existing alias

12. If you need to remove an existing alias, then just click on the **trash** icon.

And what about the **external** (advanced) type of alias I mentioned before? Let's look at it.

Aliases and pf tables

To manage an external (advanced) alias type, it is necessary to look at the pf tables directly. There are two different ways of doing that:

- **WebGUI**: Go to **Firewall | Diagnostics | Aliases**, where you will be able to see the alias (`pftable`) or edit its content:

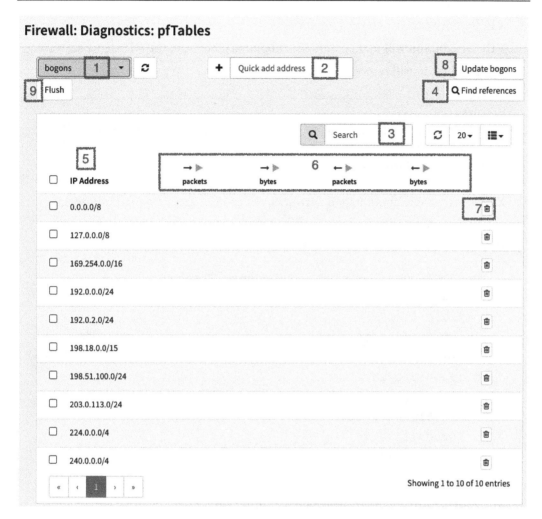

Figure 5.6 – The pfTables page

The numbers marked in *Figure 5.6* are explained here:

I.   Selected **table/alias** to manage.

II.   Add an IP address in the existing table.

III. Search for an IP address alias's content.

IV.   Search for an IP address in all existing tables.

V.   The IP addresses table's content.

VI.   When the **Statistics** option is checked in the alias, the counters will be shown in these columns.

VII. Remove a table entry.

VIII. This button will update the special **bogons** table from the OPNsense servers.

IX. This button will remove all the tables content.

> **Notes**
>
> On this page, only aliases with an IP address in their content will be listed. You can also edit their content here, but remember that changes made here are not saved in the configuration file.

X. To learn more about bogon networks, refer to `https://en.wikipedia.org/wiki/Bogon_filtering`.

- **CLI**: To manage pf tables in the CLI, we will use the `pfctl` command:

```
root@OPNsense:~ # pfctl -t bogons -T show
    0.0.0.0/8
    127.0.0.0/8
    169.254.0.0/16
    192.0.0.0/24
    192.0.2.0/24
    198.18.0.0/15
    198.51.100.0/24
    203.0.113.0/24
    224.0.0.0/4
    240.0.0.0/4
```

The command syntax to manage tables is as follows:

```
pfctl -t <table> -T <action>
```

In this book, we will be covering just three `pfctl` actions:

- `add`: To add a new IP address entry to an existing table or create a new one.

- `del`: To remove an existing table's IP address entry.

- `show`: To list the table's content. The complete actions list can be found at `https://www.freebsd.org/cgi/man.cgi?query=pfctl&sektion=8`.

Now that you have learned all about aliases and tables, it's time to learn how to use them in the firewall rules.

The firewall rules

One of the most important and useful features in OPNsense is the **firewall rules**. With them, OPNsense can control network traffic, and block, allow, or forward packets based on the firewall ruleset.

Before we start creating firewall rules, let's learn about some rule concepts used in OPNsense.

The rule processing order

OPNsense uses the pf with the quick parameter set by default, which means that the matched rule will be processed immediately, on a first-match basis. Otherwise, if we leave the **Quick** option unchecked in the rule, the last-match basis will be used, which means that all the rules will be processed.

OPNsense divides rules by network interfaces, except for the floating rules, which permit creating rules on any interface and are processed before the rules defined in the interfaces ruleset.

The rule processing order is as follows:

Figure 5.7 – The rule processing order

To see the complete list of rules created in your OPNsense using the CLI, you can run the `pfctl sa` command:

```
root@OPNsense:~ # pfctl -sa | less
TRANSLATION RULES:
no nat proto carp all
nat on em0 inet from (em1:network) to any port = isakmp ->
(em0:0) static-port
nat on em0 inet from (lo1:network) to any port = isakmp ->
(em0:0) static-port
...
```

```
FILTER RULES:
scrub on em1 all fragment reassemble
scrub on lo1 all fragment reassemble
scrub on em2 all fragment reassemble
scrub on em0 all fragment reassemble
block drop in log on ! em1 inet from 192.168.56.0/24 to any
...
pass out log quick inet6 proto ipv6-icmp from (self)
to fe80::/10 icmp6-type echorep keep state label
"acdbb900b50d8fb4ae21ddfdc609ecf8"
```

It will output the complete ruleset, including the translation rules **Network Address Translation (NAT)**.

If you want to only see the firewall rules, type `pfctl -sr`.

Let's now look at the rule actions.

Rule actions

There are three types of rule actions available:

- **Pass**: This will just allow the traffic.
- **Block**: This will drop the traffic, denying it without any response to the source. This causes TCP timeouts, so you can slow down your network! It is not recommended on **Local Area Network (LAN)** and **Demilitarized Zone (DMZ)** networks.
- **Reject**: This will deny the traffic and send a packet to the source to let it know about the block. In the case of TCP traffic, it will return an **Reset (RST)** flagged packet to the source. For UDP packets, an ICMP unreachable packet is sent to the source.

Let's see how to create a firewall rule in *WebGUI*:

Creating the first rule

Let's create our first rule to practice the concepts learned in previous sections:

1. To create your first firewall rule, go to **Firewall | Rules | LAN**:

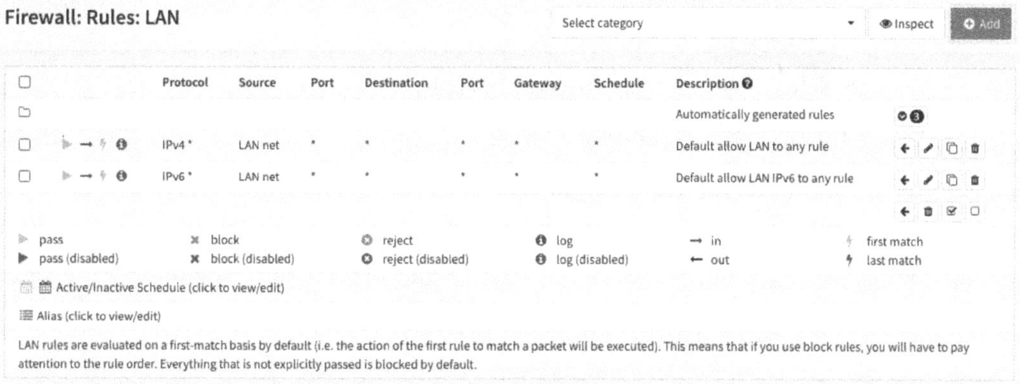

Firewall: Rules: LAN

		Protocol	Source	Port	Destination	Port	Gateway	Schedule	Description ❓	
☐										
☐									Automatically generated rules	
☐	▶ → ⚡ ❶	IPv4 *	LAN net	*	*	*	*	*	Default allow LAN to any rule	← ✎ ▣ 🗑
☐	▶ → ⚡ ❶	IPv6 *	LAN net	*	*	*	*	*	Default allow LAN IPv6 to any rule	← ✎ ▣ 🗑
										← 🗑 ☑ ☐

▶ pass	✖ block	⊘ reject	❶ log	→ in	⚡ first match
▶ pass (disabled)	✖ block (disabled)	⊘ reject (disabled)	❶ log (disabled)	← out	⚡ last match

📅 Active/Inactive Schedule (click to view/edit)

☰ Alias (click to view/edit)

LAN rules are evaluated on a first-match basis by default (i.e. the action of the first rule to match a packet will be executed). This means that if you use block rules, you will have to pay attention to the rule order. Everything that is not explicitly passed is blocked by default.

Figure 5.8 – The LAN's firewall rules page

2. Click on the **Add** button to create a new rule.

    A new page will be opened, with the following options:

    - **Action**: This is the rule action. Choose **Pass**, **Block**, or **Reject**.

    - **Disabled**: Check this option to disable the current rule.

    - **Quick**: Enabled by default, this will apply the first match criteria to the ruleset. If this rule matches, it will stop to evaluate the ruleset and will apply the selected action.

    - **Interface**: The network interface that this rule applies to.

    - **Direction**: As we have learned before, this will select which direction the rules will process, **in** or **out**; the default is to filter **incoming traffic**.

    - **TCP/IP Version**: Select which version of TCP and IP protocols this rule will process – **IPv4**, **IPv6**, or **IPv4+IPv6** (both).

    - **Protocol**: Select which IP protocol this rule will process. OPNsense supports a lot of protocols in this option, but the most common ones to select here are **TCP**, **UDP**, or both.

- **Source/Invert**: If checked, this option will invert (negate) the source in this rule. **!** will be added before the source to identify it as an inverted source (after the rule is created). For example, if checked, a rule with the `192.168.1.1` source IP address will apply to all other addresses except this specified one.

- **Source**: Click on the **Advanced** button to show the source and source port range options:

  - **Source**: Select the source address of this rule.

  - **from**: The first port number of a range.

  - **to:** The last port number of a range.

> **Note**
>
> The source port range option will be only available if you selected a compatible protocol in the **Protocol** option – for example, TCP or UDP.

- **Destination / Invert**: If checked, this option will invert (negate) the sense of destination in this rule. **!** will be added before the source to identify it as an inverted source (after the rule is created). For example, if checked, a rule with the destination of an alias with the RFC1918 networks (private networks) in it will apply this rule only to public IP addresses, because it will not match all private networks (compliant with RFC1918).

> **Note**
>
> If you never heard about RFC1918, refer to `https://datatracker.ietf.org/doc/html/rfc1918`.

- **Destination**: Select the destination address that this rule will apply to:

  - **Destination port range**: Select the source address to this rule.

  - **from**: The first port number of a range.

  - **to**: The last port number of a range.

> **Note**
>
> The destination port range option will be only available if you selected a compatible protocol in the **Protocol** option – for example, TCP or UDP.

- **Log: Log packets that are handled by this rule**: Check this option to enable the log for this rule. By default, OPNsense only logs the default system rules.

- **Category**: Categories are a useful resource added recently to OPNsense to help manage firewall rules (especially the larger ones). If you type text here, it will create a new category if it doesn't already exist, and the rule will be grouped in this category. After creating a category, you can select it on the **Rules** page, as shown in the following screenshot:

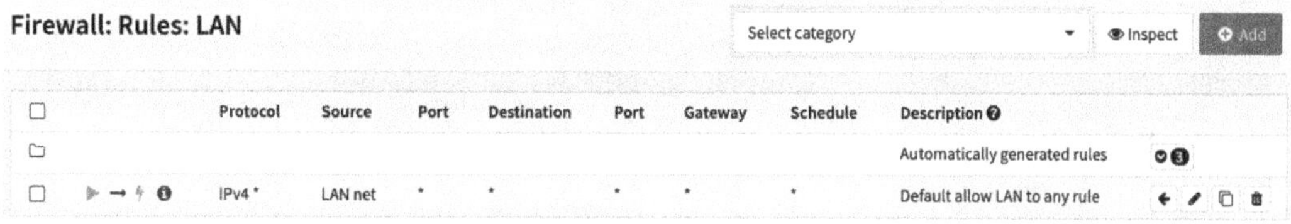

Figure 5.9 – The Rules category selection option

- **Description**: You can type a description of the rule in this option. Add some description that will make sense to others; it will help to keep your firewall ruleset clean and easy to understand!

Now that we have described each option, let's create a LAN rule:

1. As the LAN default rule is to allow all traffic, let's create a new rule with the **Block** action:

2. Create a new rule and select the following options (maintain the default values for the options not mentioned here):

   I.   **Action: Block.**

   II.   **Protocol: ICMP.**

   III. **Destination: LAN address.**

   IV.   **Category**: You can type LAB here.

   V.   **Description**: My first rule.

VI. Click on the **Save** button and check out the results:

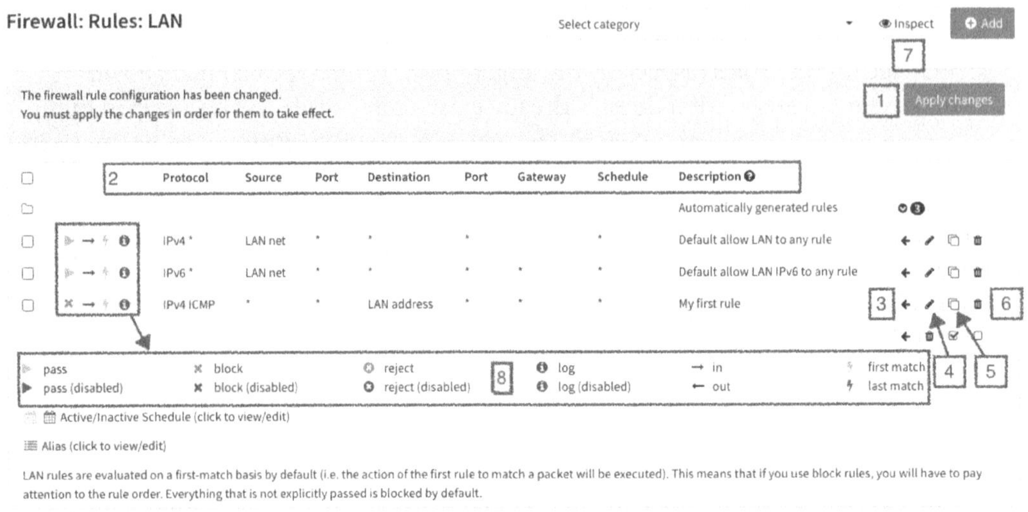

Figure 5.10 – Firewall rules options

The numbers in the preceding screenshot refer to various options:

1. After creating or editing rules, this dialog will show to make changes effective, and to do so, just click on **Apply changes**.

2. The **Rule** columns: They will show the options as the rules are configured.

3. The **Move rule** button: To move a rule, you need to first select the rule you want to move; after that, click on this button in a rule that holds the position that you want to move the selected rule to. By doing this, the selected rule will be moved before the rule you have clicked on the **Move rule** button.

4. The **Edit rule** button: Click on this button to edit the selected rule.

5. The **Clone rule** button: To clone a rule, just click on the clone button in a rule you want to clone.

6. The **Delete rule** button: If want to remove a rule, just click on this button.

7. The **Inspect** button: This will show the rule statistics by changing the columns to the options, as shown in the following screenshot:

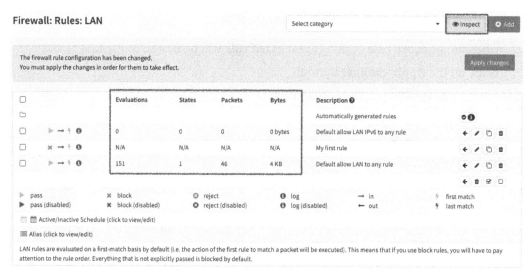

Figure 5.11 – Toggle rule inspection

8. **Legend**: This is the rule's icon legend.

After the rule is created, try to run a `ping` command to your OPNsense LAN's IP address (in my case, the LAN is `192.168.56.3`; change it to your OPNsense LAN's IP address):

```
$ ping -c 3 192.168.56.3
PING 192.168.56.3 (192.168.56.3): 56 data bytes
64 bytes from 192.168.56.3: icmp_seq=0 ttl=64 time=71.023 ms
64 bytes from 192.168.56.3: icmp_seq=1 ttl=64 time=1.205 ms
64 bytes from 192.168.56.3: icmp_seq=2 ttl=64 time=1.326 ms
--- 192.168.56.3 ping statistics ---
3 packets transmitted, 3 packets received, 0% packet loss
round-trip min/avg/max/stddev = 1.205/24.518/71.023/32.884 ms
```

The rule you have just created isn't blocking packets yet because the ICMP packets are matching the default `allow` rule created before the new rule. Let's try to move it to the first position in the LAN's ruleset and see what happens:

		Protocol	Source	Port	Destination	Port	Gateway	Schedule	Description ❷	
☐										
☐									Automatically generated rules	◐❸
☐	▶ → ⚡ ❶	IPv6 *	LAN net	*	*	*	*	*	Default allow LAN IPv6 to any rule	← ✏ ⧉ 🗑
☐	▶ → ⚡ ❶	IPv4 *	LAN net	*	*	*	*	*	Default allow LAN to any rule	← ✏ ⧉ 🗑
☑	✖ → ⚡ ❶	IPv4 ICMP	*	*	LAN address	*	*	*	My first rule	← ✏ ⧉ 🗑
										← 🗑 ☑ ⬚

Figure 5.12 – Moving the rule

To move the rule position in the ruleset, follow these steps:

1. Click on the **move rule** button (the selected rule will be moved to the first position).

2. Click on the **Apply changes** button.

3. Now, try to run the `ping` command again:

```
ping -c 3 192.168.56.3
PING 192.168.56.3 (192.168.56.3): 56 data bytes
--- 192.168.56.3 ping statistics ---
3 packets transmitted, 0 packets received, 100% packet
loss
```

4. As you can see, the rule now is blocking ICMP packets! Congratulations! Your first rule is working! You just created and moved a rule in OPNsense's *WebGUI*.

5. Now that you know how to create and manage rules in OPNsense, try to create other rules using different protocols and actions to improve your firewall's rule skills! We will use a lot of firewall rules till the end of this book, so this small exercise was to just introduce you to managing firewall rules in *WebGUI*.

Now, let's look at the firewall settings and tweaks.

Firewall settings

In the firewall settings options, you see how to adjust some firewall configurations such as **optimization**, **firewall's behavior** and maximum values to settings such as **states** and **table entries**.

In the following section, we will learn how each option can change the firewall settings and behavior globally in the system:

1. To start, head to **Firewall | Settings | Advanced**.

2. On this page, the first option is related to IPv6 traffic – **IPv6 Options**.

3. **Allow IPv6**: The default action is to block all the IPv6 traffic; if you don't want to block it, just check this option.

4. **NAT**: The following options are related to NAT. In the next chapter, we will learn how to create NAT rules in OPNsense; for now, we'll just look at some settings that can adjust the way some NAT rules work in OPNsense.

- **Reflection for port forwards**: If OPNsense has NAT rules for forwarding ports using external (internet) addresses and you want your local networks to reach these ports, enable this option. This option will create rules to permit local hosts to reach ports forwarded using external addresses to hosts that rely on local networks.

- **Reflection for 1:1**: The same option mentioned in the preceding section applies here but to the NAT 1:1 rule. In the next chapter, we'll explore the types of NAT rules that OPNsense is capable of.

- **Automatic outbound NAT for Reflection**: This is another NAT-related option; as explained in the previous option, this will set the reflection to outbound NAT rules.

- **Bogon Networks**: Bogon networks are IP addresses that are reserved and should not be seen in internet routing traffic. OPNsense frequently updates these IPs, and the following option adjusts the frequency that it is done.

- **Update Frequency**: The available options are **Monthly**, **Weekly**, and **Daily**.

5. **Gateway Monitoring**: Each OPNsense monitor created a gateway, sending ICMP (ping – ICMP request) packets to check whether the gateway host is alive. The following options are related to this gateway monitoring process:

   - **Kill states – Disable State Killing on Gateway Failure**: When a gateway goes offline, OPNsense will flush states related to this gateway by default. If you want to change this behavior, just click on this option and it will not flush the related states.

   - **Skip rules – Skip rules when gateway is down**: When a gateway specified in a rule is down (which we'll discuss in *Chapter 9, Multi-WAN – Failover and Load Balancing*), the rule is not created (skipped) when this option is checked.

6. **Multi-WAN**: OPNsense includes multi-**Wide Area Network** (**WAN**) and failover capabilities, and the way it works can be configured in the following options:

   - **Sticky connections – use sticky connections**: When using a load balancer with multiple WANs, OPNsense will send packets on a round-robin basis, with connections being sent to all available gateways. This behavior can break some connections and cause problems to users accessing certain services. To avoid that, you can set this option to make connections stick to the same gateway, based on the source address. This way, connections will only have a change of gateway after they expire or when a gateway goes down.

- **Shared forwarding Use shared forwarding between packet filter, traffic shaper, and captive portal**: Rules with a specific gateway skip the processing of its packets for a traffic shaper and a captive portal. The reason this happens is that the packet filtering system used by these two services (`ipfw`) is different from the one used by the firewall in OPNsense (pf). To make them work with the same rules from the firewall, check this option.

> **Note**
>
> To learn more about the available pf systems on FreeBSD, refer to `https://docs.freebsd.org/en/books/handbook/firewalls/`.

- **Disable force gateway – Disable automatic rules which force local services to use the assigned interface gateway**: By default, OPNsense will use the interface's gateway to outgoing packets. If you want it to use the system routing table instead, check this option.

7. **Schedules**: Firewall rules can be scheduled, specifying a date and time. You can set a schedule on the **Firewall | Settings | Schedules** page. An example of a schedule is shown in the following screenshot:

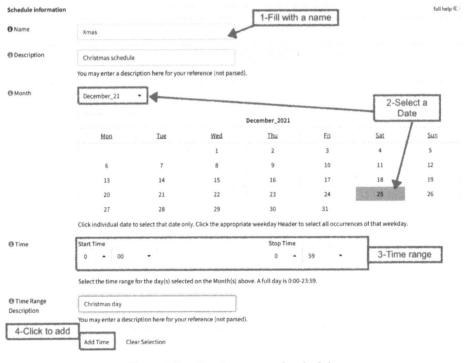

Figure 5.13 – Creating a new rule schedule

8. Following these steps, the schedule will be added; you can add more dates or times to a schedule, but in this example, we'll just add one:

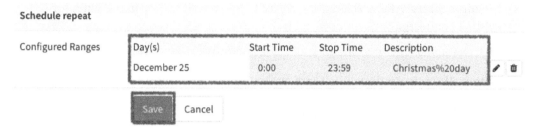

Figure 5.14 – Saving the new schedule

9. On the same page, to save the newly created schedule, just click on the **Save** button:

Figure 5.15 – Selecting the created schedule in a firewall rule

10. To assign the created schedule to a rule, just edit an existing firewall rule, select the schedule in the **Schedule** option, and then click on the **Save** button:

Figure 5.16 – A schedule assigned to a firewall rule

11. After saving the rule, the schedule can be visualized in the **Schedule** column.

> **Note**
> The ruleset is reloaded every 15 minutes, so this will be the period of time a schedule can take to become active. Another detail to consider while using schedules is that an active connection will stay active until it closes, or until the state is killed on the firewall!

Congratulations! We just learned how to create a scheduled rule! Let's get back to the **Firewall | Settings | Advanced** page to move on with the schedule options:

- **Schedule States**: After a schedule is expired, the firewall will clean the states for existing connections; if you don't want it to do so, check this option.

- **Miscellaneous**: The firewall's miscellaneous options are described as follows:

  - **Firewall Optimization**: OPNsense has four firewall optimization profiles:

    - **normal**: This is the default option, which will apply the default optimization algorithm.

    - **high-latency**: This will increase the idle connection timeout. Links with high latency may need to select this option.

    - **aggressive**: This will decrease the idle connection timeout. This way, they expire quicker, saving system resources. The cost of selecting this option is that it may drop some legitimately idle connections.

    - **conservative**: The most system resource-consuming option, this will try to maintain legitimate idle connections longer while increasing CPU and memory utilization.

  - **Firewall Rules Optimization**: Select which rules optimization algorithm OPNsense will apply to the ruleset:

    - **none**: This will disable the firewall rules optimizer.

    - **basic**: This algorithm will perform some actions to try to optimize the ruleset such as removing duplicated rules, reordering rules when it can improve the performance, combining some rules when it is an advantage, and removing rules that are a subset of other ones.

    - **profile**: This uses the currently loaded ruleset as a feedback profile to tailor the ordering of quick rules to actual network traffic.

  - **Bind states to interface**: By default, the connection states in OPNsense are floating, which means that they will not bind to an interface. To bind the connection states to the relevant interface, check this option. In the following screenshot, you can see the difference between when this option is turned on and off (OPNsense defaults) in **Firewall | Diagnostics | States**:

Figure 5.17 – The firewall states dump – a comparison

- **Disable Firewall – Disable all packet filtering**: This option will disable the pf system, which means no more packet filtering or NAT! With this option checked, OPNsense will act as a network router.

> **Note**
>
> If you need to temporarily disable the firewall and NAT, you can run on the CLI the `pfctl -d` command, and to re-enable it, just run `pfctl -e`. When OPNsense is intended to be used as just a network router, this option may increase the throughput. Be careful running those commands in a production environment!

- **Firewall Adaptive Timeouts**: To prevent the state table from getting full quickly in a state's burst, for example, you can define two values here: the **start** value, which will trigger a scaled timeout factor for the connections in the state table, and the **end** value, which will be used as the scale factor. When it has reached the connections, the timeout will become zero, removing all entries immediately, which isn't a good idea, so try to maintain this value high enough to not be reached. The scaling process will work based on this formula: (*adaptive end value – current number of states) / (adaptive end value – adaptive start value*).

12. Here is an example:

```
end value: 1000
current states number: 750
start value: 500
(1000 - 750) / (1000 - 500) = 0,5
```

In this case, when the state's number reaches 500, it will start to apply scaling. In the preceding example, the scale factor (with 750 current states) will be 50% (0,5), which means that the default timeout value will be decreased by 50%. The default value is 0.

13. **Firewall Maximum States**: These are the maximum connections entries OPNsense will keep in the state table. The default value will vary based on the system's memory (RAM). Each state entry will consume 1 KB of RAM.

14. **Firewall Maximum Fragments**: This is the maximum number of packet fragments (used for reassembly) that the system will hold.

15. **Firewall Maximum Table Entries**: This is the maximum number of entries that each alias (pf table) will hold. The default value will vary based on the system's memory (RAM).

16. **Static route filtering**: While using static routes in OPNsense, enabling this option will bypass packet filtering for packets that are incoming and outgoing in the same interface. This option can be used when there are multiple subnets connected to the same interface.

17. **Disable reply-to**: By default, while using multi-WAN, OPNsense will keep the same path for traffic arriving by a certain interface, preventing asymmetric routing behavior. For example, in some cases, when using two or more interfaces in bridge mode or with local hosts using another gateway IP, rather than the one configured on the WAN interface, it may be necessary to disable **reply-to** to make the traffic path flow properly.

18. **Disable anti-lockout**: OPNsense, by default, automatically adds rules to allow system administration through WebGUI- and **Secure Shell Protocol** (**SSH**)- configured ports. Check this option to disable these automatic rules. Be warned that checking this option can lock your OPNsense administrative access; be sure of what you are doing before doing it.

19. **Aliases Resolve Interval**: When using hostnames inside aliases, OPNsense must resolve them periodically; here, you can set the intervals in which it does that. The default is 300 seconds (5 minutes).

20. **Check certificate of aliases URLs – Verify HTTPS certificates when downloading alias URLs**: While using URL/URL table aliases with HTTPS addresses, OPNsense will check for a valid certificate before it starts downloading the alias' content. If this option is checked, only fetch content from the server if the certificate is valid.

21. **Dynamic state reset – Reset all states when a dynamic IP address changes**: If some interface configured with a dynamic IP has changed, the whole state table will be flushed to reflect the new IP (which is only used for IPv4 addresses).

That's all the available options on the firewall advanced settings page. Now, let's move on and look at some of the traffic normalization options available in WebGUI.

Go to **Firewall | Settings | Normalization**. On this page, it will be possible to change some traffic normalization (also known as scrubbing) settings and create new scrub rules for very specific cases when needed. Traffic normalization is useful to normalize packets that may be malformed or created intentionally – for example, for attacks.

The options available in **General settings** are as follows:

- **Disable interface scrub**: Checking this option will disable the normalization rules to all interfaces.

- **IP Do-Not-Fragment**: If a protocol generates fragmented packets with **don't fragment bit** set, OPNsense by default will drop these packets. To avoid this behavior, check this option; this way, the packets will pass through the filter that will just clean the **don't fragment bit** instead of dropping these packets. If you want to deep-dive into IP fragmentation, refer to `https://en.wikipedia.org/wiki/IP_fragmentation`.

- **IP Random id**: When checked, this will replace the IP identification of the non-fragmented packet's field.

- **Detailed settings**: You can optionally specify your own packet normalization rules by clicking on the + **Add** button at the top-right corner of the page. There, you will be able to specify custom rules for several parameters for traffic normalization. As this is an extensible topic, we'll not cover it here. If want to learn more about scrub rules in pf, a good starting point is the pf man page: `https://www.freebsd.org/cgi/man.cgi?query=pf.conf&sektion=5&n=1#TRAFFIC%09NORMALIZATION`.

Now, we'll explore a little bit about firewall diagnostics and troubleshooting in the next section.

Diagnostics and troubleshooting

Truth has no answer – against facts, it's difficult to argue, so always count on logs and diagnostics tools to help you solve issues related to firewalls. Based on facts, your troubleshooting will be as sharp as a katana (a samurai's sword). As an airplane pilot, aviation experience taught me that things must be checked using a checklist based on facts, and as a firewall administrator, experience has taught me that a firewall will be blamed for network problems in 99.9% of cases. So, don't guess a problem – check it! OPNsense has a lot of tools that can help you with your troubleshooting. In time, users will start relying on just your word and not blame the firewall for any network (or maybe entire internet) problems. This is my advice for you based on a couple of decades of personal experience.

Starting at **Diagnostic tools**, go to **Firewall** | the **Diagnostics** menu:

- **pfInfo**: This page shows different information based on the `pfctl` command output. The available options are as follows:

  - **Info**: This is the output of the `pfctl -si` command, which shows information about interface statistics, the state table, and counters.

  - **Memory**: This will show the `pfctl -sm` output and displays the memory hard limits for pf.

  - **Timeouts**: This is the `pfctl -st` output and shows the global timeout values.

  - **Interfaces**: This outputs the `pfctl -sI -vv` command. It displays the interface statistics available to pf.

  - **Rules**: This shows the current rules loaded in the memory. This is the output of the `pfctl -sr -vv` command.

  - **NAT**: This is the same as the preceding but applies to the NAT rules. It outputs the `pfctl -sn -vv` command.

- **pfTop**: This page displays the `pftop` command output. It also displays pf states and can be filtered or sorted by the available options – **View type**, **Sort type**, and **Number of States**.

- **pfTables**: We already explored this earlier in the chapter; please refer to the *Alias and pf tables* section.

- **States Dump**, **States Reset**, and **States Summary**: We also explored these before; please refer to the *Stateful firewall* section if you want to review these pages.

Let's move on to the log options; head to **Firewall** | **Log Files** | **Live View** to start.

On this page, we will be able to see in real time all events generated by rules with the log option enabled or by system default rules, such as the default `deny` rule. This is a useful page and is the starting point for most firewall troubleshooting:

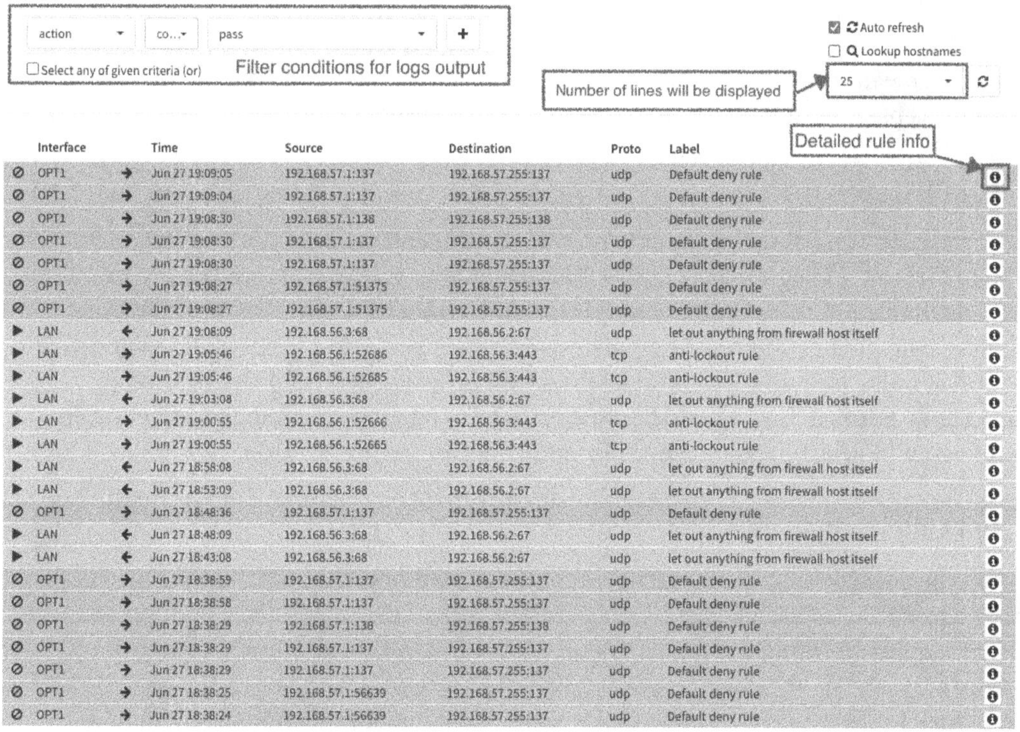

Figure 5.18 – The firewall logs live view page

To see the raw log output, go to **Firewall | Log Files | Plain view**. As you can see in the preceding screenshot, the output of the live view is formatted and much easier to understand. The red lines are blocked traffic and the green lines are the allowed ones. OPNsense has a page that outputs graphics of the firewall logging; you can explore these graphics by going to **Firewall | Log Files | Overview**. Let's now explore some troubleshooting cases to see how to use diagnostic tools and log files to find firewall issues.

Troubleshooting

Let's look at some real use cases of issues relating to firewalls and see how we can try to solve each one using OPNsense tools.

The firewall is blocking a connection

In cases like this, a good start is looking at the firewall logs, specifically on the **Live View** page; this will help to filter the information about the connectivity issue. It is important to know what the source or destination address is (both is better), and the port number will help a lot but isn't essential to know in most cases. If you can't see connections being blocked, it means that the firewall is passing the traffic, or it isn't arriving in the firewall. If the traffic is passing, let's move on to the next item to be checked.

Firewall states dump

If a connection is passing through the firewall, it will generate a connection state. To check it, go to **Firewall | Diagnostics | States Dump** and filter for connections using a source or destination address, or even the port number. Here's a hint – you can filter for port numbers by typing `:port` (For example, `443` to filter port `443`) in the **Filter expression** textbox. If traffic isn't arriving in the firewall, you probably will need to check other connectivity issues in the network to find out what is happening. Going to **Interfaces | Diagnostics | Packet Capture** may help you.

Rule reviewing

If traffic is blocked, then check the ruleset of the related interface. Maybe you'll need to create a new rule, adjust an existing one, or even change the rule order. Common issues related to rules are wrong protocols or port numbers, typos in addresses (source or destination), and so on; when reviewing a rule, be focused. I have seen good analysts take a long time to solve a firewall issue because they were not paying enough attention to the rule reviewing process.

In very rare cases, it's necessary to double-check whether a rule is applied to the pf reviewing the `/tmp/rules.debug` file in the CLI. Another way to check it is by running the `pfctl -sr` command. It's very rare in OPNsense to see rules not being loaded, but if you are in doubt, check it.

You can also check whether a rule has been matched with traffic by clicking on the **Inspect** button on the **Rules** page.

To end this topic, the last piece of advice is to always use your knowledge and experience. The OSI model is a good guide to follow when troubleshooting connectivity issues. If you don't know the OSI model yet, take a break from this chapter and start studying it; it will save you a lot of time. You can start here: `https://en.wikipedia.org/wiki/OSI_model`.

Summary

In this chapter, we learned about the OPNsense packet filtering system (the pf). We also learned what a stateful firewall is and how to check states on OPNsense, how to manage aliases and rules, how they are processed in OPNsense, creating rules' schedules, changing firewall settings, and explored firewall diagnostics and troubleshooting. Now you understand how OPNsense's firewalling works, how to create aliases and rules, how to adjust firewall settings, and how to troubleshoot firewall issues. Understanding firewalling is important in preparation for moving on to the next chapters.

In the next chapter, we'll continue the firewalling saga by exploring NAT rules!

6
Network Address Translation (NAT)

Following the last chapter, where we explored firewalling, **Network Address Translation (NAT)** is a simple way to have an entire local network using a single public IP address. Since NAT is an IPv4-exclusive method, we will not explore IPv6 and its equivalent NPTv6 (also known as NAT66). This book aims to provide valuable and practical information. Unfortunately, these days, IPv6 and NPTv6 aren't the predominantly used technology in local networks.

> **IPv6 Adoption**
>
> According to Google's IPv6 statistics page (in July 2021), IPv6 adoption was 35% globally (`https://www.google.com/intl/en/ipv6/statistics.html`).

Coming back to our chapter's topic, we will learn about NAT concepts, which types OPNsense supports, and how to create NAT rules on WebGUI. By the end of this chapter, you will be able to manage NAT rules on OPNsense and will understand how the NAT method works.

The following are the topics we'll explore in this chapter:

- NAT concepts
- Outbound port forwarding
- One-to-one NAT

Technical requirements

This chapter requires that you have a clear understanding of firewall concepts and the TCP/IP network stack, and feel comfortable running commands on the CLI. You will need OPNsense running to follow some of this chapter's steps. I suggest you use the virtual machine we set up earlier in this book, in *Chapter 2*, *Installing OPNsense*.

NAT concepts

Before we start to talk about NAT concepts, let's understand what kind of problems it solves. Let's think of a small company with a network of 10 devices and all of them need to be connected directly to the internet. We are talking about a small company with a limited budget, so the available WAN connection has only one public **Internet Protocol (IP)** address and there is no possibility to get an upgrade to a service that provides an entire IP network range. Sound familiar? In Brazil and Portugal, this scenario is very common. So, how do we connect all those devices using a single public IP address? If you answered *using NAT*, you're right! This is one of the most common scenarios for using NAT. Another one is when you need to provide, let's say, web services, but you have more web servers than available public IP addresses. In this case, NAT can help too, by using different ports of the same IP address. But what is the difference between these two examples? The direction. The first example uses **outbound NAT** and the second one uses **inbound NAT**, also known as **port forwarding**. There is a third supported type of NAT in OPNsense – *one-to-one*. This kind of NAT can be used in both directions. A special term used when talking about one-to-one NAT is when you have networks of the same size and need to apply NAT in both directions. We call this **BINAT** in OPNsense. We will explore all these three types of NAT in the following sections.

Port forwarding

Using our previous example, let's consider a small company with three web servers but with just one public IP address and a lot of users needing to access them from the internet. How can we solve this problem using just firewall features? By creating an inbound NAT rule! We will refer to this type of NAT in this book in the same way as OPNsense: **port forwarding**. It will forward a port or a port range from the public interface to an internal host such as, for example, a web server. At the same time, the port number/range can be changed.

Using the preceding example, let's take a look at the following topology:

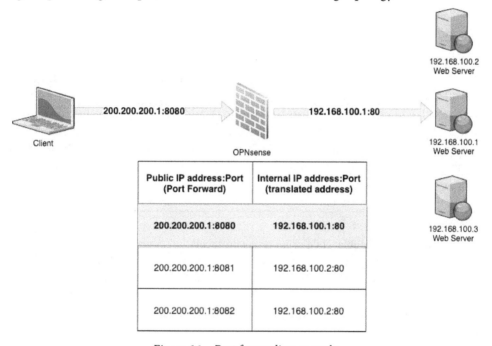

Public IP address:Port (Port Forward)	Internal IP address:Port (translated address)
200.200.200.1:8080	192.168.100.1:80
200.200.200.1:8081	192.168.100.2:80
200.200.200.1:8082	192.168.100.2:80

Figure 6.1 – Port forwarding example

In the preceding figure, you can see a client requesting access from the internet to the public IP 200.200.200.1 on port 8080. When this request arrives in our firewall, it will look for a NAT entry that forwards port 8080 in its public IP address to an internal address and port – 192.168.100.1 on port 80 in this case. As OPNsense is a stateful firewall, the reply from the web server will follow the same path back to the internet client, leaving the internal network with the translated address 200.200.200.1, instead of 192.168.100.1. Otherwise, the packet will not be routable to the internet since its internal IP address is an RFC1918 private IP address and couldn't be routed on the internet. Using port forwarding allows us to assign thousands of ports to internal services/ports with just one public IP address.

> **Notes**
>
> A common issue related to port forwarding NAT is when the internal host has a default gateway set to another router/host rather than the firewall that is forwarding the inbound NAT rules. This will break the connection and will not work, since the outgoing packets (replies) are leaving the internal network from another gateway.
>
> For more about RFC1918 address allocation for private internets please refer to `http://www.faqs.org/rfcs/rfc1918.html`.

Let's now look at caveats, which is another important NAT concept.

Caveats

Port forwarding will not always save the day. Some situations will require more sophisticated solutions, such as reverse proxies. An example is when you have just one public IP address available and multiple web servers that require serving on port 443 (HTTPS), which will not be possible with only one IP address available. While using NAT, you must configure a different port for each server. In this case, a reverse proxy could help. OPNsense has some plugins that may help in this case, such as **HAProxy** and **NGINX**, but for logging the real client's IP addresses, they may demand that the application supports reverse proxy protocols to work as expected.

> **Note**
>
> If you have some trouble while using proxy protocols with reverse proxies, check out this awesome project that could help you: `https://github.com/cloudflare/mmproxy`. Thanks to Fabian Franz for this tip!

Now that we have learned about the port forwarding basics, let's look at how to create a rule of this type.

Creating a port forwarding rule

OPNsense WebGUI simplifies a lot of how to create a NAT port forwarding rule. Let's see the steps we will need to create our first port forwarding rule:

1. Log into WebGUI and go to **Firewall | NAT | Port Forward**:

Figure 6.2 – NAT Port Forward rules page

2. Click on the **+ Add** button to create a rule.

A new page will be opened, with the following options:

- **Disabled**: Check this option to disable the current rule.

- **No RDR (NOT)**: If this option is marked, OPNsense will NAT the matching packets. There are a few special cases where this is necessary and to explore it, we would need to cover the **port-forwarding (pf)** in detail, which is not our goal in this book. Please refer to Packt's books about this subject to learn more, or to the pf manual: `https://www.freebsd.org/cgi/man.cgi?pf.conf(5)`.

- **Interface**: The network interface that this port forwarding rule applies to.

- **TCP/IP Version**: Select which version of TCP and IP protocols this rule will process: **IPv4**, **IPv6**, or **IPv4+IPv6** (both).

- **Protocol**: Select which IP protocol this port forwarding rule will match. OPNsense supports a lot of protocols in this option, but the most common ones to select here are TCP and UDP.

- **Source**: Click on the **Advanced** button to show the source and source port range options.

- **Source / Invert**: If checked, this option will invert (negate) the sense of the source in this rule. A **!** will be added before the source to identify it as an inverted source (after the rule is created). For example, if checked, a rule with the *source* 192.168.1.1 will apply to all other addresses *except* this specified one.

- **Source**: Select the source address for this port forwarding rule.

- **Source port range**:

  - **from**: The first port number of a range

  - **to**: The last port number of a range

- **Destination / Invert**: If checked, this option will invert (negate) the sense of destination in this port forwarding rule. An **!** will be added before the source to identify it as an inverted source (after the rule is created). For example, if checked, a rule with the **destination** of an alias with the RFC1918 networks (private networks) in it will apply this rule only to public IP addresses, because it will *not match* all private networks (compliant with RFC1918).

- **Destination**: Select the destination address that this port forwarding rule will apply to. You can select previously created aliases, virtual IPs, interface addresses, or choose the **Single host or Network** option to type an IP address.

- **Destination port range**: Select the destination port range for this rule.

  - **from**: The first port number of a range

  - **to**: The last port number of a range

- **Redirect target IP**: Enter the internal IP address that the packets will redirect to.

- **Redirect target port**: Enter the port of the internal IP address.

- **Pool Options**: The pool options can be used when you define a host alias with two or more addresses. The following pool options are available:

  - **Default**: This option will not apply any NAT algorithm to this NAT rule, and is the preferred option while using a single redirect IP port forwarding rule.

  - **Round Robin**: This algorithm will send packets for each available IP address sequentially. An example of usage is balancing traffic to web servers.

  - **Random**: As the name suggests, the option will randomize the available addresses in the pool (alias).

  - **Round Robin with Sticky Address / Random with Sticky Address**: The last two options can be combined with the sticky address option, which means that this port forwarding rule will map connections based on the source address. This is quite useful for applications and protocols that are session-based, such as TCP, for example.

  - **Source Hash**: Similar to the sticky address, will use a hash based on the source address to determine the pool's IP address to use.

  - **Bitmask**: This will map the source address using its subnet mask to define the translated address. For example, if the source address is `192.168.0.10/24` and the pool's subnet is `172.16.0.0/24`, then the translated IP address will be `172.16.0.10`.

- **Log: Log packets that are handled by this rule**: Check this option to enable the log for this port forward rule.

- **Category**: Like the firewall rules, you can define categories for NAT rules. Please refer to *Chapter 5*, *Firewall*, to review more details about firewall rule categories.

- **Description**: Describe what this port forwarding rule is doing as clearly as possible. It will help your teammates!

- **Set local tag**: Tags are a useful pf resource to manage complex rulesets when you need to check something more than source/destination addresses. You can set a tag to mark packets processed by this rule to be checked later by other rules.

- **Match local tag**: Once you have a ruleset using tags, you can set a tag here to match it.

- **No XMLRPC Sync**: Check this option while using OPNsense in a high-availability configuration, to prevent this rule from being automatically synced to the other firewall members.

- **NAT reflection**: This option is useful when you have an external NAT created but with internal clients accessing it. When enabled, it will redirect internal clients to the internal IP address of this rule. The available options are **Enabled**, **Disabled**, and **Use system default**. This last one can be set as you learned in *Chapter 5*, *Firewall*, in the *Settings and tweaks* section.

- **Filter rule association**: When you create a port forwarding rule, it is still necessary to add a firewall rule to permit its traffic. In this option, you can select **None** to not create a firewall rule or add one manually later. **Add associated filter rule** adds a non-editable rule associated with this NAT rule. If the NAT rule is deleted, the rule will also be removed. The opposite occurs when you select the **Add unassociated filter rule** option. In this case, the NAT rule and the created filter rule can be removed, not depending on each other. The **Pass** option will not create a filter rule. Instead, it will permit traffic using only the NAT rule. *Caution*: the **Pass** option will not work properly with multi-WAN installations and will demand an interface with a default gateway configured in it.

Now that we have explored the port forwarding rules page, let's practice a little bit.

Proposed exercise

To see a working port forwarding rule, we can redirect a new port to WebGUI. To follow these steps, we'll need a working OPNsense installation. In this exercise, we will assume that you are using the virtual machine we set up earlier in this book.

Create a new port forwarding rule that will forward the TCP port 8443 of our LAN network interface address to the loopback address at port 443, the WebGUI default port (the options not mentioned keep the default values):

1. **Interface**: Select **LAN**.

2. **Destination**: Select **LAN address**.

3. **Destination port range**: Select the **(other)** option and type 8443 (both in the **from** and **to** options).

4. **Redirect target IP**: Select the **Single host or Network** option and type the following in the textbox that will show below (after selecting the **Single host or Network** option): 127.0.0.1.

5. **Redirect target port:** Select the **(other)** option and type 443 in the textbox that will show below (after selecting the **(other)** option).

6. **Description**: NAT port forwarding test (suggestion).

7. **Filter rule association**: Select **Add associated filter rule**.

8. Click on the **Save** button.

9. Click on the **Apply** button.

After creation, your rule will look like this:

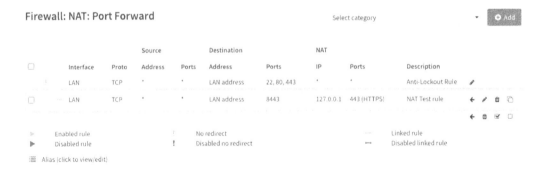

Figure 6.3 – Example port forwarding rule

Now try to access OPNsense WebGUI using the LAN IP address but using port 8443, for example, `https://<OPNsense_LAN_IP_Address>:8443`.

If your rule is correct, WebGUI will load using port 8443. It's done! Your first port forwarding rule is working!

If you want to check that the rule just added is working, go to **Firewall | Diagnostics | States Dump**, and you should see the translated connection states as in the following screenshot:

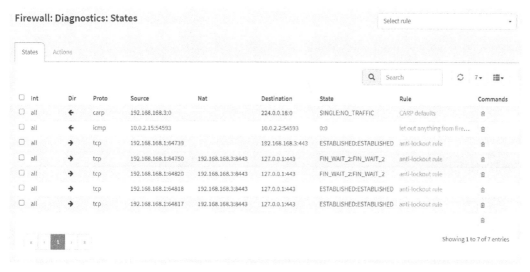

Firewall: Diagnostics: States

Int	Dir	Proto	Source	Nat	Destination	State	Rule	Commands
all	←	carp	192.168.168.3:0		224.0.0.18:0	SINGLE:NO_TRAFFIC	CARP defaults	🗑
all	←	icmp	10.0.2.15:54593		10.0.2.2:54593	0:0	let out anything from fire...	🗑
all	→	tcp	192.168.168.1:64739		192.168.168.3:443	ESTABLISHED:ESTABLISHED	anti-lockout rule	🗑
all	→	tcp	192.168.168.1:64750	192.168.168.3:8443	127.0.0.1:443	FIN_WAIT_2:FIN_WAIT_2	anti-lockout rule	🗑
all	→	tcp	192.168.168.1:64820	192.168.168.3:8443	127.0.0.1:443	FIN_WAIT_2:FIN_WAIT_2	anti-lockout rule	🗑
all	→	tcp	192.168.168.1:64818	192.168.168.3:8443	127.0.0.1:443	ESTABLISHED:ESTABLISHED	anti-lockout rule	🗑
all	→	tcp	192.168.168.1:64817	192.168.168.3:8443	127.0.0.1:443	ESTABLISHED:ESTABLISHED	anti-lockout rule	🗑

Showing 1 to 7 of 7 entries

Figure 6.4 – Translated connection states example

As we can see, the IP address inside the parentheses (192.168.168.3:8443) is the *destination address* and *port* set in the rule, and the address after the <- is the *redirect target IP*. As you can see, even with this unusual example, port forwarding works very well!

> **Note**
> I tried to use a simple example that does not demand an entire network lab setup, but feel free to add some virtual machines in your laboratory to practice other port forwarding examples.

Now that we have learned about port forwarding basics and created our first rule, let's move on with outbound NAT.

Outbound NAT

Back to our examples, as we discussed at the beginning of this chapter, let's use the example of a small company with 10 computers and just a single public IP address in its WAN connection. Moving on in this scenario, we have the goal to connect all those computers to the internet just using firewall capabilities. How do we achieve that? By creating an outbound NAT! Let's see how things work. The following is an example topology of outbound NAT traffic:

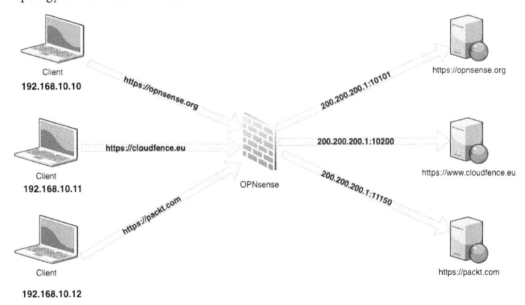

source Public IP address:Port (Outbound NAT)	source Internal IP address	Website
200.200.200.1:10101	192.168.10.10	https://opnsense.org
200.200.200.1:10200	192.168.10.11	https://cloudfence.eu
200.200.200.1:11150	192.168.10.12	https://packet.com

Figure 6.5 – Outbound NAT example

As we can see in the preceding figure, three clients are each accessing a website. Let's pick the host `192.168.10.11`: it is accessing the `https://cloudfence.eu` website, but to the CloudFence web server, the source IP address is the public IP of OPNsense firewall `200.200.200.1` with source port `10200`. So what is happening here? The outbound NAT rule is translating from the internal source IP to a public IP address, so from a TCP perspective, the client for the web server is the OPNsense public IP. In this way, all the internal clients can access different internet hosts at the same time, sharing the same public IP address, and as we can observe in the preceding table, the mapping is based on the internal source IP and public IP source port. Cool, huh?

> **Note**
>
> I have used a translation from an internal to a public IP address in this example, but it is common to use outbound NAT in scenarios that demand a translation from an internal to another internal IP address. An example of doing that is while using VPN tunnels.

Let's go to OPNsense WebGUI and see how to create NAT outbound rules.

NAT outbound modes

To create a new rule, we need to first set OPNsense to the correct NAT outbound mode. There are four possible modes, and you can set them by going to **Firewall | NAT | Outbound**:

- **Automatic outbound NAT rule generation**: The default OPNsense configuration, this mode will automatically generate the NAT outbound rules based on configured networks, but it does not permit adding manual rules. It is like autopilot – if it is engaged, the pilot becomes a passenger!

- **Manual outbound NAT rule generation**: This option is recommended when you want to create your own rules with 100% control of the outbound NAT behavior. In this mode, no automatic rule is generated by OPNsense.

- **Hybrid outbound rule generation**: This is a good option for most scenarios. With it, you can count on the automatic rules generated by OPNsense and can also create your own rules. The processing order, in this case, is first manual rules are processed and then comes the automatic rule. In this way, the manual rules will always prevail.

- **Disable outbound NAT rule generation**: This option simply turns outbound NAT off.

To create a new outbound rule, I'll suggest you set the **Hybrid outbound NAT rule generation** mode to on. To do that, you will need to do the following:

Figure 6.6 – Firewall: NAT: Outbound page

Let's now look at how to add an outbound NAT rule.

Adding an outbound NAT rule

To add a new outbound NAT rule, first, select a compatible mode – **Hybrid outbound NAT rule generation** or **Manual outbound NAT rule generation** – then click on the **Save** button. As I said before, I suggest using the hybrid mode.

Figure 6.7 – Firewall: NAT: Outbound page

As you can see in the preceding screenshot, **Manual rules** is displayed before **Automatic rules**, following the processing rules order.

After selecting a compatible mode, next click on the **+ Add** button to create a new rule. The following options will be shown:

- **Disabled**: Check this option to disable the current rule.

- **Do not NAT**: If this option is marked, it will disable NAT for this outbound rule.

- **Interface**: The network interface that this NAT outbound rule applies to.

- **TCP/IP Version**: Select which version of TCP and IP protocols this rule will process: **IPv4** or **IPv6**.

- **Protocol**: Select which IP protocol this port forwarding rule will match. OPNsense supports a lot of protocols for this option, but the most common ones to select here are TCP and UDP.

- **Source invert**: If checked, this option will invert (negate) the sense of the source in this rule. A *!* will be added before the source to identify it as an inverted source (after the rule is created). For example, if checked, a rule with the source `192.168.1.1` will apply to *all other addresses except* this specified one.

- **Source address**: Select the source address for this rule.

- **Source port**: The source port that this rule must match with.

- **Destination invert**: If checked, this option will invert (negate) the sense of destination in this rule.

- **Destination address**: Select the destination address that this rule will apply to.

- **Destination port**: Select the destination port for this rule.

- **Translation / target**: Enter the IP address that this outbound rule will translate the source address(es) for. The default option is **Interface address**.

- **Log: Log packets that are handled by this rule**: Check this option to enable the log for this port forwarding rule.

- **Translation / port**: Enter the port that will be translated as the source port of this rule.

- **Static-port**: Check this option if you want to always use the same port as the source port after the NAT outbound translation.

- **Pool Options**: The pool options can be used when you define a host alias with two or more addresses. The following pool options are available:

  - **Default**: This option will not apply any NAT algorithm to this NAT rule. This is the preferred option while using a single redirect IP port forwarding rule.

  - **Round Robin**: This algorithm will send packets for each available IP address sequentially. An example of usage is balancing traffic to web servers.

  - **Random**: As the name suggests, the option will randomize the available addresses in the pool (alias).

  - **Round Robin with Sticky Address / Random with Sticky Address**: The last two options can be combined with the sticky address option, which means that this port forwarding rule will map connections based on the source address. This is quite useful for applications and protocols that are session-based, such as TCP, for example.

  - **Source Hash**: Similar to the sticky address, this will use a hash based on the source address to determine the pool's IP address to use.

- **Bitmask**: This will map the source address using its subnet mask to define the translated address. For example, if the source address is `192.168.0.10/24` and the pool's subnet is `172.16.0.0/24`, then the translated IP address will be `172.16.0.10`.

- **Set local tag**: You can set a tag to mark packets processed by this rule to be checked later by other rules.

- **Match local tag**: Once you have a ruleset using tags, you can set a tag here to match it.

- **No XMLRPC Sync**: Check this option while using OPNsense in a high-availability configuration, to prevent this rule from being automatically synced to the other firewall members.

- **Category**: Defines which category this rule will be grouped in (optional).

- **Description**: The rule description.

Manually created outbound NAT rules are usually used in multi-WAN scenarios and while using high availability. We will explore both features later in this book, and we will have the opportunity to create outbound NAT rules when the right time arrives. Usually, the automatic mode will fit most small networks that need to share an internet connection with the internal network hosts.

Let's now move on to the last type of NAT available in OPNsense WebGUI — one-to-one.

One-to-one NAT

So far, we have learned about NAT types that allow us to map one-to-many IP addresses, so the main difference of this type of NAT is that it will map one IP to another one in a one-to-one manner. Every port will be forwarded to the internal IP or network, and if all traffic is permitted by the filter rule, this can mean an internal IP is exposed to the internet, so be careful using this type of NAT.

In my personal experience, I have seen a few instances of OPNsense using NAT one to one in corporate networks. A common situation I will mention is when you need to connect two remote sites using an IPsec tunnel and the internal networks overlap between those sites. In this case, one-to-one BINAT usually helps a lot!

Next, we'll see how to add a one-to-one rule.

Adding a one-to-one NAT rule

To add a rule, go to **Firewall | NAT | One-to-One** and click on the **+ Add** button. A new page will be opened with the following options:

- **Disabled**: Check this option to disable the current rule.

- **Interface**: The network interface that this NAT rule applies to.

- **Type**: There are two one-to-one NAT types:

  - **BINAT**: The default option, this mode will map networks with the same netmask (CIDR) in a bidirectional way. If you are mapping just one IP to another, it will work well too.

  - **NAT**: If you need to map networks that don't have the same netmask, this option is the one you need to set.

- **External network**: Enter the IP address or subnet that will start the translation mapping. For example, if you select a WAN interface in the **Interface** option, you'll probably want to set its IP address here.

- **Source / Invert**: If checked, this option will invert (negate) the source in this rule. A *!* will be added before the source to identify it as an inverted source (after the rule is created).

- **Source**: The source address for this rule. For example, if you are using this rule for a **BINAT** from an external to an internal address (and vice versa), you need to set the internal IP address.

- **Destination / Invert**: If checked, this option will invert (negate) the destination in this port forward rule. A **!** will be added before the source to identify it as an inverted source (after the rule is created).

- **Destination**: Set the destination address this rule will apply to. This is most used when you set a **NAT** type rule.

- **Category**: Defines which category this rule will be grouped in (optional).

- **Description**: The rule description.

- **NAT reflection**: This option is useful when you have an external NAT created but with internal clients accessing it. When enabled, it will redirect internal clients to the internal IP address of this rule. The available options are **Enabled**, **Disabled**, and **Use system default**. This last one can be set as you learned in *Chapter 5, Firewall*, in the, *Settings and tweaks* section.

- To add the rule, just click on the **Save** button.

> **Note**
> To permit traffic, you need to also add a firewall rule once. One-to-one rules don't automatically add rules in the firewall ruleset.

To illustrate a scenario using BINAT, let's consider a network with a Linux host using the **Secure Shell** (**SSH**) protocol for remote connections (incoming traffic NAT), and the same host using OPNsense as the default gateway to access the internet (outgoing traffic NAT):

External IP
200.1.2.3
Incoming NAT
BINAT
Outgoing NAT
Windows host
Internal IP
192.168.10.10
OPNsense
Internet

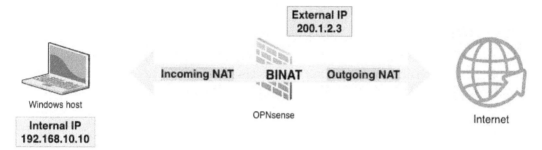

Figure 6.8 – BINAT example

The created one-to-one BINAT rule for the preceding scenario will look like this:

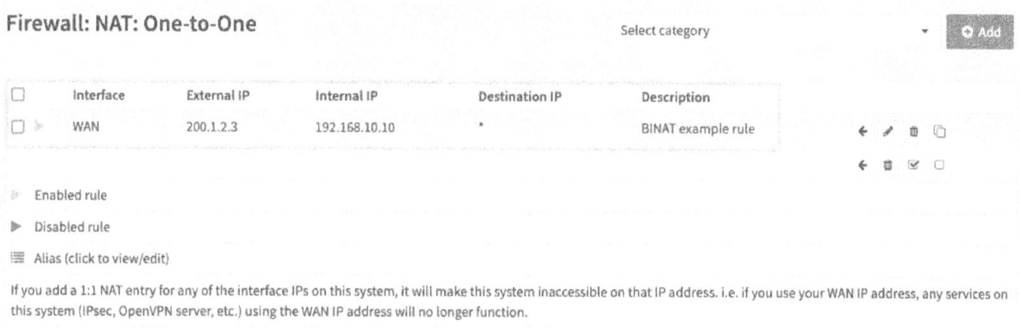

Firewall: NAT: One-to-One
Select category
Add
Interface
External IP
Internal IP
Destination IP
Description
WAN
200.1.2.3
192.168.10.10
*
BINAT example rule
Enabled rule
Disabled rule
Alias (click to view/edit)
If you add a 1:1 NAT entry for any of the interface IPs on this system, it will make this system inaccessible on that IP address. i.e. if you use your WAN IP address, any services on this system (IPsec, OpenVPN server, etc.) using the WAN IP address will no longer function.

Figure 6.9 – BINAT example rule

Another example using BINAT is using it with IPSec tunnels. We will explore VPN tunnels in the next chapter, but if you are curious about this BINAT implementation using IPSec, you can take a look at `https://docs.opnsense.org/manual/how-tos/ipsec-s2s-binat.html`.

> **Note**
>
> I just used the SSH protocol as an example; it is not a good idea to allow SSH access to the public internet. Instead, always prefer accessing it through VPN tunnels.
>
> To allow both incoming and outgoing traffic, you must create the proper firewall rules in the respective network interface.

As you can see, one-to-one NAT rules are an option in some specific situations that NAT port forwarding and NAT outbound rules will also fit, but the way you set it is, in most cases, simpler than other NAT-type rules.

Summary

In this chapter, we learned about the types of NAT OPNsense supports, how they work, and when to use each one. Of course, not all possible examples can fit in just one chapter, or maybe even one book, but I tried to cover the most common usages. Now you can understand and create port forwarding, outbound, and one-to-one rules using OPNsense WebGUI.

In the next chapter, we'll learn about traffic shaping and how to use OPNsense to create rules using it. See you there!

7

Traffic Shaping

This chapter presents the basics of traffic shaping and **Quality of Service (QoS)**. You'll learn what the traffic shaper rules can do in OPNsense, how to create and apply QoS in a local network, and how to monitor and test it.

The following are the topics we'll explore in this chapter:

- Introduction to traffic shaping
- Possible scenarios
- Creating rules
- Monitoring

Technical requirements

To follow this chapter, we'll need an understanding of the TCP/IP network stack and how to manage filter rules in OPNSense's WebGUI. It is optional to have an OPNsense instance running with a host in a local network, but this is advisable in order to be able to test some of the steps presented in this chapter.

Introduction to traffic shaping

Let's think about car traffic – on a bustling road, if an ambulance, fire truck, or police vehicle needs to pass, they have priority, right? The same situation also applies in the network packets world, where some packets need to be treated with higher priority to keep the protocols operating smoothly. With all sorts of different packets passing through the firewall, it is necessary to classify them to choose which traffic requires higher priority to keep an application, such as a **voice over IP** (**VoIP**) application. For example, if there are 20 VoIP packets and 1 for HTTP, the firewall or the streaming apps may work smoothly. The term QoS is also often used to refer to traffic shaping, and in this chapter, it will be used to refer to the classification and prioritization of packets.

To apply QoS to network traffic, OPNsense uses dummynet and ipfw, independent of the packet filter rules (which use pf). Ever heard of dummynet and ipfw before? Let's clarify them in the following sections.

dummynet and ipfw – a brief introduction

The OPNsense core team chose to not support or use the pf QoS (*ALTQ*) system in favor of the combination of dummynet and ipfw. This means basically that *under the hood*, packet filtering, which is done by pf (as we learned above), is independent of the traffic shaping system, so you can disable packet filtering and keep the traffic shaping in operation. dummynet is a FreeBSD system facility that allows classifying packets to the ipfw to manage later. Just in case you don't remember ipfw, it is one of the available packet filtering options in FreeBSD, but in OPNsense it is only used for QoS and Captive Portal, so you can't create firewall rules using ipfw on OPNsense.

Before moving on to the common scenarios in which we can use QoS, let's take a look at common terms used in the traffic shaping context in the WebGUI.

Pipes, queues, and rules

A **pipe** is basically used to define bandwidth in our traffic shaper rules. In more advanced usage, it is possible to define propagation delays, queue sizes, and packet loss rates in a pipe.

A **queue** is combined with a pipe to define a weight, which doesn't mean priority, but defines how the available bandwidth will be shared between the packets in the same defined pipe.

A **rule** is where we can select pipes or queues to apply traffic shaping, defining an interface, source, destination, protocols, and ports.

Let's now check out some examples of when you can use traffic shaping.

Possible scenarios

Some possible scenarios are necessary or considered as contexts where it is a good practice to employ QoS in a network firewall. Let's explore two common scenarios.

Controlling hosts' and users' bandwidth usage

When sharing the same WAN connection across more than one host, it is a good idea to keep some control to avoid a single user consuming all the available bandwidth. A simple way to do that is by applying rules that will limit the maximum bandwidth that each user will have available.

Protocol prioritization

With the massive usage of VoIP and other streaming technologies, it is becoming rare to see someone use a PBX extension with a regular voice line. So, it is very important to prioritize packets for these kinds of application. For example, if there are 20 VoIP packets and 1 for HTTP, the firewall may send out 10 VoIP packets, the HTTP packet, and the rest of the VoIP packets. With all sorts of packets passing through our OPNsense firewall, we need to guarantee that protocols that need real-time processing are first in our packet line, and a good way to do that is by creating traffic shaper rules.

With the massive usage of internet connections in the present world, it is so important to have a good packet filtering ruleset to create and maintain an accurate traffic shaper ruleset. Let's see how to do this in OPNsense.

Creating rules

To start and create a new traffic shaping rule, log in to the WebGUI and go to the **Firewall | Shaper** menu.

As an example, we will create a rule that defines the *maximum download bandwidth usage* for each user in our local network. Let's take the example of a local network with a 100 Mbps WAN connection shared across a few hosts.

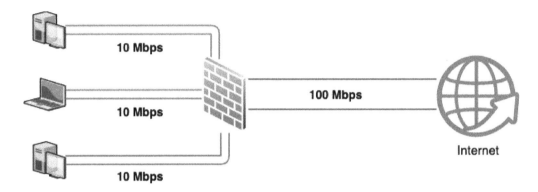

Figure 7.1 – Limiting maximum bandwidth per host

To start creating the rule, go to **Firewall | Shaper | Pipe**:

1. Click on the + icon to add a new pipe.
2. Make sure the **Enabled** option is checked.
3. In the **Bandwidth** field, type 10.
4. In the **Bandwidth Metric** option, select **Mbit/s**.
5. In the **Mask** option, select **destination**. This way, each destination will have 10 Mbps available. In our example, with this setting, you can have a maximum of 10 hosts: *10 hosts x 10 Mbps = 100 Mbps*.
6. Leave the **Enable CoDel** and **Enable PIE** schedulers unchecked.
7. For the **Description** field, enter something like Download-10Mbps.
8. Click on the **Save** button.

> **Note**
> To learn more about CoDel and PIE schedulers, please take a look at the ipfw man page: https://www.freebsd.org/cgi/man.cgi?query=ipfw&sektion=&n=1

After being created, your pipe will look like this:

Firewall: Shaper

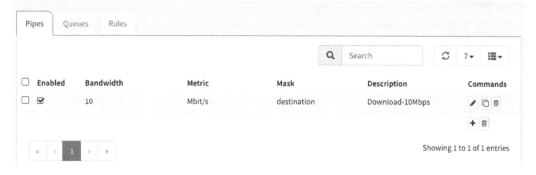

Figure 7.2 – Download pipe example

9. Now go to **Firewall | Shaper | Rules**.

10. Click on the + icon to add a new rule.

11. Make sure the **Enabled** option is checked.

12. In **Sequence**, leave the default value (1).

13. **Interface**: For this example, you may select **WAN**.

14. **Proto**: Leave the default protocol **ip** selected.

15. **Source**: Leave the default.

16. **Invert source**: As in the case of the firewall rules, you can apply a negate sign here to invert the source selection. Leave this unchecked.

17. **Src-port**: In this example, we will use the default: **any**.

18. **Destination**: Fill this field with the LAN network address (example: `192.168.0.0/24`).

19. **Dst-port**: Leave the default: **any**.

20. **Target**: Select the previously created pipe.

21. **Description**: Enter something like `Download rule`.

22. Click on the **Save** button.

23. After creating the rule, click on the **Apply** button. The created rule will look like this:

Firewall: Shaper

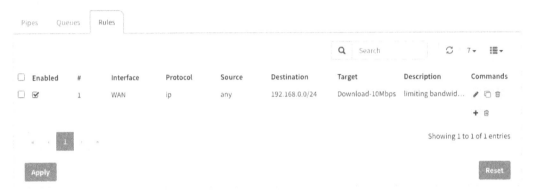

Figure 7.3 – Traffic shaper rule added

To test it, you will need a host connected to your OPNsense LAN. If you try to run a speed test on this host, you might see results like this:

Figure 7.4 – Download results example after traffic shaper rule applied

Notice that only the download was limited. Want to limit the upload too? Pretty easy with OPNsense! Let's do it.

Just *clone* the created rule, changing **Source** to the *LAN network address* and **Destination** to **any**. I'll also advise you to set the **Description** field to something like Upload rule, as this will be useful for clarity in the later *Monitoring* section in this chapter. Lastly, click on **Save** and then **Apply**. Try another speed test:

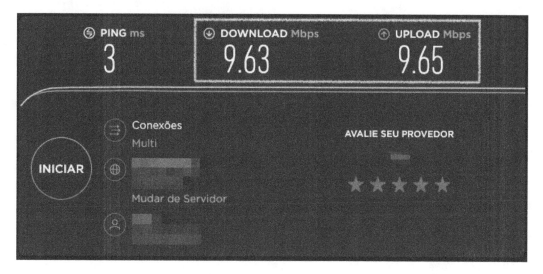

Figure 7.5 – Download and upload results after modifying the traffic shaper rule

> **Note**
> Download and upload tests were done using https://speedtest.net.

Great! We have created our first traffic shaping rule, limiting both download and upload bandwidth. Let's now check how to monitor created rules.

Monitoring

To check if the traffic shaper rules are working, you can go to the **Firewall | Shaper | Status** page. On this page, you will be able to check each rule's statistics, active flows, the number of packets, and the bytes matched by each one.

As an example, I took a couple of screenshots while the speed tests were running:

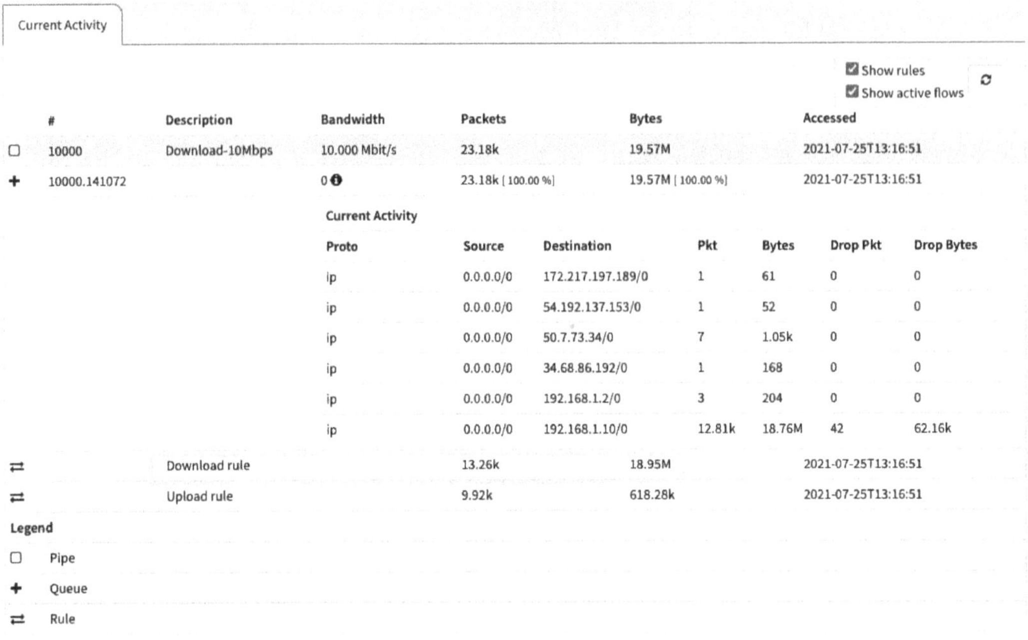

#	Description	Bandwidth	Packets	Bytes	Accessed
□ 10000	Download-10Mbps	10.000 Mbit/s	23.18k	19.57M	2021-07-25T13:16:51
+ 10000.141072		0 ⓘ	23.18k [100.00 %]	19.57M [100.00 %]	2021-07-25T13:16:51

Current Activity

Proto	Source	Destination	Pkt	Bytes	Drop Pkt	Drop Bytes
ip	0.0.0.0/0	172.217.197.189/0	1	61	0	0
ip	0.0.0.0/0	54.192.137.153/0	1	52	0	0
ip	0.0.0.0/0	50.7.73.34/0	7	1.05k	0	0
ip	0.0.0.0/0	34.68.86.192/0	1	168	0	0
ip	0.0.0.0/0	192.168.1.2/0	3	204	0	0
ip	0.0.0.0/0	192.168.1.10/0	12.81k	18.76M	42	62.16k

| ⇄ | Download rule | | 13.26k | 18.95M | | 2021-07-25T13:16:51 |
| ⇄ | Upload rule | | 9.92k | 618.28k | | 2021-07-25T13:16:51 |

Legend

□ Pipe

+ Queue

⇄ Rule

Figure 7.6 – Statistics while testing download speed

Note that the **Packets** and **Bytes** values in the preceding screenshot are higher for the download rule than the upload rule.

The following screenshot shows some more traffic shaping rule statistics:

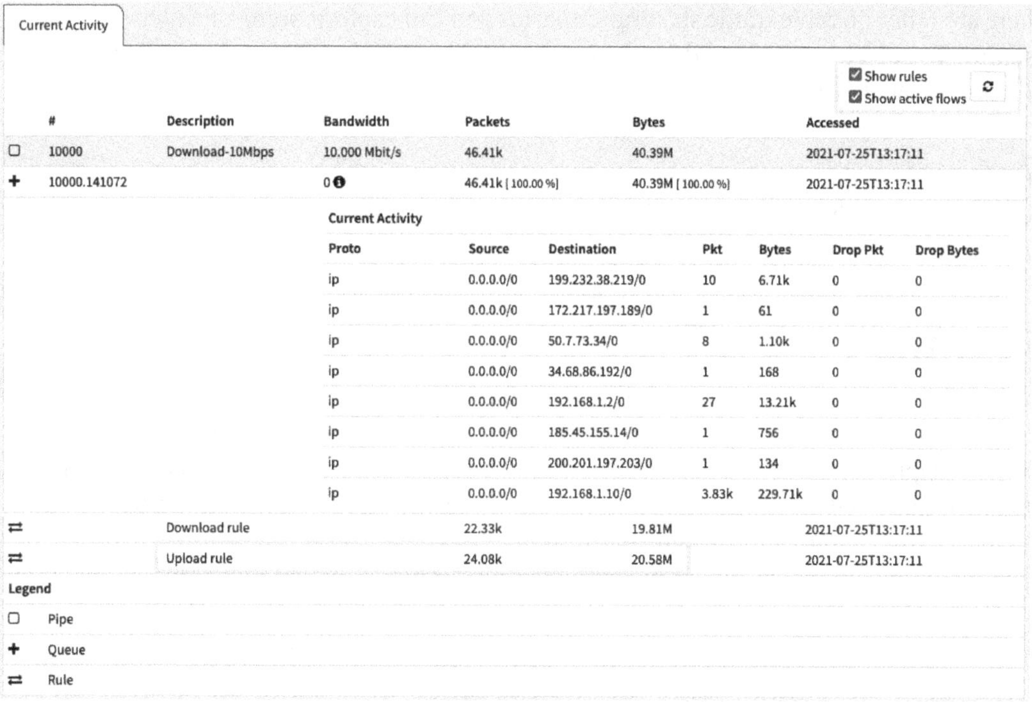

	#	Description	Bandwidth	Packets	Bytes		Accessed	
☐	10000	Download-10Mbps	10.000 Mbit/s	46.41k	40.39M		2021-07-25T13:17:11	
✚	10000.141072		0 ❶	46.41k [100.00 %]	40.39M [100.00 %]		2021-07-25T13:17:11	

Current Activity

Proto	Source	Destination	Pkt	Bytes	Drop Pkt	Drop Bytes
ip	0.0.0.0/0	199.232.38.219/0	10	6.71k	0	0
ip	0.0.0.0/0	172.217.197.189/0	1	61	0	0
ip	0.0.0.0/0	50.7.73.34/0	8	1.10k	0	0
ip	0.0.0.0/0	34.68.86.192/0	1	168	0	0
ip	0.0.0.0/0	192.168.1.2/0	27	13.21k	0	0
ip	0.0.0.0/0	185.45.155.14/0	1	756	0	0
ip	0.0.0.0/0	200.201.197.203/0	1	134	0	0
ip	0.0.0.0/0	192.168.1.10/0	3.83k	229.71k	0	0

		Description		Packets	Bytes		Accessed
⇄		Download rule		22.33k	19.81M		2021-07-25T13:17:11
⇄		Upload rule		24.08k	20.58M		2021-07-25T13:17:11

Legend

☐	Pipe
✚	Queue
⇄	Rule

Figure 7.7 – Statistics while testing upload speed

Note that the upload rule's values increased compared to *Figure 7.6*.

> **Note**
>
> I checked both **Show rules** and **Show active flows** before taking these screenshots.

Another way to check bandwidth utilization in real time is on the **Reporting | Traffic** page, but note that you will be presented with the overall network interface statistics and not specific traffic shaping stats – this is just another visual way to see the maximum bandwidth utilization.

There are other possible traffic shaping scenarios you can explore, some of which you can find in OPNsense's official documentation. There, you'll find some step-by-step "*how to*" guides that are easy to follow. You can access them at `https://docs.opnsense. org/manual/shaping.html`.

Due to the limited number of pages available in this chapter, it will not be possible to explore each scenario individually, but the basics you've learned here are sufficient for you to follow any OPNsense traffic shaping tutorial available online.

Summary

In this chapter, you learned about QoS and traffic shaping basics, how to create pipes, queues, and rules in the OPNsense WebGUI, and how to test them. Now you can design a traffic shaping policy based on your network requirements and create rules to apply the policy using OPNsense features. In the next chapter, we will begin our exploration into the world of virtual private networking.

8
Virtual Private Networking

This chapter will explore the **Virtual Private Network** (**VPN**) technologies available on the OPNsense core: IPsec and OpenVPN. With OPNsense, you can connect securely to remote sites and users using site-to-site and remote user deployments, and we'll learn how to do it in this chapter. After reading this chapter, you will be able to deploy VPNs using OPNsense, connecting remote networks and users in a secure manner.

We will explore the following topics in this chapter:

- OPNsense core VPN types
- Site-to-site deployments using IPSec
- VPN deployments using OpenVPN

Technical requirements

For this chapter, you need to know how to create firewall rules on OPNsense and read logs using a **Command-Line Interface** (**CLI**). Practicing the example configurations will require knowledge about creating virtual machines on VirtualBox, or other virtualization environments, to connect to OPNsense. Basic knowledge about certificates and cryptography is recommended.

OPNsense core VPN types

Virtual private networking enables remote networks to connect through WAN connections using cryptography to protect data exchanged inside a tunnel. It is a fundamental concept of VPNs, but we can extend it in scenarios that demand data protection using cryptographic mechanisms.

OPNsense in a stock installation supports two types of VPN technologies: **IPsec** and **OpenVPN**.

IPSec

The **Internet Protocol Security** (**IPsec**) protocol implements authentication and encryption of data over the IP protocol. It is a prevalent protocol when compared with other VPN technologies such as OpenVPN, for example. So, while connecting OPNsense with other firewall technologies, especially the closed source ones, IPsec will be the first option you should consider. This type of VPN is most commonly used to connect remote networks utilizing a site-to-site deployment. We will explore the possible deployments on OPNsense later in this chapter.

OpenVPN

OpenVPN is an open source protocol that uses the OpenSSL library to implement cryptography. It is a popular solution for open source projects, so with OPNsense, it isn't different than other open source firewall projects. As a flexible protocol, it is based on **Transport Layer Security** (**TLS**) and is commonly used in both site-to-site and remote user deployments.

IPsec versus OpenVPN

If you are not familiar with these technologies, you are probably thinking, *When should I choose one or another VPN technology?* Let's explore this by getting to know the pros and cons of each protocol.

The pros of the IPsec protocol are as follows:

- **As fast as it gets**: An IPsec tunnel doesn't demand powerful hardware to get a good connection performance once it runs on the IP protocol (layer 3). This protocol will work well once both VPN tunnel sides have a WAN connection with decent speed and network latency.
- **Compatibility**: Most modern firewalls support IPsec, so it is a widespread protocol for VPNs.

The cons of the IPsec protocol are as follows:

- **Tricky configuration**: It isn't rare to spend hours and hours getting an IPsec tunnel up and running. It's common for it to happen while trying to connect different firewall technologies.

- **NAT unfriendly**: Some devices don't support IPsec traversing NAT, so when traffic goes through connections using NAT, you will probably face issues while using IPsec. For this reason, connecting remote users using IPsec will likely be more complex than using OpenVPN. This is actually a NAT issue that breaks transported protocols by manipulating packet headers.

- **Different Internet Key Exchange (IKE) versions (v1 and v2)**: If both devices aren't using the same IKE version, getting the tunnel up and running will not be possible.

> Note
>
> You can read more about IPsec NAT issues here: `https://www.uninet.edu/6fevu/text/IPSEC-NAT.SGML.html`.

Now, let's look at OpenVPN's pros and cons.

OpenVPN's pros are as follows:

- **Flexibility**: OpenVPN can operate in most WAN connection scenarios without issues no matter what connection technology is being used, given that it has a decent speed and latency. Even with several NAT rules being applied to the traffic, it is a good option, especially while connecting remote users. The client-server design permits better support to multi-WAN connections or those that don't have public IPs on one of the VPN endpoints.

- **Better control features**: OpenVPN has more control features while using remote users' connections than IPsec has. The client-server design used by OpenVPN allows controls to be applied on the client (user) side. It is possible to push routes and other commands to clients without any user-required action; this is a considerable advantage while managing hundreds, or even thousands, of users.

- **Cost-effective and better security**: There are free OpenVPN clients for the most common operating systems (Windows, Linux, macOS, FreeBSD, and mobile platforms such as Android and iOS). So, if you need to implement a VPN solution with excellent security at a low cost, OpenVPN will probably be the best option. For remote user (also known as road warrior) deployments, OPNsense even supports **Two-Factor Authentication (2FA)** natively.

> **Note**
>
> An excellent example of fine-grained control of the client side using OpenVPN is blocking Windows users from using an external DNS server while connected to an OpenVPN tunnel. Details about this feature and how it was implemented on OPNsense can be found here: `https://github.com/opnsense/core/issues/4422`.

OpenVPN's cons are as follows:

- **Hardware consumption on high loads**: OpenVPN needs decent hardware with some crypto acceleration onboard to perform well on high-speed demanding tunnels. You must pay this price to have OpenVPN as a VPN solution in larger network environments; otherwise, if you connect small networks with low-speed demanding traffic on a tunnel, any modern hardware will work very well.

- **Requires additional software installation**: Compared to IPsec, which has a native client on Windows, Android, macOS, and iOS, for example, OpenVPN demands software installation to work, which can become an issue at the deployment phase with dozens of users; it may require a lot of work and planning to deploy access to all users.

As we can see, there are pros and cons for both VPN technologies, and based on my personal experience, I can tell you that you'll need both running on your OPNsense installations most of the time so you will be able to choose which one better suits your network demands.

Now, let's look at the possible deployments we can do using these two protocols.

Site-to-site deployments using IPsec

This deployment type is commonly used to connect remote networks: for example, a branch office to a head office. In the past, it was common to use private lines/WAN connections based on **Multiprotocol Label Switching** (**MPLS**) and frame relay, for example. Nowadays, with large offers from ISPs of high-speed WAN connections, it is cost-effective to use a site-to-site VPN solution rather than contracting a private line service.

An example of multiple site-to-site VPN tunnels connecting branch offices to a head office is shown in the following diagram:

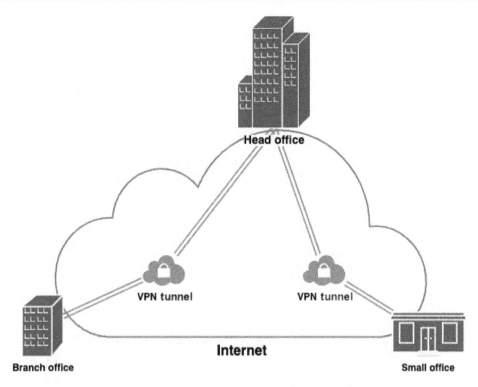

Figure 8.1 – Site-to-site VPN topology example

As we can see in the preceding diagram, the communication between the company offices is made by using the internet but protected by a VPN tunnel.

On OPNsense, we can use IPSec or OpenVPN to create a site-to-site tunnel. Let's see what the options are while creating an IPsec tunnel using the webGUI.

> **Note**
>
> We won't cover a complete lab setup here, but you can create two OPNsense virtual machines using *VirtualBox* or try any other public cloud service to practice it. Many cloud providers offer a free trial period, which you can use to create a lab and practice some of the steps shown in this book.

The OPNsense official documentation has many recipes for configuring site-to-site and remote users (called Road Warriors in OPNsense documentation). You can find it here: `https://docs.opnsense.org/manual/vpnet.html#ipsec`.

A typical IPsec site-to-site topology will look as follows:

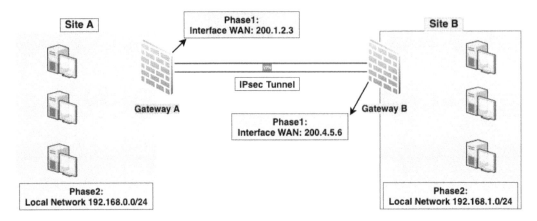

Figure 8.2 – Typical site-to-site IPsec topology

Before starting to investigate the OPNsense webGUI options, let's review some essential concepts about IPsec tunnels:

- **Phase 1**: In this first phase, the **IKE** protocol will exchange messages between the two gateways (also known as peers) to ensure an encrypted channel to start the phase 2 negotiation. On the OPNsense, webGUI will be in the phase 1 parameters that the gateways IP addresses will be defined.

- **Phase 2**: After phase 1 is successfully completed, the IPsec gateways start to negotiate phase 2, where the gateways define which networks will be allowed to send traffic into the tunnel, and how to authenticate and encrypt traffic, for example.

Now that we have explored the IPsec basics, let's see how to set a tunnel using the OPNsense webGUI:

1. Go to **VPN | IPSec | Tunnel Settings**.
2. Click on the + button to create a new IPsec phase 1 tunnel. After clicking on it, a new page will be shown with the tunnel settings.
3. **Disabled**: Click on this option to disable this phase 1 tunnel.
4. **Connection method**: Select one of the following options:

   - **Default**: This, by default, will behave like the **Start on traffic** option, as explained later in this list.

   - **Respond only**: This option will load the phase 1 connection without starting it. This option is recommended while using the **Common Address Redundancy Protocol (CARP)**.

- **Start on traffic**: This option tells the IPsec service to load this phase 1 connection and as soon as traffic is detected between gateways, the connection is established.

- **Start immediate**: When set, this will start up the tunnel immediately, ignoring the connection.

5. **Key Exchange Version**: Set the IKE protocol version that will be used in this connection. You can select **auto**, which will use IKEv2 (**V2**) as default, **V1** to force the use of IKEv1, or **V2**, which will set it to IKEv2.

6. **Internet Protocol**: Select which IP protocol family will be used in this phase 1 tunnel: **IPv4** or **IPv6**:

   - **Interface**: Select which network interface or virtual IP will be used for this IPsec tunnel.

   - **Remote gateway**: Type the IP address or hostname of the remote gateway the tunnel is connecting to. You will probably want to choose the OPNsense WAN interface here.

   - **Dynamic gateway**: Check this option when another tunnel endpoint is using a dynamic IP address. This option will usually be required when the remote gateway option is set with a dynamic DNS hostname.

7. **Description**: Provide a description of this IPsec tunnel.

8. **Phase 1 proposal (Authentication)**: The following options are authentication-related. Pay attention to specify precisely the same on both tunnel gateways:

   - **Authentication method**: Select which authentication method will be used by both tunnel sides. For site-to-site tunnels, you'll want to choose a mutual **Pre-Shared Key** (PSK) here. For better security, prefer using certificates instead.

   - **My identifier**: This option will be used to identify the peer and associate it with this tunnel on the remote side.

   - **Peer identifier**: This option will define how the remote peer will be identified in this tunnel by OPNsense. It is common to use the default option, **Peer IP address**, but it will depend on the configuration used by both tunnel sides.

   - **Pre-Shared Key**: This is the string used by both tunnel sides to authenticate, so choose something strong to keep this tunnel safe. It is a good practice to exchange this string out of band (by phone call, for example).

> **Tip**
> You can use the OPNsense CLI to generate a strong PSK; an example is using
> the `openssl rand -base64 48 | tr -dc '[:alnum:]\n'`
> OpenSSL command.

Phase 1 configuration

Phase 1 proposal (Algorithms) contains the following options that will define which encryption algorithms will be used in the tunnel proposal phase. Always check on both tunnel endpoints whether the options are precisely the same:

1. **Encryption algorithm**: This option must specify which encryption algorithm will be used by both tunnel endpoints. It is recommended to use strong algorithms such as **Advanced Encryption Standard** (**AES 256**) instead of deprecated ones such as **Triple Data Encryption Algorithm** (**3DES**).

2. **Hash algorithm**: Select which hash algorithms will be used in this tunnel. You can select multiple algorithms. Using more robust hash algorithms such as **SHA256** or higher is recommended to have a good level of security.

> **Note**
> Avoid using deprecated hash algorithms such as MD5 and SHA-1. More details can be found at the following page:
>
> 3DES deprecation NIST: `https://csrc.nist.gov/news/2017/update-to-current-use-and-deprecation-of-tdea`

Consider not using deprecated algorithms to avoid data leaking in your IPsec tunnels. Let's carry on with the phase 1 options.

3. **DH key group**: The **Diffie-Hellman** (**DH**) key exchange groups that will be used in this authentication phase. Consider using higher bit numbers to have a solid key to increase the security level on the key exchange process.

4. **Lifetime**: The **IKE Security Association** (**IKE SA**) lifetime in seconds; after that time, IKE SA will expire, starting the rekeying process.

5. **Advanced Options**: The following options define advanced parameters that will be necessary for specific scenarios. Some IPsec tunnels won't be connecting using the same operating system or firewall appliance, so sometimes you need to fine-tune the tunnel using these options:

- **Install policy**: This option will be checked by default, which means that the IPsec daemon (`charon`) will execute the tunnel routes installation in the kernel. Otherwise, if you need to use a routed IPsec tunnel (IPsec **Virtual Tunnel Interface (VTI)**), uncheck this option.

- **Disable Rekey**: Disable the IKE SA rekey process.

> Note
>
> An SA is a set of shared attributes (such as a cryptographic algorithm and encryption key) between the two IPSec gateway nodes.

- **Disable Reauth**: Disable the IKE SA reauthentication process on IKEv2, while in IKEv1 the reauthentication is always enabled by default.

- **Tunnel Isolation**: This option should be required with some firewall vendors such as Fortinet. It will create a tunnel for each phase 2 tunnel when the IKEv2 is in use.

- **NAT Traversal**: IPsec uses the **Encapsulating Security Payload (ESP)**, and a NAT implementation will break this if the NAT traversal technique isn't used. So, if this tunnel traverses devices that implement NAT, you'll probably need to change this option to **Enable** or **Force**. To learn more about NAT traversal and how it is related to IPsec, go to `https://en.wikipedia.org/wiki/NAT_traversal`.

> Note
>
> Remember to create rules allowing traffic for port 4500/UDP on both sides.

- **Disable MOBIKE**: The **Mobility and Multihoming protocol (MOBIKE)** is supported by IKEv2, and it helps IPSec manage IKE SAs efficiently. An example is when an IPsec endpoint has multiple IP addresses or changes frequently due to user mobility. If you don't need MOBIKE, check this option to disable it.

- **Dead Peer Detection**: Also known as **DPD**, it implements a periodic check to determine whether the tunnel is still alive.

- **Inactivity timeout**: Specifies in seconds how much time to wait before closing the tunnel due to inactivity.

- **Keyingtries**: Specifies how many times the IPsec tunnel will try to be negotiated. The default value is 3 (if left blank), and -1 will try forever.

- **Margintime**: Specifies how much time (in seconds) to wait for IKE SA to expire and start a rekeying process.

- **Rekeyfuzz**: Based on the Margintime defined time (seconds), define a percentage number in this option to increase it (Margintime) randomly. This is used to avoid both tunnel sides rekeying while creating many SAs that can break tunnel communication. Please refer to the following site: `https://linux.die.net/man/8/ipsec_pluto`.

After you have filled in all the necessary options to bring your tunnel up, just click on the **Save** button. You should see your newly created tunnel as in the following figure:

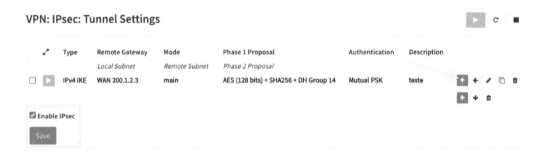

Figure 8.3 – IPSec phase 1 tunnel example

While using IPsec, be sure **Enable IPsec** is checked. Due to the position of this option on the page, sometimes it can lead some users to leave the service disabled by mistake.

Phase 2 configuration

It is in phase 2 where it is defined which networks will communicate using the IPSec tunnel. To start the phase 2 configuration, click on the + button, as indicated in the preceding screenshot.

A new page will be shown with the following options:

1. **General information**: Following the general setting of Phase 2, a single Phase 1 can have several Phase 2 settings:

   - **Disabled**: Check this option to disable this Phase 2.

- **Mode**: Select which mode this Phase 2 will operate. For deployments using IPv4 select **Tunnel IPv4**, or **Tunnel IPv6** for IPv6-based networks. **Route-based** is used while configuring IPSec VTI-based tunnels. **Transport Mode** is often used for protecting non-encrypted protocols such as **Generic Routing Encapsulation (GRE)** tunnels or transfer data without an additional hop (direct connection between two networks).

- **Description**: Fill with a description for this Phase 2 network.

- **Local Network**: Specifies Phase 2's local network options.

- **Type**: You can choose the available **local network interface subnet** options or select **Network** to type a custom network address. You can also select **Address** for a tunnel including just a host (/32 CIDR).

- **Address**: Fill with the network or address, respectively, if defined as **Address** or **Network** in the preceding option.

- **Remote Network**: The same Phase 2 options, but for the remote site network.

- **Type**: For the remote network, just two options are available, **Network** and **Address**; you can choose and fill it as defined by the remote side.

- **Address**: Fill with the network or address, respectively, depending on what is defined in the previous option.

- **Phase 2 proposal (SA/Key Exchange)**: The following options will define the proposal parameters for this Phase 2:

  i.   **Protocol**: Select between **ESP** for an encrypted tunnel and **AH** for authentication only. In most tunnels, you'll want to leave this option in the ESP protocol.

  ii. **Encryption algorithms**: Select which encryption algorithms this Phase 2 will support. As we explored in Phase 1, remember to select robust algorithms for a good security level.

  iii. **Hash algorithms**: Select hash algorithms will be supported by Phase 2. The algorithms' good practices apply here too.

  iv. **PFS key group**: **Perfect Forward Secrecy** (PFS) will guarantee that an already used key will not be used again. By default, this option is disabled, but you can select which group and how many bits will be employed while using PFS. Notice that the bigger the number of bits, the more CPU power is needed.

  v.   **Lifetime**: How much time (in seconds) before the Phase 2 SA expires.

  vi. **Advanced Options**: Phase 2 advanced options are described here.

vii. **Automatically ping host**: Specifies a remote host to send **Internet Control Message Protocol** (**ICMP**) packets (the `ping` command). This option might be helpful when an IPsec tunnel keeps going down for inactivity. Some firewalls close the tunnel due to traffic inactivity. It is useful for keeping the state alive; this is a common issue while using NAT.

viii. **Manual SPD entries**: **Security Policy Database** (**SPD**) is created using the network defined in this Phase 2 tunnel. Sometimes you need to send traffic through the tunnel from a network that is not specified or allowed on the remote side. You can add these networks or addresses here for a workaround on situations like that. This option is required while using IPSec with NAT/BINAT scenarios.

To add a new Phase 2 tunnel, we finish by clicking on the **Save** button:

Figure 8.4 – Newly added Phase 2

As we can see, the newly added Phase 2 tunnel will be below the respective Phase 1 configuration.

> **IPSec Remote User Deployment Note**
>
> In this book, I will try to cover the most common deployments using OPNsense VPN-supported protocols. So, for the IPsec protocol, we will cover only the site-to-site deployment; the remote users scenario will be explored in the OpenVPN topic. In my personal experience of managing OPNsense and pfSense firewalls, I can tell you that a customer never asked me to set an IPsec remote user's deployment, using OpenVPN instead after comparing both protocols' pros and cons. I am not saying that IPSec shouldn't be used on a remote user's deployment, but I can't just ignore years of experience and advise you to use it.

There are two relevant points we need to explore before finishing this site-to-site topic using IPSec: **BINAT** and **VTI**.

IPSec BINAT

Sometimes, you receive a call from a customer or a work colleague asking you to connect two remote networks using IPsec with overlapping addresses, which can become a big headache if you don't use OPNsense as the network's firewall.

Let's explore an example through the following diagram:

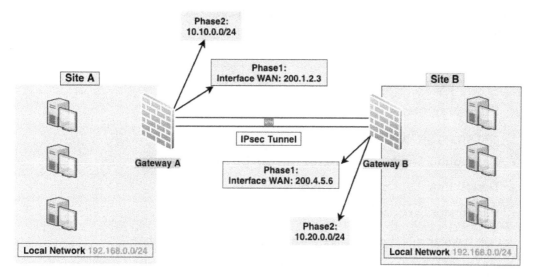

Figure 8.5 – IPSec BINAT example

As shown in the preceding diagram, both **Site A** and **Site B** have the same local network address, `192.168.0.0/24`, which is a problem while connecting both sites using an IPsec tunnel. How could we fix this issue and connect these networks without changing the network address on one of the sites? The short answer is we can apply a NAT rule before packets enter the tunnel. Let's see how to do that.

To achieve our goal, we first need to define two new network addresses that don't exist on both sides. In our example, these are as follows:

- **Site A**: `10.10.0.0/24`
- **Site B**: `10.20.0.0/24`

These two new network addresses will be only used in Phase 2 and NAT configuration. Let's go through the steps needed to make it happen.

> **Note**
> Let's assume here we are doing all the steps on **Site A**.

Let's move on to create our first NAT rule.

Creating a NAT one-to-one rule

To masquerade the traffic from the local network to the IPsec tunnel, we need to create a one-to-one NAT rule.

To create a new NAT one-to-one rule, go to **Firewall | NAT | One-to-One** and click on the + button. Fill the new rule with the following options:

- **Interface: IPsec**.

- **Type: BINAT**.

- **External network**: 10.10.0.0/24.

- **Source: Single host or Network** – 192.168.0.0/24.

- **Destination: Single host or Network** – 10.20.0.0/24.

- **Description**: IPSec BINAT example.

- Click on the **Save** button.

After that, the new rule is created as shown in the following screenshot:

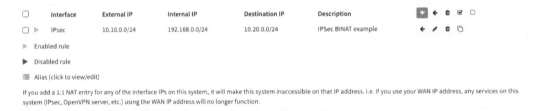

Figure 8.6 – NAT one-to-one BINAT example rule

As we can see in the preceding screenshot, the new network is configured as **External IP** while the actual network address is **Internal IP**. Now let's take steps to add the configuration on the IPSec tunnel.

Creating a new IPSec tunnel

Next, we'll create an IPsec tunnel to work with the previously created BINAT rule through the following steps:

1. Go back to **VPN | IPSec | Tunnel Settings** to create new Phase 1 and Phase 2 configurations.

2. Add a new Phase 1 configuration. I will provide some example configuration options, but you can change or adapt to your lab or network environment settings (leave the not mentioned options with default values or empty):

  - **Connection method: default.**

  - **Key Exchange version: auto.**

  - **Internet Protocol: IPv4.**

  - **Interface: WAN.**

  - **Remote gateway**: Type the Site B/other firewall's IP address.

  - **Description**: Site A Phase 1 configuration.

  - **Authentication method: Mutual PSK.**

  - **My identifier**: My IP address.

  - **Peer identifier**: Peer IP address.

  - **Pre-Shared Key**: Create a strong PSK here.

  - **Encryption algorithm: AES 256.**

  - **Hash algorithm: SHA 256.**

  - **DH key group: 14 (2048 bits).**

  - **Lifetime**: 28800.

  - **Install policy**: Checked.

  - To finish, click on the **Save** button.

3. Add a new Phase 2 configuration (leave the not mentioned options with default values or empty):

  - **Mode Tunnel: Tunnel IPv4**

  - **Description**: Site A Network

  - **Local Network:**

    i.   **Type: Network**

    ii. **Address:** 10.10.0.0/24

  - **Remote Network:**

    i.   **Type: Network**

    ii. **Address:** 10.20.0.0/24

- **Protocol**: ESP

- **Encryption algorithms**: AES auto

- **Hash algorithms**: SHA256

- **Lifetime**: 3600

- **Manual SPD entries**: 192.168.0.0/24

> **Note**
>
> The **Manual SPD entries** option ensures that the previously created NAT rule works with the IPsec tunnel.

3. Click on the **Save** button to finish adding this Phase 2 configuration.

After creating both phases, the new IPSec configuration should look as follows:

Figure 8.7 – IPsec example configuration

The preceding screenshot shows the **Site A** configuration. To create the **Site B** configuration, repeat the preceding steps, swapping the networks and gateway-related options on firewall B to get it working.

As you can see, OPNsense has powerful features to help with tricky scenarios such as in the preceding example. To finish our IPsec site-to-site expedition, let's look at the routed IPsec tunnel.

IPsec routed tunnel (VTI)

The routed tunnel configuration uses a **Virtual Tunnel Interface** (**VTI**), so you will probably hear people calling this IPsec configuration just VTI. The main difference with this kind of configuration is that instead of using a policy based on routes managed by the IPsec daemon and the kernel, it will use installed routes on the operating system using the virtual interface. If you already use OpenVPN tunnels, it will sound familiar to you; the working principle is almost the same.

Following the steps from the IPsec BINAT example, you can alter a few steps to change it from a policy-based to a route-based tunnel. You don't need to follow the NAT rule steps and change the network address to the existing local networks configured in Site A and B firewalls.

Let's go through the steps:

1. Follow the steps for Phase 1 as discussed in the *Creating a new IPSec tunnel* section, leaving the following options *unchecked*:

   - **Install policy**
   - **Disable Rekey**
   - **Disable Reauth**
   - **Dead Peer Detection**
   - **Disable NAT Traversal**: **NAT Traversal: Disabled**

2. In Phase 2, we must create a tunnel IP for each tunnel side instead of defining the local network as we did before while configuring a policy-based tunnel. Follow the steps and change the following options (assuming Site A as an example):

   - **Local Network**:

     i. **Type: Address**

     ii. **Address**: 10.254.1.1

   - **Remote Network**

     i. **Type: Network**

     ii. **Address**: 10.254.1.2

   Leave the **Manual SPD entries** option empty.

3. Go to **System** | **Gateways** | **Single** and create a new gateway:

   - **Name:** IPSecGW.
   - **Interface:** Choose the IPSec interface (the tunnel must be up and running).
   - **IP address:** 10.254.1.1.
   - **Far Gateway:** Checked.

   Click on the **Save** button to add the new IPsec gateway.

4.   Go to **System | Routes | Configuration** to add the tunnel route.

Click on the + button and fill in the options as follows:

- **Network Address**: Site B's network (for example, `192.168.1.0/24`)

- **Gateway**: `IPSecGW`

- **Description**: `IPSec Site B route`

Click on the **Save** button to finish adding the new route. Congratulations! You just set your new IPSec routed tunnel!

Now that we have explored IPSec site-to-site configurations, let's look at some diagnostics tools that we can use while checking everything is OK with our tunnels.

IPSec diagnostics

Sometimes IPSec issues can be tricky, but OPNsense has some helpful diagnostic resources to help to identify and solve IPSec issues. If you need to check whether a tunnel is up and running, go to **VPN | IPsec | Status Overview**:

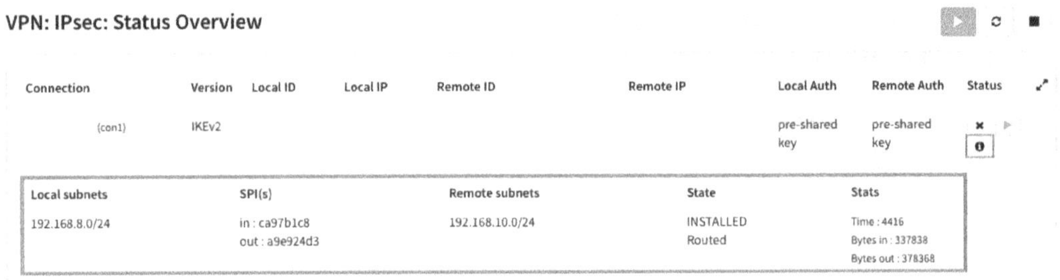

Figure 8.8 – A running IPsec tunnel status

As shown in the screenshot, you can check that Phase 1 is up (green play icon). To check Phase 2, you must click on the **i** button, and the defined local subnets will be shown with some extra information such as statistics, state, and remote subnets.

To check log files on the webGUI, go to **VPN | IPsec | Log file**. On this page, you can check all logs generated by the IPsec daemon. You can also check the log file using the CLI:

```
#tail -f /var/log/ipsec.log
```

To list all current IKE SAs, you can go to **VPN | IPsec | Security Association Database**.

To list the **SPD**, check **VPN | IPsec | Security Policy Database**.

To manage the IPsec daemon on the CLI, use the `ipsec` command. Some examples of its usage are the following:

To list *routed connections* and *security associations*, use the following command:

```
#ipsec status
```

If you want to see a specific tunnel output, add the connection name (`con`, for example) and use the following command:

```
#ipsec status con1
```

To list the daemon's *listening IP addresses, connections, routed connections,* and *security associations*, use the following command:

```
#ipsec statusall
```

To list all the available options, type the following:

```
#ipsec --help
```

As you may notice, the IPsec topic has content that could fill an entire book, so I tried to cover the most relevant subjects related to it in this chapter. As mentioned earlier in this chapter, I recommend you take a look at the official OPNsense documentation to explore other IPsec topics: `https://docs.opnsense.org/manual/vpnet.html#ipsec`.

Firewall rules

Finally, remember to create all the firewall rules to allow traffic to pass through each firewall. In the previous examples, we'll need to add rules on local networks and IPsec interfaces. The official OPNsense documentation recommends creating rules on WAN interfaces to allow *ESP, ISAKMP*, and *NAT-T* traffic. Still, I have seen many OPNsense firewalls without those rules, working with IPsec tunnels. So why should you create those rules? First, it's good practice! Another reason is that with OPNsense's evolution, this behavior can change in the future, and your tunnels could stop working after an update. You can check the rules out here: `https://docs.opnsense.org/manual/how-tos/ipsec-s2s.html#firewall-rules-site-a-site-b-part-1`.

Now let's move on with OpenVPN and see how to employ it in different VPN deployments.

VPN deployments using OpenVPN

As discussed previously, OpenVPN is a versatile VPN protocol and works well both in site-to-site and remote user deployments. Let's assume a similar scenario to the one we used in the IPsec site-to-site topic as an example to start this section.

Site-to-site deployment

In the following topology, you will notice that the tunnel network has a `10.10.10.0/30` address with just two usable addresses, one to each firewall:

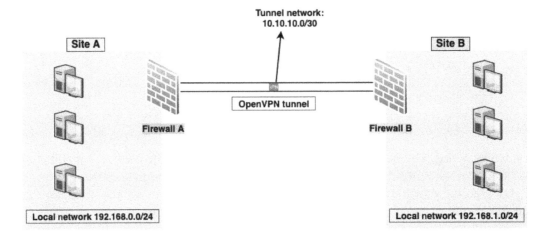

Figure 8.9 – OpenVPN site-to-site topology example

We will configure this tunnel using a *shared key* to stay on the same track as our previous IPsec example. OpenVPN supports authentication using certificates for site-to-site deployments, although this chapter will look at it later, in the *remote user deployment* topic.

OpenVPN uses a client-server approach, so to define which side of the tunnel should be set as a server or client, you should take these things into consideration:

- **MultiWAN OPNsense as OpenVPN client**: Sometimes, a site is set as a client even if it is the only one with the multi-WAN configuration. The reason? A well-configured multi-WAN OPNsense firewall will switch the default gateway quickly if a WAN link goes down. Otherwise, if it is set as the server on the OpenVPN deployment, changing from one downlink to another online can take longer once the client side waits for a timeout before reconnecting to another configured WAN link.

- **Dynamic IP firewall/NATed IP address**: In this case, you'll probably want to set this firewall as the OpenVPN client, especially if the other firewall has a static public IP configured. The client side needs to find the server to bring the tunnel up, hence, it is better to have the client side using a dynamic or NATed IP address; otherwise, setting the tunnel wouldn't be possible.

Considerations made, let's see how to configure a site-to-site tunnel using OpenVPN:

1. Go to **VPN | OpenVPN | Servers**.
2. Click on the + icon to create a new OpenVPN server.

   A new page will open with the following options. Please refer to the preceding example topology to fill in these options:

   - **General information**: This section refers to general tunnel information.

   - **Disabled**: Leave this unchecked to enable this server.

   - **Description**: Fill it with something that describes this server.

   - **Server Mode**: This option will set which operation mode the server will be configured in. The following are the operation modes:

     i. **Peer to Peer (SSL/TLS)**: Choose to set a site-to-site tunnel using a certificate as the authentication method.

     ii. **Peer to Peer (Shared Key)**: Choose to set a site-to-site tunnel using a shared key string. *We'll select this for our example.*

     iii. **Remote Access (SSL/TLS)**: This option is used for remote user tunnels using a certificate as the authentication method.

     iv. **Remote Access (User Auth)**: When selected, this will configure a remote user tunnel using only a username and password as the authentication method.

     v. **Remote Access (SSL/TLS + User Auth)**: Sets a remote user tunnel utilizing a combination of both previous options, that is, **certificate and username/ password** authentication. It is the safer option for remote user tunnels but isn't easy to configure while using an external authentication backend such as RADIUS or LDAP (Active Directory, for example).

   - **Protocol**: The available options are **UDP4**, **TCP4** for *IPv4* and **UDP6**, or **TCP6** for *IPv6*. To use both IP versions, select one of the **UDP** or **TCP** options. *For our example, we'll choose UDP*. Using TCP could impact the tunnel's throughput performance.

- **Device Mode**: Then, you can choose which operation mode the virtual network will use:

    i.   **tun**: Select for a route-based tunnel. *Select this one.*

    ii. **tap**: Select for both a route-based and a capable bridge tunnel (layer 2).

- **Interface**: Select which interface OpenVPN will listen to for connections. To enable all, select **any**. *Leave the default option: any.*

- **Local port**: Choose which port the server will listen to for incoming connections. *Leave the default 1194.*

- **Cryptographic Setting**: All related cryptographic settings will be set in this section, some of which are discussed here:

    i.   **Shared Key**: Check **Automatically generate a shared key** to generate a shared key, or uncheck this option to paste a shared key if you already have one. *Leave the default: checked.*

    ii. **Encryption algorithm**: Select which encryption algorithm you want to use. For our example, select **AES-256-CBC**.

    iii. **Auth Digest Algorithm**: Select which authentication digest algorithm to use in this tunnel. For our example, select **SHA 256**.

- **Tunnel Settings**: In this section, the tunnel's networks and routes will be configured:

    i.   **IPv4 Tunnel Network**: Defines the tunnel's virtual network for the peer-to-peer connection. We will fill it with `10.10.10.0/30`.

    ii. **IPv6 Tunnel Network**: Define the network, if you want to set the tunnel's virtual network using IPv6.

    iii. **Redirect Gateway**: Check this option to force all clients' traffic to be routed to the tunnel. Use it with caution!

    iv. **IPv4 Local Network**: Defines which networks will be routed through the tunnel. You can add more than one network, filling in the addresses separated by a comma. For our example, fill it with `192.168.0.0/24`.

    v.   **IPv6 Local Network**: The same as the previous option, but using IPv6 networks instead.

    vi. **IPv4 Remote Network**: Defines which remote networks will be reachable through the tunnel. You can add more than one network by filling in the addresses separated by a comma. For our example, fill it with `192.168.1.0/24`.

vii. **IPv6 Remote Network**: The same as the previous option, but using IPv6 networks.

viii. **Concurrent connections**: Defines the maximum number of concurrent clients connected in this tunnel.

- **Compression**: Defines which compression algorithm will be used in this tunnel. Leave **No preference** for no compression.

- **Type-of-Service**: Check this option to enable TOS on traffic going through the tunnel. This may be desirable while using VoIP traffic inside the tunnel.

> Note
> Using the **Type-of-Service** option can leak information about the traffic.

- **Duplicate Connections**: If this option is checked, the OpenVPN server will allow clients to connect with the same **Common Name** (**CN**). It is not recommended!

- **Disable IPv6**: Checking this option means it doesn't send IPv6 through the tunnel.

- **Client Settings**: This section will define client-related options.

- **Dynamic IP**: Check this option for better stability for roaming users with frequent IP address changes.

- **Address Pool**: This option will automatically set the IP addresses pool to be leased to clients. *In our example, you can uncheck it.*

- **Topology**: If checked, this will assign one IP instead of a /30 subnet per connected client. Using a topology can save IP addresses from the tunnel network pool once all usable addresses from the subnet have been allocated. According to the OpenVPN official documentation, Windows clients with old OpenVPN versions (before *2.0.9*) will not work with the **Topology** option enabled, so take care before enabling it.

- **Client Management Port**: Select this option to change the default client's management port (166).

- **Advanced configuration**: From this point forward, you must go with caution. Advanced options, when set without proper knowledge, can break things!

- **Advanced**: You can set OpenVPN server-compatible options separated by a semicolon (;). Everything you type here will be written on the server's configuration file. Caution!

- **Verbosity level**: Select which log verbosity level you want to set. The levels are as follows (extracted from the OPNsense webGUI):

    i.   **0**: No output except fatal errors.

    ii.   **1**: Startup information, connection-initiated messages, and non-fatal encryption and network errors.

    iii. **2** and **3**: Show TLS negotiations and route information.

    iv.   **4**: Normal usage range.

    v.   **5**: Output *R* and *W* characters to the console for each packet read and write. Uppercase is used for TCP/UDP packets and lowercase is used for TUN/TAP packets.

    vi.   **6-11**: Debug information range.

- **Force CSO Login Matching**: **Client-Specific Overrides** (**CSO**) is a way to set custom configurations to clients based on CN. To use the username to match a client instead of CN, check this option.

- To add the new OpenVPN server configuration, click on the **Save** button.

After creating the Site A OpenVPN server, we'll need to set the Site B OpenVPN client:

1. On the Site B firewall, go to **VPN | OpenVPN | Clients** and add a new client configuration, clicking on the + button.

2. Configure the options as follows (leave not mentioned options as the default):

    - **Server Mode: Peer to Peer (Shared Key)**

    - **Protocol**: **UDP**

    - **Device mode: tun**

    - **Interface: any**

    Remote server:

    - **Host or address**: Fill with the OpenVPN server address (probably the WAN interface address).

    - **Port**: Fill with the configured server's port (1194 in our example).

3. In **Cryptographic Settings**, *uncheck* the **Shared Key** option **Automatically generate a shared key**, go back to the OpenVPN server settings, and copy/paste the shared key here. In the other options, select the same configuration as the server.

4. In **Tunnel Settings**, fill **IPv4 Tunnel Network** with `10.10.10.0/30`, and fill **IPv4 Remote Network** with the Site A local network address: `192.168.0.0/24`.

5. Click on the **Save** button to add the new OpenVPN client configuration.

> **Note**
>
> OPNsense must have a firewall rule on the WAN interface permitting traffic on *UDP* port `1194` to allow connections on the new OpenVPN tunnel on the server side.

6. To check whether the new OpenVPN tunnel is up, go to **VPN | OpenVPN | Connection Status**.

7. You will see something like the following:

VPN: OpenVPN: Connection Status

Client Instance Statistics

Name	Remote Host	Virtual Addr	Connected Since	Bytes Sent	Bytes Received	Status	
S2S DC Matrix UDP	200.201.	10.15.16.2	2021-08-12 20:29:46	27.15 MB	93.68 MB	up	

Figure 8.10 – OpenVPN Connection Status page

In the preceding screenshot, we can see the status on the OpenVPN client side with the tunnel up. If you did everything right, you should see the tunnel up on both sides of your OPNsense firewalls.

Remote user deployment

The remote user deployment is often used to provide secure access for remote employees and their companies. In this topic, we'll explore an example connecting remote users to a company's local network using user authentication. Let's go through the steps to set up our remote user deployments.

The following topology shows a typical client-to-site OpenVPN scenario, in which each user can have their own rules and routes controlled by the OPNsense firewall. We'll go through the steps to configure an OpenVPN tunnel based on the screenshot in *Figure 8.10*.

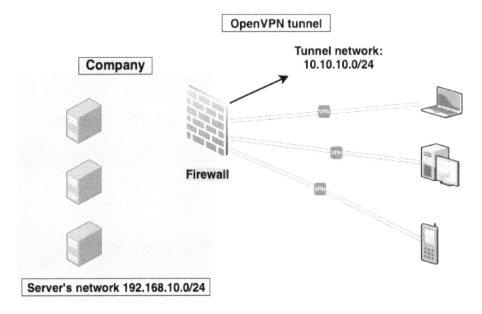

Figure 8.11 – OpenVPN remote users topology example

The first step is to create a **Certificate Authority** (**CA**) that will be used in an OpenVPN tunnel configuration. We'll need it to create a certificate later. To start, go to **System | Trust | Authorities** and click on the + button.

Fill in the options as follows:

- **Descriptive name:** OpenVPN CA.

- **Method:** Select **Create an internal Certificate Authority**.

- **Key Type:** Select **RSA**.

- **Key length (bits):** Select **4096**.

- **Digest Algorithm:** Select **SHA512**.

- **Lifetime (days):** You can leave the default value as **825**.

> **Note**
>
> It is recommended to create a CA with a longer lifetime. Otherwise, you would have to exchange all VPN connections that rely on certificates based on it.

- **Country Code**: Select your country; in this example, we'll choose **Portugal**.
- **State or Province**: Fill with yours; for the example, we'll use `Lisboa`.
- **City**: Fill with yours; for the example, we'll use `Lisboa`.
- **Organization**: Fill with yours; for the example, we'll use `Cloudfence`.
- **Email Address**: Fill with your company department or server administrator address; for the example, we'll use `soc@cloudfence.pt`.
- **Common Name**: `internal-openvpn-ca`.
- To finish creating the **CA** click on the **Save** button.

Next, we need to create the certificate that we'll also use in our OpenVPN tunnel configuration. It will be required to sign the client certificates. Follow the given steps:

1. To create the certificate, go to **System | Trust | Certificates** and click on the + button.

    Use the following points to guide you through creating a certificate, which will be required to set up your OpenVPN tunnel:

    - **Method**: Select **Create an internal Certificate**.

    - **Descriptive name**: `OpenVPN certificate`.

2. In the **Internal Certificate** section, set the options as follows:

    - **Certificate authority**: *Select the previously created CA*.

    - **Type**: Select **Server certificate**.

    - **Key Type**: Select **RSA**.

    - **Key length (bits)**: Select **4096**.

    - **Digest Algorithm**: Select **SHA512**.

    - **Private key location**: **Save on this firewall**.

3. In the **Distinguished name** section, set the options as follows:

    - **Country Code**: Select your country; in this example, we'll choose **Portugal**.

    - **State or Province**: Fill with yours; for the example, we'll use `Lisboa`.

    - **City**: Fill with yours; for the example, we'll use `Lisboa`.

    - **Organization**: Fill with yours; for the example, we'll use `Cloudfence`.

- **Email Address**: Fill with yours; for the example, we'll use `soc@cloudfence.pt`.

- **Common Name**: `OpenVPN server certificate`.

4. To finish creating the certificate, click on the **Save** button.

The following are the steps for creating the OpenVPN server:

1. Go to **VPN | OpenVPN | Servers** and click on the + button. Set the options as described in the following. (Options that are not mentioned here can be left as the default.)

    On the **General information** page, set the options as follows:

    - **Description**: For example, `Remote users server`.

    - **Server Mode**: Select **Remote Access (SSL/TLS + User Auth)**.

    - **Backend for authentication**: Select **Local Database**.

    - **Protocol**: Select **UDP**.

    - **Device Mode**: Select **tun**.

    - **Interface**: Select **any**.

    - **Local port**: Leave the default value; for this example, it is `1195`.

2. Next, we'll define the **cryptographic settings**:

    - **TLS Authentication**: Both options must be checked – **Enable authentication of TLS packets** and **Automatically generate a shared TLS authentication key**.

    - **Peer Certificate Authority**: Select the CA we created before.

    - **Server Certificate**: Select the server certificate we created before.

    - **DH Parameters Length**: Select **4096 bit**.

    - **Encryption algorithm**: Select **AES-256-CBC**.

    - **Auth Digest Algorithm**: Select **SHA512**.

    - **Certificate Depth**: Select **One (Client+Server)**.

    - **IPv4 Tunnel Network**: `10.10.10.0/24`.

    - **Disable IPv6**: Check this option.

    - To finish the OpenVPN server creation, click on the **Save** button.

3. Now, we need to create the users to connect to this OpenVPN server. Go to **System | Access | Users** and click on the + button to create a user.

4. Fill in the **Username** and **Password** fields, check the **Certificate** option **Click to create a user certificate**, and click on the **Save and go back** button to add the user.

5. A new page will open to create the user's certificate. Change the **Method** option to **Create an internal Certificate** and check whether the CA we have created is selected. It might be selected by default; if not, do it.

6. Click on the **Save** button to create the user's certificate.

> **Note**
>
> OPNsense must have a firewall rule on the interface that will receive the OpenVPN connections permitting traffic on UDP port 1195 (set in **Local Port**) to allow connections to the newly created OpenVPN tunnel on the server side.

7. To test our tunnel, you will need to export the client configuration. To do that, go to **VPN | OpenVPN | Client Export**.

8. **Remote Access Server**: Select the tunnel you have created.

9. **Export type**: Depending on the client's operating system and OpenVPN client version, you can choose the right option. For example, on a Windows-based client, you can select the **File Only** or **Archive** option.

10. **Hostname**: Type in your firewall's WAN address or an FQDN hostname (published on the internet).

11. **Port**: Use the port of the OpenVPN server; in our example, it is 1195. You can set some extra options, including custom options, in the **Custom config** field. For this example, we'll use the default configuration. To download the user *config* file, click on the download button as shown in the following screenshot:

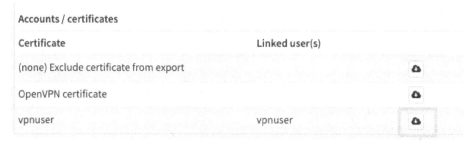

Figure 8.12 – Exporting OpenVPN client configuration

12. Create a firewall rule on the OpenVPN server permitting all the traffic for testing purposes.

13. After downloading the client's configuration, you can connect to the newly created OpenVPN tunnel using your favorite client:

  - **Windows**: `https://openvpn.net/community-downloads/`

  - **macOS**: `https://tunnelblick.net/downloads.html`

  - **Linux**: Most distributions have the OpenVPN package as default. You will be able to install it without significant problems. Some Linux distributions allow you to configure using the network manager, which will ease the usage a lot. For example, after a connection is made to a hotel WLAN, autostart the VPN.

  - **Android and iOS**: Search on Google Play and App Store for OpenVPN, and you'll find the client for your favorite mobile platform.

14. To configure each OpenVPN client, I suggest you consult the official documentation of the client you'll use. The OPNsense official documentation has some guides for each platform as a starting point: `https://docs.opnsense.org/manual/how-tos/sslvpn_client.html#step-3-export-client-configuration`.

15. Once you have downloaded the OpenVPN client and imported the configuration, you can try connecting to your new OpenVPN server. Once you do so from **VPN | OpenVPN | Connection Status**, your client's connection will look like this:

VPN: OpenVPN: Connection Status

Common Name	Real Address	Virtual Address	Connected Since	Bytes Sent	Bytes Received	
ademir.rodrigues	45.160.107.152:13345	10.111.111.6	2021-08-17 10:06:01	664.99 MB	88.09 MB	✕

Figure 8.13 – OpenVPN connected user example

If you can see something like the preceding screenshot on your OPNsense firewall, congratulations! Your remote user connection is working! If it isn't working yet, repeat the steps you took to configure and check your OpenVPN client logs to try to find a clue. If you already did that and it is still not working, try the following topic.

OpenVPN diagnostics

For both site-to-site and remote user scenarios, when things are not working as expected, we can try using some valuable tools offered by OPNsense to find possible issues. Let's look at some examples.

OpenVPN is connected but the traffic is not reaching the tunnel's destinations

I have tried to list the more common issues related to OpenVPN troubleshooting based on my personal experience here:

- The first step is to check your firewall rules and look for packets that have been blocked. Especially check the OpenVPN firewall rules and check all necessary rules have been created.

- On the client side, it can happen that the operating system is not creating the proper routes. This issue can be related to the user's permissions or some local network problem while creating the tunnel routes. Another common cause is conflicting routes (with existing interfaces or even with an IPsec tunnel).

- Always check the log file on both sides (server and client). On OPNsense, you can check logs by accessing the **VPN | OpenVPN | Log** file menu, or on the CLI by reading the openvpn.log file:

```
#tail -f /var/log/openvpn.log
```

This is a short list, but you might face issues that will have a mix of them. You can also explore the OPNsense forum to get some more help when you need it.

OpenVPN client is not connecting to the server/a site-to-site tunnel doesn't become up

This might be related to firewall rules on the configured tunnel interface (the **Interface** option) or even a wrong interface selected in this option.

A single user cannot connect

Most user connection issues are related to the operating system configuration or user privileges, so it is good to start looking at the user's computer for the following:

- In this case, a common issue is when some users forget or use the wrong credentials. The best place to look is in the OpenVPN log file. It can happen too when there is a problem with the user's connection or OpenVPN client. A good alternative is to ask the user to change the internet connection (from Wi-Fi to a cable or cellular network). On Windows hosts, reinstalling the OpenVPN client sometimes can help.

- For all situations, always keep your eye on the logs; they can help a lot!

- The tunnel connection port might be blocked in the user network (for example, hotels, airports, and other guest networks).

Unfortunately, we cannot list all the issues related to OpenVPN here. Still, you can always count on OPNsense's community, especially on the forum, to understand what could be happening with your OpenVPN issues.

Summary

In this chapter, you have learned about **Quality of Service** (**QoS**) and traffic shaping basics, how to create pipes, queues, and rules in the OPNsense webGUI, and how to test them. Now you can design a traffic shaping policy based on the network requirements and create rules to apply it using OPNsense features. In the next chapter, we will begin our exploration of the VPN world.

In this chapter, we learned about IPsec and OpenVPN, the pros and cons of each one, how to employ IPsec and OpenVPN to connect remote networks using a site-to-site approach, and how to use OpenVPN to connect remote users. We have explored the diagnostics tools OPNsense provides and some common issues you can face while using them. In the next chapter, we will learn about OPNsense's multi-WAN capabilities.

9
Multi-WAN – Failover and Load Balancing

This chapter will explore some multi-**Wide Area Network** (**WAN**) strategies such as load balancing and failover using the policy-based routing concept. You will also explore some common issues and how to solve them.

By the end of this chapter, you will be able to understand and configure the following multi-WAN related topics:

- Failover and load balancing
- Policy routing
- Troubleshooting

Technical requirements

You will need a running OPNsense and a host to practice this chapter's steps. Knowledge of how to create and change network settings on VirtualBox is required. A good understanding of how to create/edit firewall rules on OPNsense is essential for this chapter.

Failover and load balancing

In the past, we used to need a dedicated network appliance to deal with multiple internet connections and guarantee good availability of internet access. One of the best features of a modern firewall is working with multiple WAN connections. OPNsense has incredible features that we'll explore in this chapter so that it can be configured with various internet connections.

Failover

The first scenario we'll explore is the failover configuration; with two or more WAN connections, it is possible to configure OPNsense to change the active internet connection to a backup one automatically. An example of an OPNsense configured with two internet connections is shown in the following topology:

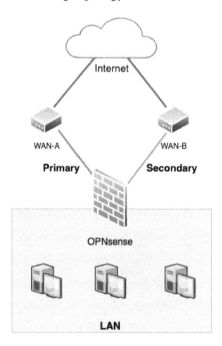

Figure 9.1 – Multi-WAN example scenario

As we can see in the preceding diagram, two WAN connections, A and B, are both connected to the internet and configured on OPNsense. *WAN-A* is configured in OPNsense as the primary connection and *WAN-B* as the secondary. If the primary fails, OPNsense will automatically change the **Local Area Network's** (**LAN**) outgoing connections to the secondary WAN in a failover manner.

> **Important Note**
>
> For simulating another WAN connection in a lab environment, you can add a new network interface and configure it on your OPNsense **virtual machine (VM)**.

To configure the OPNsense VM to work in a failover manner, as proposed in the previous diagram, we need to adjust the VirtualBox network settings of our OPNsense VM:

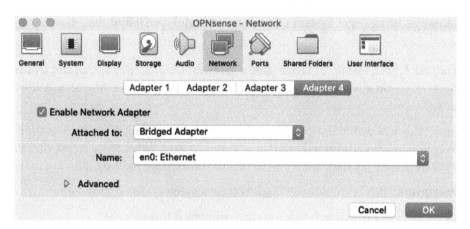

Figure 9.2 – Additional WAN interface, VirtualBox's example configuration

As you can see in the preceding screenshot, I have configured an additional interface using the **Bridged Adapter** option and selected my internet-connected network interface (**en0**). These network options can be accessed by clicking on the **Settings** button of your VirtualBox's VM. The configuration in your OPNsense will depend on the network's settings connected to this new WAN interface. If you are not sure, try to configure it as **Dynamic Host Configuration Protocol (DHCP)** first.

With your OPNsense configured with two WAN interfaces, it is time to go through the steps to set the failover configuration:

1. On the webGUI, go to **System | Gateways | Single** and check whether both WAN interfaces have configured gateways:

System: Gateways: Single

Name	Interface	Protocol	Priority	Gateway	Monitor IP	RTT	RTTd	Loss	Status	Description	
▶ WAN_A_DHCP (active)	WAN_A	IPv4	254 (upstream)	10.0.2.2		~	~	~	Online	Interface WAN_A_DHCP Gateway	✎ 🗑 ⧉
▶ WAN_B_DHCP	WAN_B	IPv4	254	192.168.1.1		~	~	~	Online	Interface WAN_B_DHCP Gateway	✎ ⧉

Figure 9.3 – System | Gateways | Single page

In the preceding screenshot, you'll note that the primary WAN (**WAN_A**) has the gateway address 10.0.2.2 (VirtualBox's gateway) and the secondary WAN (**WAN_B**) has the gateway configured as 192.168.1.1. As I mentioned previously, using VirtualBox, you can set one WAN using a **Network Address Translation** (**NAT**) adapter (**WAN_A**) and another one using a **bridged** adapter (**WAN_B**).

Let's check the options available on the gateway configuration page by clicking on the edit button (pencil icon):

On the gateway editing page (**System** | **Gateways** | **Single**), you'll find the following options:

- **Disabled**: Check this option to disable a gateway. *Caution: this option can lock you out of OPNsense if you access it from a network that depends on this gateway to be reached.*

- **Name**: This will define the gateway's name. If you are setting it for the first time, it's clever to name it with something that will make sense. *Think twice before choosing a name; OPNsense does not allow you to rename an existing gateway.*

- **Description**: This is an optional field. You set a gateway description here.

- **Interface**: Select which network interface will reach this gateway.

- **Address Family**: Select which IP version this gateway address will be set, IPv4 or IPv6.

- **IP address**: The gateway IP address.

- **Upstream Gateway**: This option defines whether a gateway can be set as the default gateway. *For our example, this option must be selected.*

- **Far Gateway**: This option will allow a gateway that addresses outside the network interface's subnet.

- **Disable Gateway Monitoring**: If this option is checked, this monitor daemon (dpinger) will not try this gateway to determine if it is alive. Consequently, it will always be considered online. *Uncheck this option to our failover example configuration.*

- **Monitor IP**: By default, OPNsense monitors the gateway's IP address. If you want to set a different IP address, we'll need to put it here. For example, if the gateway is blocking pings, or if you want to check not only whether the gateway is up but also whether the internet is accessible.

- **Mark Gateway Monitoring**: Check this option if it is a gateway to be forced to the **Offline** state (considered offline).

- **Priority**: This option allows you to set the priority for the gateway, where a higher priority will be selected as the default gateway (*a lower number means higher priority*).

- **Advanced**: Click on this button to expand the gateway's advanced configuration.

- **Weight**: This will set which weight the gateway will have while configuring in a gateway group. This setting can be used in a load balance configuration. For instance, if one link has 100 Mbps of bandwidth and the second has 25 Mbps, you may set the first one (100 Mbps) with weight 4 and the second one with weight 1; therefore, the larger link will get four times more traffic than the second one.

- **Latency thresholds**: Defines the range from the low to high latency thresholds (in milliseconds). The default values are **From**: 200 and **To**: 500. The **From-To** range will be set in a *Warning* state; any value below the **From** configured number will be considered *Online*, and values above the **To** configured number will be changed to the *Offline* state.

- **Packet Loss thresholds**: Like the preceding option, the **From-To** range will set the gateway in a *Warning* state. The default values are **From**: 10 and **To**: 20 (in percent). Any value lower than the **From** set value will be *Online*, and above the **To** value will be considered *Offline*.

- **Probe Interval**: This is the time interval expressed in seconds that each **Internet Control Message Protocol** (**ICMP**) probe will be sent. The default value is 1 second.

- **Alert Interval**: The time interval (in seconds) before triggering a gateway condition change.

- **Time Period**: The time frame in which the results will be averaged. The default value is 60 seconds.

- **Loss Interval**: The time frame (in seconds) in which the packets will be considered lost.

- **Data Length**: Which size the ICMP probes will have (in bytes).

To save the gateway configuration, click on the **Save** button, then click on the **Apply changes** button.

After saving the gateway, you might notice two significant changes on the **System | Gateways | Single** page:

Figure 9.4 – Gateways with gateway monitoring enabled

As you can see in the preceding screenshot, the monitoring daemon will start to measure the gateway's latency, represented on the page by the **round-trip time** (**RTT**) column and the standard deviation, **RTTd**. The **Loss** column is how many packets (in percent) are being lost.

Remember to uncheck the **Mark Gateway Monitoring** option in the other gateway (**WAN_B_DHCP**).

> **Important Note**
>
> The RTTd, or RTT standard deviation is, in a simple manner of speaking, how much it varies over time.
>
> To learn more about ping/ICMP standard deviation, check this link:
> `https://newbedev.com/what-does-mdev-mean-in-ping-8`.

The `dpinger` monitor daemon can be noted as running in the top-right corner of the page. You can also check whether it is running on **Lobby | Dashboard**:

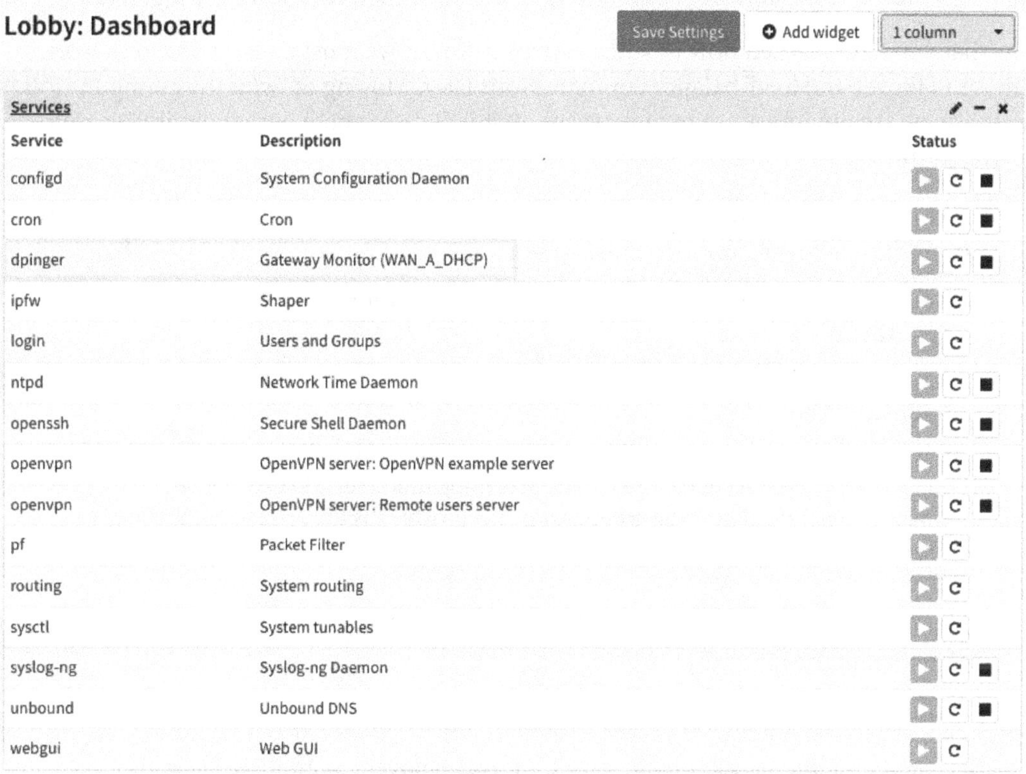

Figure 9.5 – The dpinger service running on the services widget (dashboard)

The preceding screenshot shows the `dpinger` service running on the service widget. You can check all the running services in this widget. On the **Command Line Interface (CLI)**, you can check it using the `pluginctl -s` command.

Creating gateway groups

A gateway group can contain several system gateways inside it and the way we set each will define how this group will behave when some predefined condition is triggered.

Now that you have learned how to add a system gateway, we must create a group to work with it in failover and load-balance configurations later. Moving on with the configuration steps, let's now configure a group of gateways that will define how our configuration will behave, in this case, in a failover manner:

1. Go to **System | Gateways | Group** and click on the + button to add a new gateways group. The following options must be set:

2. **Group Name**: Set a name for the gateways group, for example, `Failover_WANA_WANB`.

> **Important Note**
>
> It is a good practice to name your gateways group by writing the primary gateway before the secondary (as in the preceding example). Thus, it will be easier to understand how the group is organized (which is the primary and secondary gateway).

3. **Gateway Priority**: This defines each gateway priority considering a *Tier level* as explained in the following:

    I. **Never**: The gateway will not be used on this gateway group.

    II. **Tier 1**: The higher priority gateway. For our failover example, we'll set the **WAN_A** gateway by selecting this option.

    III. **Tier 2** to **5**: The following priorities define in which order each gateway will be triggered (*the lowest number equals higher priority*). For our example, we'll set the **WAN_B** gateway as **Tier 2**.

4. **Trigger Level**: The trigger in which this gateway group will be ruled. The following options are available for selection:

    I. **Member Down**: A gateway will be considered down when it has 100% packet loss. The monitor daemon (`dpinger`) will monitor gateways using ICMP packets to define if it is alive or not. *For our example, select this option.*

    II. **Packet Loss**: Will be triggered when each gateway's defined packet loss thresholds are reached in advanced settings.

    III. **High Latency**: Will be triggered when each gateway's defined latency loss thresholds are reached in advanced settings.

IV. **Packet Loss or High Latency**: Will be triggered if one of these conditions is satisfied.

V. **Description**: Fill with a description of this gateway group. For example, `Failover group`.

To finish, click on the **Save** button and then on the **Apply changes** button.

So far, we have configured an additional WAN interface, configured the basics of each WAN gateway, and created a gateway group to work in a failover manner. Now, it's time to learn about how policy routing works on OPNsense.

Policy-based routing

Unlike the static routes added to the system, policy-based routes will be created through firewall rules on OPNsense. In *Chapter 5, Firewall*, we explored firewall concepts and rules, but nothing related to using a gateway on rules, so now it's time to learn how to do that.

Before starting, to follow these steps, we'll need a host connected to OPNsense's LAN. If you are using VirtualBox as your lab platform, with an additional VM installed, follow these steps to connect it to the OPNsense LAN:

1. To attach the VM to OPNsense's LAN, change the network settings to **Host-only Adapter** by editing the VM settings on VirtualBox:

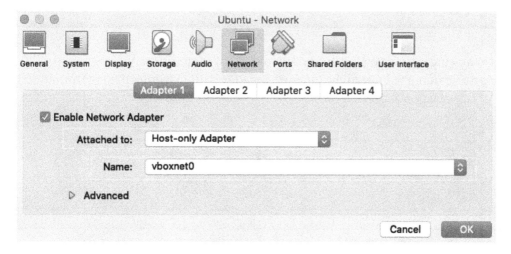

Figure 9.6 – Changing VM network settings to connect on OPNsense's VM LAN

2. As we can see in the preceding screenshot, we need to select **Host-only Adapter** in the **Attached to** option. Choose the same network interface your OPNsense's VM uses as the LAN interface in the **Name** option. Following the steps we took in the earlier chapters, it should be **vboxnet0**.

> **Important Note**
>
> I'll use an Ubuntu VM host; feel free to choose your preferred operating system but pay attention to the commands demonstrated, as they will only work on Ubuntu (and probably most Linux distributions)!

3. Start your VM, which is connected to the OPNsense LAN. From now on, I'll refer to the Ubuntu VM just as the *host*.

4. Test the communication by running a ping test: `ping <OPNsense LAN address>`. If the ping replies, it works! Otherwise, check your VirtualBox network settings to ensure everything is configured as described in the previous steps.

5. After checking that the communication is working, it's time to change the default gateway on the host to our OPNsense VM LAN's address. On Ubuntu, you can run the following:

```
sudo ip route del default
sudo ip route add default via <OPNsense LAN address>
```

Test whether the host is reaching the internet using OPNsense as the default gateway:

```
ping 8.8.8.8
```

With ping running on the host, check on OPNsense to ensure that the traffic is passing from the LAN to the internet; go to **Firewall | Diagnostics | States** and filter the destination address, `8.8.8.8`:

Figure 9.7 – Firewall | Diagnostics | States page showing the traffic from our LAN's host

6. In the preceding screenshot, you will notice that the LAN's host address (192.168.56.4 in my lab) is using OPNsense as the default gateway to reach the public IP address 8.8.8.8.

With this, we finished the required steps to connect an additional VM as a LAN host to OPNsense. Now, it is time to move on with the policy-based routing rule to make the failover work!

Creating a firewall rule to enable the failover configuration

To enable the previously configured gateways settings, we need to add a firewall rule specifying the gateway group to enable the failover configuration. To do this, follow the given steps:

1. Let's start editing the default *allow all* rule on the LAN; go to **Firewall | Rules | LAN** (the rule description is **Default allow LAN to any rule**).

2. On the host, run traceroute to check the path the packets are following:

```
opnsense@ubuntu:~$ traceroute 8.8.8.8
traceroute to 8.8.8.8 (8.8.8.8), 30 hops max, 60 byte
packets
  1  _gateway (192.168.56.3)  0.817 ms  0.720 ms  0.806 ms
  2  10.0.2.2 (10.0.2.2)  6.692 ms  6.660 ms  6.631 ms
  3  * * *
  4  192.168.15.1 (192.168.15.1)  12.545 ms  12.644 ms
13.841 ms
```

> **Important Note**
>
> In my lab, **WAN_A_DHCP** has the IP address 10.0.2.2, which is the OPNsense default gateway.
>
> If your Ubuntu host doesn't have traceroute installed, you can install it by running apt install traceroute.

3. To change the outgoing gateway from **WAN_A_DHCP** to **WAN_B_DHCP**, edit the same rule, and in the **Gateway** option, select the **WAN_B_DHCP** gateway, save, and apply the changes.

> **Important Note**
>
> Policy-based routing is only supported by rules with **Direction** set to **in** (inbound rules). For this reason, you need a LAN host connected to the OPNsense LAN interface to see things working.

4. Back to the host, and run `traceroute` again:

```
opnsense@ubuntu:~$ traceroute 8.8.8.8
traceroute to 8.8.8.8 (8.8.8.8), 30 hops max, 60 byte
packets
 1  _gateway (192.168.56.3)  0.784 ms  0.997 ms  0.974 ms
 2  192.168.1.1 (192.168.1.1)  2.408 ms  2.388 ms  2.378
ms
 3  192.168.15.1 (192.168.15.1)  8.664 ms  8.638 ms
8.453 ms
 4  * * *
```

5. As you can see, the path was changed. Now, the packets are using the **WAN_A_DHCP** gateway (which is `192.168.1.1` in my lab). We just made the policy-based routing work! Let's now adjust the rule to achieve our goal: the failover!

6. Edit the rule again, and now set the **Gateway** option to the gateway group we added previously: **Failover_WANA_WANB**.

7. Repeat the `traceroute` step to ensure that **WAN_A_DHCP** (OPNsense's default gateway) is in use:

```
opnsense@ubuntu:~$ traceroute 8.8.8.8
traceroute to 8.8.8.8 (8.8.8.8), 30 hops max, 60 byte
packets
 1  _gateway (192.168.56.3)  2.238 ms  2.972 ms  4.418 ms
 2  10.0.2.2 (10.0.2.2)  5.921 ms  10.674 ms  10.626 ms
 3  * * *
```

8. The traffic is back to the **WAN_A** gateway.

9. Now, let's simulate that **WAN_A** is down by disconnecting its cable. In VirtualBox, you can do that by unselecting the network interface, as shown in the following:

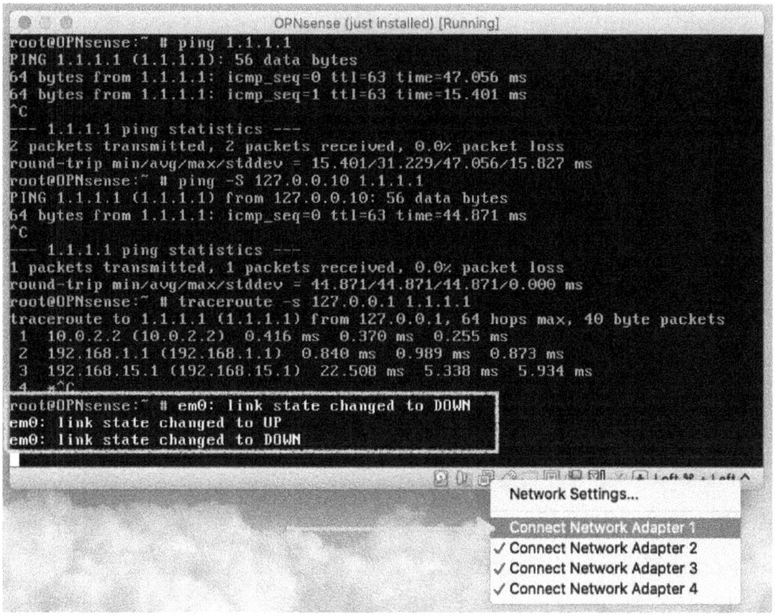

Figure 9.8 – Disconnecting a network adapter in VirtualBox

10. After disconnecting the network interface, check on webGUI if the gateway changed to the **Offline** state; go to **System | Gateways | Single** to do that.

11. Try `traceroute` again on the host:

```
opnsense@ubuntu:~$ traceroute 8.8.8.8
traceroute to 8.8.8.8 (8.8.8.8), 30 hops max, 60 byte
packets
 1  _gateway (192.168.56.3)  2.450 ms  2.203 ms  2.151 ms
 2  192.168.1.1 (192.168.1.1)  4.363 ms  4.308 ms  3.954
ms
 3  192.168.15.1 (192.168.15.1)  8.288 ms  9.896 ms
9.852 ms
 4  * * *
```

12. The traffic path has changed! Now, it is using the **WAN_B** gateway. The failover is working!

Even with this simple example, you can see that OPNsense works very well in a failover configuration. Users barely notice when a WAN link is down in a production environment, even with dozens of different protocols and thousands of hosts. I have excellent experience with failover scenarios (CloudFence's customers) that have five or more WANs with thousands of users; OPNsense is fantastic with that!

Now we have explored the failover configuration, let's look at the outbound load balance.

Load balance

The load balance configuration differs slightly from the failover, and it can also act as a failover when a gateway goes offline. The main idea of a load balance configuration is to send packets through the gateways in an alternate manner. How the packets will alternate the gateway will depend on some configurations:

- The gateway's *tier level* in the group
- The **Sticky connections** option in the **Firewall | Settings | Advanced** (the **Multi-WAN** section) – see *Chapter 5, Firewall*
- **Firewall rule**: Depending on the packet's match with the ruleset

To configure load balance, follow the given steps:

1. Create a new gateway group using the same steps you have followed in the failover topic, changing only the **Gateway Priority** option for both gateways to **Tier 1**. Save and apply your changes.

2. We will disable the **Sticky connections** option in **Firewall | Settings | Advanced** (the **Multi-WAN** section) only for testing purposes. Once this option will stick the connections based on the source host, and we have only one in our lab, it is safe to disable it to save time in our testing. In a production environment, it is not recommended to do that while using load balance, as this can break some internet connections (those that have source address check, for example, **Transmission Control Protocol** (**TCP**) and partially **User Datagram Protocol** (**UDP**)).

3. Go to **Firewall | Rules | LAN** (the rule description is **Default allow LAN to any rule**) and alter the **Gateway** option to the new load balance gateway group we just created. Save and apply the changes.

4. To do a test on the Ubuntu host, try `traceroute 8.8.8.8`:

```
opnsense@ubuntu:~$ traceroute 8.8.8.8
traceroute to 8.8.8.8 (8.8.8.8), 30 hops max, 60 byte
packets
 1  _gateway (192.168.56.3)  2.277 ms  4.733 ms  4.707 ms
 2  10.0.2.2 (10.0.2.2)  4.685 ms  4.548 ms  4.512 ms
 3  * * *
 4  192.168.15.1 (192.168.15.1)  13.798 ms  14.349 ms
14.316 ms
```

5. In the preceding output, we can see the traffic leaving the firewall using the **WAN_A** gateway.

6. Now, repeat `traceroute` to another destination (`1.1.1.1`, for example):

```
opnsense@ubuntu:~$ traceroute 1.1.1.1
traceroute to 1.1.1.1 (1.1.1.1), 30 hops max, 60 byte
packets
 1  _gateway (192.168.56.3)  7.422 ms  7.342 ms  7.305 ms
 2  192.168.1.1 (192.168.1.1)  16.747 ms  16.715 ms
10.0.2.2 (10.0.2.2)  16.682 ms
 3  192.168.15.1 (192.168.15.1)  54.729 ms  58.178 ms *
```

Notice that the path changed! Our load balance configuration is working!

These were quite simple examples, but in a production environment, a lot of complexity might be added, and some issues can appear. Let's now see some examples of how to troubleshoot them.

Troubleshooting

Let's look at some of the common issues while configuring load balance and failover configurations:

- **Failover/load balance isn't working**: The tier 1 WAN line goes down, but the traffic is still trying to leave the firewall through it. OPNsense, by default, will set the gateway's IP address as the monitor IP address; for instance, let's suppose the WAN line was interrupted somewhere between the customer and the **Internet Service Provider** (**ISP**). The router/modem will be still alive and responding to ICMP requests. This can happen because the gateway's IP address is the local network interface of the router/modem, and it won't be down in this case, so the OPNsense monitoring daemon will consider it online; therefore, the condition to change to another WAN will not be triggered. To avoid this issue, always set the monitor IP address to an ISP WAN's cloud address; this way, when the communication between the ISP's router/modem is interrupted, the failover will work as expected. For internet gateways, ensure the **Upstream Gateway** option (**System** | **Gateway** | **Single**) is checked in each configured gateway in the group.

- **The monitor IP address is offline, even with the WAN line as Online**: This is the opposite of the preceding condition. Before setting an alternate monitor IP address to a gateway, test whether it replies to ICMP packets. Another common issue is when a host stops responding to ICMP packets, so choosing the right host as the monitor IP address is essential to avoid future headaches.

- **OPNsense has internet access, but hosts using the failover/load balance rule don't**: This can happen because OPNsense and all services running on it will use the default gateway to access the internet. The option that will configure OPNsense to change the default gateway in a multi-WAN configuration is **Gateway switching** (in **System** | **Settings** | **General**). Once a gateway group is set to follow the tier level of each configured gateway, a policy-based routing rule may send traffic to a different gateway that isn't working well. So, in cases like that, it is crucial to compare the gateway group's configuration and OPNsense's in-use gateway.

- **Always check the logs**: The `dpinger` daemon has a log file that can help you troubleshoot issues related to gateways. You can view this log file on webGUI at **System** | **Gateways** | **Log File**.

These are some examples of failover/load balance issues we face daily while working with OPNsense. Sometimes a problem can be a combination of other ones. It will depend on the complexity of the OPNsense configuration. You can always count on the community's support in the forum to help you!

Summary

In this chapter, we have explored the failover, load balance, and policy-based concepts. Now, you can understand, create, and manage gateways, groups of gateways, and firewall rules using them. You also learned how to troubleshoot common issues involving failover, load balance, and gateways on OPNsense. In the next chapter, we will go through the reporting features available on OPNsense!

10
Reporting

In this chapter, you will learn how to read system graphs, which is an essential part of managing a firewall. We will explore each available graph and use it to identify possible unexpected behaviors in the network or to check our firewall's health.

By the end of this chapter, you will have learned about the reporting tools that are available on OPNsense's webGUI and how to use each one to get the most out of OPNsense.

In this chapter, we will cover the following topics:

- System health graphs
- Understanding Netflow and how to use it
- Exploring real-time traffic
- Troubleshooting common problems in the network using Netflow and graphs

Technical requirements

For this chapter, you may wish to have a running version of OPNsense to follow along, though this isn't mandatory. All the concepts presented in this book will be enough for you to follow this chapter's steps and examples. No additional knowledge will be necessary.

System health graphs

Like a pilot that keeps monitoring an airplane's instruments so that it continues flying safely, a firewall administrator (or a firewall pilot, if you like) must monitor each aspect of its firewall to keep the network secure. Instead of flight instruments, in OPNsense, we have graphics that help us know how the system is working so that we can make decisions based on each graph that's read. Add logs to this, and you will know everything about your firewall, especially during troubleshooting. If you want to be known as someone who solves issues fast, my advice is this: pay attention to the logs and graphs and always read the documentation!

OPNsense provides several graphs that you can use to monitor a firewall system. Let's explore each graph and learn how to use them to help us keep our firewall *flying*.

Our quest begins with accessing the **Reporting | Health** menu. After accessing this menu, the following page will open:

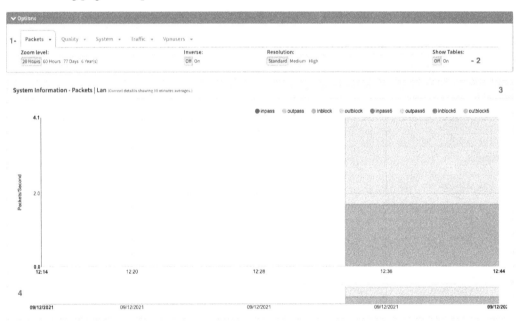

Figure 10.1 – Accessing the Reporting | Health menu

Referencing the numbers specified in the preceding screenshot, we can see the main parts of this page:

1. **Graphs menu**: This shows the available graphs in the following order: **Packets**, **Quality**, **System**, **Traffic**, and **Vpnusers**.

2. **Graphs options**: These options will change each graphic's visualization. The available options are as follows:

I.   **Zoom level**: With this option, you can select how much time the graph will show data, from **20 Hours** to **6 Year(s)**.

II. **Inverse**: This option is helpful while you're using graphs that have incoming and outgoing flows where each direction will be plotted in a positive/negative manner.

III. **Resolution**: This will change the graph's resolution. There are three options: **Standard**, **Medium**, and **High**. Each resolution will change the graph's data points, increase or decrease them, and change the average values.

IV. **Show Tables**: This turns the detailed tables **On** or **Off**:

Current View - Overview

item	min	max	average
inpass	0	110.5958306775	31.276428571335124
outpass	-5221.103489901359	0	-1105.895476201366
inblock	0	0	0
outblock	0	0	0
inpass6	0	0	0
outpass6	0	0	0
inblock6	0	0	0
outblock6	0	0	0

Current View - Details

Toggle Timeview:
Timestamp Full Date & Time   ⬇ Download as CSV

#	timestamp	inpass	outpass	inblock	outblock	inpass6	outpass6	inblock6	outblock6
1	1631459640	0	0	0	0	0	0	0	0
2	1631460240	1.9199999999989998	-1.093333333326	0	0	0	0	0	0
3	1631460840	1.92	-1.093333333335	0	0	0	0	0	0
4	1631461440	110.5958306775	-5221.103489901359	0	0	0	0	0	0
5	1631462040	75.4791693219069	-2455.6131768418113	0	0	0	0	0	0
6	1631462640	11.7033333333339998	-22.641666666659997	0	0	0	0	0	0
7	1631463240	17.316666666599996	-39.72333333307	0	0	0	0	0	0

Figure 10.2 – Detailed tables turned on

As you can see, the **Show Tables** option shows a table and its graph data (shown below the graph). With this table, you can export data in CSV format to create a more complex datasheet that includes graph data, for example.

3. **Main graph area**: This is where the graph will be plotted. Passing your mouse cursor over it will give you more details.

4. **Graph zoom area**: Selecting the graph in this area will allow you to zoom into it.

With the page layout explained, let's explore each available graph, starting with the **Packets** menu.

Clicking on the **Packets** menu will open a submenu containing various graphs. Each graph is a configured network interface in OPNsense:

Figure 10.3 – The Packets menu

In the preceding screenshot, you can see the network interfaces that have been configured in my lab for OPNsense. By clicking on each interface, you can see the corresponding graph in the main graph area.

The following is an example of the **WAN** packets graph:

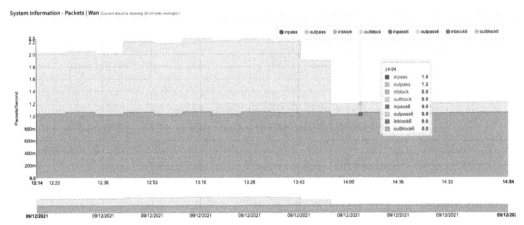

Figure 10.4 – Packets graph

In the preceding screenshot, the graph's labels are highlighted. In the packets graph, the available labels are related to **passed packets** (allowed by firewall filtering) and **blocked packets**. They are classified by IP protocol version – that is, versions 4 and 6; the last one has a **6** added to each related label and the flow direction, which is either **in**coming or **out**going. The labels in a vertical layout are shown when you pass your mouse cursor over the graph.

Important Note

You can select the label you want to be enabled in the graph by clicking on it in all the graphs.

The packets graph helps measure how many packets OPNsense is processing, classified by IP protocol version, and how much is allowed/blocked by the firewall rules. Here, you can find, for example, possible bottlenecks related to packets processing.

The next menu is **Quality**. *It will only be available if you have added a gateway* by ensuring the **Disable Gateway Monitoring** option is *unchecked* on the **System | Gateways | Single** page (for each added gateway).

Important Note

Sometimes, it can be confusing that webGUI has options that must be checked to *disable* some features and others that must be checked to *enable* them. Maybe soon, with the legacy webGUI being redesigned, this will be changed to a standard.

The submenus that are available in the **Quality** graph will depend on the number of gateways that have been added to the system. The quality graphs are generated using the ICMP packets from the dpinger monitoring daemon. As we learned in the previous chapter, this daemon does IP monitoring – that is, measuring the time it takes to get replies from the ICMP. Using quality graphs helps determine WAN link conditions, for example.

The next menu is the **System** menu. Here, you will find the following submenus:

- **Processor**: This graph will measure the CPU usage by **user**, **nice**, **system**, **interrupt**, and the number of **processes**. When you need to read the processor's utilization precisely, I recommend that you turn **processes [#]** off to generate a clear graph. The labels that are available in this graph are based on the `top` command.

- **Mbuf**: The **memory buffer** (**mbuf**) is a common concept in networking stacks. It is used to hold packet data as it traverses the stack. It also generally stores header information or other networking stack information that is carried around with the packet. More information can be found at `https://mynewt.apache.org/latest/os/core_os/mbuf/mbuf.html`:

  - The mbuf is automatically calculated by FreeBSD. Its max limit will be around 50% of the physically installed RAM on the system. You can check the percentage that's being used in the **System Information** widget by going to **Lobby | Dashboard**. This will give you a good idea of its usage.

- **States**: This graph plots the number of firewall connection states. The available labels are **pfrate** (states rate), **pfstates** (number of concurrent states), **pfnat** (NAT rules related states), and **srcip/dstip**, which show the number of active sources and destination IPs, respectively.

- **Cputemp**: This shows the CPU temperature over time, when available. Sometimes, it is necessary to enable it by going to the **System | Settings | Miscellaneous** page and selecting your CPU architecture by choosing the **Hardware** option (the **Thermal Sensors** section).

- **Memory**: This graph will plot the system's memory usage. The available labels are **active**, **inactive**, **free**, **cache**, and **wire**. As in the case of the CPU graphs, these labels are also based on the `top` command.

> **Important Note**
>
> For more information about FreeBSD's memory classes, please refer to `https://wiki.freebsd.org/Memory`.

The next menu is where we can check the network interface traffic graphs. Click on the **Traffic** menu to list all the configured interfaces as a submenu, as shown in the following screenshot:

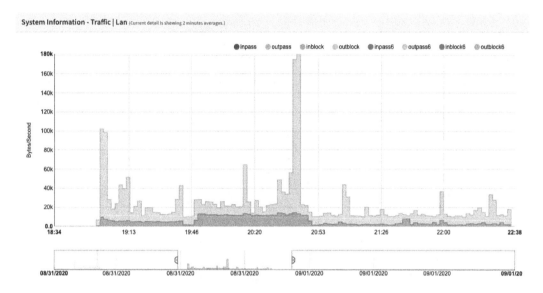

Figure 10.5 – LAN traffic graph example

As we can see, the **Traffic** graph uses the same labels as the **Packets** graphs – incoming and outgoing traffic (IPv4 and IPv6), both passing or blocked.

> **Important Note**
> The traffic graph uses *Bytes/Second*, not Bits/Second as usual. Pay attention to that to avoid interpreting this graph incorrectly.

The traffic graph helps check how high the network interface's usage is.

The last available graph menu is **Vpnusers**. Each configured OpenVPN server will be listed as a submenu in it. The graph is depicted as follows:

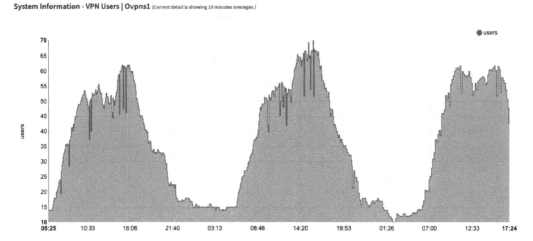

Figure 10.6 – VPN Users graph example

As shown in the preceding screenshot, the **Vpnusers** graph plots the number of connected users in the respective OpenVPN server graph. But how do we know which graph corresponds to an OpenVPN server? This is the limitation of this graph. It doesn't display the tunnel's description; instead, it only shows its virtual network interfaces. In the following steps, I'll show you how to link the graph to the respective OpenVPN server easily:

1. Go to **System | Routes | Status** and type the OpenVPN interface's name in the search bar (as shown in the VPNusers' graph). For example, type ovpns1 and look for the line that contains **UH** flags:

System: Routes: Status

		🔍 ovpns1	10 ▾	☰ ▾

Proto	Destination	Gateway	Flags	Use	MTU	Netif	Netif (name)	Expire	Action
ipv4	10.10.10.2	link#10	UH	0	1500	ovpns1			🗑
ipv4	192.168.1.0/24	10.10.10.2	UGS	0	1500	ovpns1			🗑
ipv6	fe80::%ovpns1/64	link#10	U	0	1500	ovpns1			🗑
ipv6	fe80::a00:27ff:feeb:a6ff%ovpns1	link#10	UHS	0	16384	lo0	Loopback		🗑

« ‹ **1** › »

Showing 1 to 4 of 4 entries

☐ **Name resolution**

Enable this to attempt to resolve names when displaying the tables. By enabling name resolution, the query may take longer.

🔄 Refresh

Figure 10.7 – The Routes: Status page – filtering by the OpenVPN graph's interface name

2. In the results, note the **Destination** network, as in the preceding screenshot (for example, `10.10.10.2`).

3. Go to **VPN | OpenVPN | Servers** and search for the network you noted in the previous step:

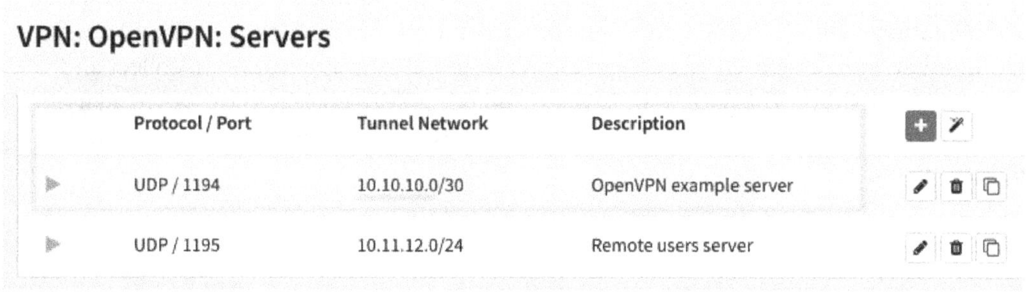

VPN: OpenVPN: Servers

	Protocol / Port	Tunnel Network	Description	
▶	UDP / 1194	10.10.10.0/30	OpenVPN example server	✎ 🗑 🗐
▶	UDP / 1195	10.11.12.0/24	Remote users server	✎ 🗑 🗐

Figure 10.8 – The VPN: OpenVPN: Servers page

Here, `10.10.10.2` refers to the tunnel using the **ovpns1** network interface. Now, you know which tunnel the graph corresponds to in the **Vpnusers** graph.

Maybe in future versions, the developers of OPNsense will implement an easy way to do this, but for now, I hope the preceding steps can help you with this task.

RRDtool and health graphs

OPNsense makes use of the round-robin database tool known as *RRDTool* to generate its health graphs. It is a very popular and helpful tool for generating network and system graphs. It saves a `.rrd` file (to the `/var/db/rrd` path) that will be used to create the graphs. RRD graphing is enabled by default, but if for some reason you need to disable it, or maybe reset or remove RRD data, you can go to the **Reporting | Settings** page. Here, you can find the options to execute the aforementioned actions. Besides this, you can also remove RRD data for each graph individually and deal with Netflow data.

Talking about Netflow, let's look at this reporting feature.

Understanding Netflow and how to use it

Introduced by Cisco back in 1996, Netflow is a protocol that's used to help analyze network traffic. Netflow has three main components: **flow exporter**, **flow collector**, and **analyzer**. An advantage of using Netflow is that it captures the packet flow, including information about the source and destination IP and port number. As OPNsense's official documentation claims, it is the only open source solution that integrates all this in a web GUI. In other words, with OPNsense, you don't need another application to collect and analyze network flows. The exception is when you have OPNsense as a firewall in a large network with a lot of traffic – here, you will need an external analyzer with a dedicated database engine.

> Important Note
> OPNsense's embedded Netflow analyzer has a local cache with a 100 MB limit (wispy for larger networks). Therefore, in large or high-throughput networks, it is highly recommended to use an external Netflow analyzer.

Next, we will learn how to configure and use Netflow in OPNsense.

Configuring Netflow in OPNsense

To enable Netflow in OPNsense, go to the **Reporting | NetFlow** page and configure the following options:

- **Listening interfaces**: Select the interfaces that you want to collect network flows.

- **WAN interfaces**: Select only the WAN interfaces here. Selecting interfaces here will prevent double network flow counting.

- **Capture local**: This option will enable the OPNsense embedded analyzer (known as **Insight**).

- **Version**: OPNsense supports Netflow versions 5 and 9. If you need IPv6 support, you need to select version 9.

- **Destinations**: If you use an external analyzer, specify the IP address and port here; for example, `192.168.0.10:2550`. If you intend to use Insight, *leave this option blank*; the webGUI will automatically fill it with the localhost address (after you click on **Apply** button).

To finish the configuration, click the **Apply** button. After saving and applying, OPNsense will start the Netflow capture. To look at Netflow's cache statistics, click the **Cache** tab. On this page, you can see network flow statistics for the configured interfaces.

If **Insight analyzer** is enabled, you can check the graphs it will plot by going to the **Reporting** | **Insight** page:

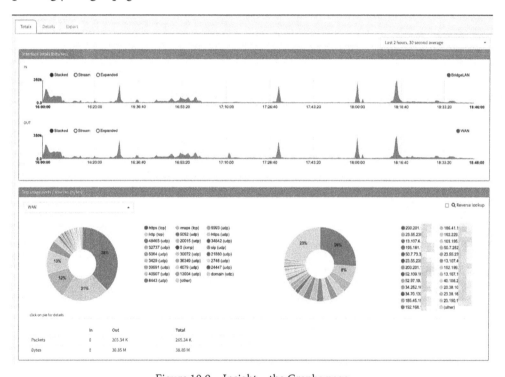

Figure 10.9 – Insight – the Graphs page

As you can see, **Insight** plots per-interface graphs and shows the most-used ports and sources:

Totals	Details	Export					

Date from	Date to	Interface	(dst) Port	(dst) Address	(src) Address		
2021-09-16 ▼	2021-09-16 ▼	WAN ▼					⟳

Service	Source	Destination	Bytes	Last seen	%
imaps (tcp)	200.	192.168.15.9	66 MB	Sep 16 18:44:29	25.21 %
3481 (udp)	52.1	192.168.15.9	22 MB	Sep 16 14:28:03	8.26 %
3480 (udp)	52.1	192.168.15.9	10 MB	Sep 16 14:28:03	3.78 %
5052 (udp)	200.	192.168.15.9	9 MB	Sep 16 18:20:14	3.55 %
50003 (udp)	185.	192.168.15.9	9 MB	Sep 16 18:29:58	3.44 %
http (tcp)	200.	192.168.15.9	6 MB	Sep 16 11:59:29	2.17 %
http (tcp)	8.24	192.168.15.9	5 MB	Sep 16 11:59:29	1.81 %
9993 (udp)	195.	192.168.15.9	5 MB	Sep 16 18:38:14	1.75 %
9993 (udp)	50.7	192.168.15.9	4 MB	Sep 16 18:38:04	1.69 %
https (tcp)	20.1	192.168.15.9	4 MB	Sep 16 16:21:17	1.69 %
9993 (udp)	103.	192.168.15.9	4 MB	Sep 16 18:38:20	1.51 %
9993 (udp)	50.7	192.168.15.9	4 MB	Sep 16 18:38:14	1.46 %
https (tcp)	200.	192.168.15.9	3 MB	Sep 16 18:44:26	1.05 %
https (tcp)	13.1	192.168.15.9	3 MB	Sep 16 18:08:55	0.96 %
https (tcp)	40.1	192.168.15.9	2 MB	Sep 16 18:44:33	0.95 %
https (tcp)	52.9	192.168.15.9	2 MB	Sep 16 17:11:42	0.90 %
https (tcp)	13.1	192.168.15.9	2 MB	Sep 16 18:45:06	0.89 %
https (tcp)	13.1	192.168.15.9	2 MB	Sep 16 18:45:10	0.88 %
0 (icmp)	192.	192.168.15.9	2 MB	Sep 16 18:39:28	0.86 %
https (tcp)	40.1	192.168.15.9	2 MB	Sep 16 18:17:00	0.83 %

Figure 10.10 – Insight – the Details page

By clicking on the **Details** tab, you can access the details page, which will list the following data in columns:

- **Service**: This column lists the top ports.
- **Source**: This column lists the source IP addresses.
- **Destination**: This column lists the destination IP addresses.
- **Bytes**: This column lists the bytes count.
- **Last seen**: This column shows the time when the traffic was seen last.
- **%**: This column shows the percentage of the respective traffic.

If you need to create a custom report from the network flow data to work in a spreadsheet, you can click on the **Export** tab and select the desired options to export.

Exploring real-time traffic

So far, we've explored traffic graphs with historical data that helps us analyze an extended time frame. But sometimes, we need to see the traffic in real time. To do so, OPNsense has a real-time traffic page on its webGUI. To access it, go to **Reporting | Traffic**:

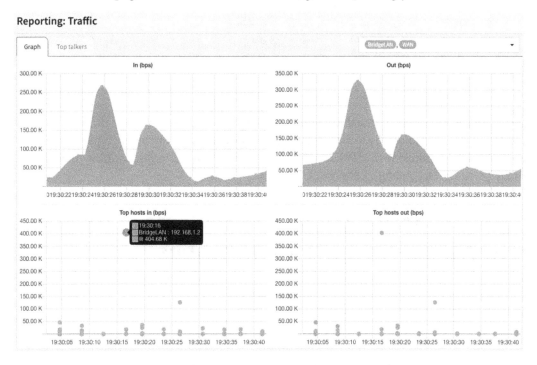

Figure 10.11 – The Reporting: Traffic page

The traffic graphs show the *input* and *output* for the selected interfaces and the *top hosts* (in bits per second). Each circle in the top host graph represents a host. If you pass your mouse cursor over it, it will display a legend (as shown in the preceding screenshot), along with information about the respective host.

By going to the **Top talkers** tab, you will find a table listing traffic information per address (in the selected interfaces):

Reporting: Traffic

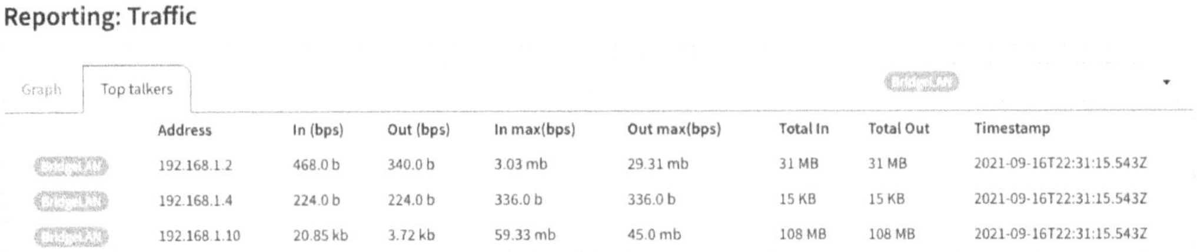

	Address	In (bps)	Out (bps)	In max(bps)	Out max(bps)	Total In	Total Out	Timestamp
Bridge(LAN)	192.168.1.2	468.0 b	340.0 b	3.03 mb	29.31 mb	31 MB	31 MB	2021-09-16T22:31:15.543Z
Bridge(LAN)	192.168.1.4	224.0 b	224.0 b	336.0 b	336.0 b	15 KB	15 KB	2021-09-16T22:31:15.543Z
Bridge(LAN)	192.168.1.10	20.85 kb	3.72 kb	59.33 mb	45.0 mb	108 MB	108 MB	2021-09-16T22:31:15.543Z

Figure 10.12 – The Top talkers page

As you can see, the selected interface (shown in the top-right corner) will be shown in front of each line, followed by the other fields. So far, we have explored the reporting resources that are available in OPNsense's webGUI, but we need to discuss how they can help us when issues come up. Let's check this out.

Troubleshooting common problems in the network using Netflow and graphs

The available OPNsense reporting tools can help us solve some problems related to performance. Based on my experience, I will give some examples of when these reporting tools can help a lot while solving issues:

- **Network Connectivity Issues**: Sometimes, you can face connectivity issues related to packet loss, latency, or both. When this kind of issue occurs – commonly on WAN interfaces, depending on the number of services enabled on OPNsense, such as **VPN** and **Proxy**, to name a few – you will receive a lot of complaints. A good starting point is checking the WAN links and quality graphs, as these are helpful resources. Even if you watch the gateway's status closely while running pings and other known network tools, the traffic and quality graphs will show a historical view of this data, which gives you a better analytical point of view.

- **Slow System/System Crashes**: When the whole system becomes slow or even crashes for no apparent reason and you don't know what's happening even after checking the logs, it might be a good idea to start looking at system graphs. The system graphs will show you historical data that will help you find the possible reasons for a slow system. Sometimes, the system slows down due to high CPU or memory usage, network throughput, and more. I have even experienced systems crashing due to a CPU's fan malfunctioning, which results in the system overheating and thus crashing or even damaging it. The CPU temperature graph helps a lot in cases like this.

Of course, the system graphs are not the only resources you need to rely on to solve problems related to OPNsense, but they can help you find auxiliary issues.

Summary

In this chapter, we dived into OPNsense's reporting resources. We learned about each available graph, the Netflow protocol, and how to use different graphs to monitor the system's health and performance. We also explored some scenarios in which various reporting tools can help solve problems in OPNsense. In the next chapter, we will discuss DHCP.

Section 3: Going beyond the Firewall

This final part will go through some extra features you can use in OPNsense to go beyond the firewall. We will explore services such as DHCP and DNS services, and how to apply web browsing controls and guestnet authorization with Web Proxy and Captive Portal. Will see how to extend Layer4 firewalling with IDS/IPS and the Sensei plugin and how to improve OPNsense availability by integrating two or more firewalls in a cluster. Finally, we will learn how OPNsense can protect websites.

This part of the book comprises the following chapters:

- *Chapter 11, Deploying DHCP in OPNsense*
- *Chapter 12, DNS Services*
- *Chapter 13, Web Proxy*
- *Chapter 14, Captive Portal*
- *Chapter 15, Network Intrusion (Detection and Prevention) System*
- *Chapter 16, Next-Generation Firewall with Zenarmor*
- *Chapter 17, Firewall High Availability*
- *Chapter 18, Website Protection with OPNsense*
- *Chapter 19, Command-Line Interface*
- *Chapter 20, API – Application Programming Interface*

11

Deploying DHCP in OPNsense

This final part of the book will go through some extra features you can add to your OPNsense system to go beyond the firewall. We will explore services such as **Dynamic Host Configuration Protocol (DHCP)** and **Domain Name System (DNS)**, applying web-browsing controls and guest net authorization with *a web proxy and a captive portal*. We will see how to extend layer 4 firewalling with **intrusion detection systems/intrusion prevention systems (IDSs/IPSs)** and Zenarmor plugins and improve OPNsense availability by integrating two or more firewalls in a cluster.

Providing **Internet Protocol (IP)** addresses to network hosts is one possible firewall duty. In this chapter, we will understand DHCP concepts used by OPNsense and how to use them to perform dynamic IP address leasing. By the end of this chapter, we will be able to understand and set up DHCP using OPNsense, going through the following topics:

- DHCP concepts
- DHCP server
- DHCP relay
- Diagnostics

Technical requirements

This chapter requires a running OPNsense instance to follow some steps, but this isn't mandatory. Knowledge of IP addressing is required.

DHCP concepts

Let's suppose you are in charge of setting up a new small local network with 30 hosts, and let's assume **IP version 4** (**IPv4**) addressing in this example. Now, think how much time setting up the IP address in each host will take. A lot, right? Setting it up in a server or gateway on the network will save a lot of time managing its IP address. So, to solve this problem, we can count on DHCP. OPNsense has a DHCP server that can be configured to deploy IP addresses to the network's hosts. DHCP works in a client/server manner, and it uses the **User Datagram Protocol** (**UDP**) to work, specifically on port 67. We will explore both IPv4 and IPv6 in this chapter. The process of a host taking a DHCP lease, which is how the IP addressing lease is named, is represented in the following diagram:

Figure 11.1 – DHCP addressing process

Next, we'll see how to configure the DHCP service on OPNsense for IPv4 addressing.

DHCP server

Each static IP-configured network interface can have a DHCP server configuration on OPNsense. To start configuring, go to **Services** | **DHCPv4** | *[Network interface name]*; for our example, I will assume a previously configured interface **local area network** (**LAN**).

The following options will be shown on the DHCP server configuration page:

- **Enable | Enable DHCP server on the LAN interface**: Checking this option will enable the DHCP service on this interface (LAN in our example).

- **Deny unknown clients**: This option, when checked, will restrict DHCP leases to hosts added to the **DHCP Static Mappings for this interface** section. Using this option is a good way of preventing unknown hosts from getting a network address and talking with OPNsense, but it will not guarantee that if someone tries to set up a static IP address instead of using DHCP on a host, they won't start accessing other hosts on the network.

> **Important Note**
>
> Using a sniffer tool (such as `tcpdump` or Wireshark, for example), finding other hosts on the network will be possible. Enabling this option is an additional protection method but not a bulletproof one.

- **Ignore Client UIDs**: A **unique identifier** (**UID**) is used on DHCP for device tracking. Sometimes, a device can send multiple UIDs to the DHCP server, and you can check this option to avoid a host doing that to take multiple leases.

- **Subnet**, **Subnet mask**, and **Available range** labels: These labels will show information based on the network interface configuration. We can use this information to decide which DHCP range will be configured in the following options:

  - **Range**: Provide here the first available address in the range in the **from** textbox and the last address in the range in the **to** textbox. In a 100% DHCP-based network, it will define how many DHCP-managed hosts the network can have.

  - **Additional Pools**: In some networks, it may be necessary to split the DHCP range, for example, and in such cases, you can use this option to add a pool. Each additional pool will have its own options, which can be useful while setting a pool for specific hosts such as **Voice over IP** (**VoIP**) telephones.

  - **WINS servers**: Old Windows versions (Windows **eXPerience** (**XP**) and earlier) may need the **Windows Internet Name Services** (**WINS**) protocol. This option allows us to configure up to two servers.

  - **DNS servers**: If you need to send DNS servers' addresses differently from OPNsense-configured servers, fill the two textboxes in this option, each with a DNS server IP address for primary and secondary servers, respectively.

> **Important Note**
>
> In the **Networking** section, you can check which DNS servers are configured on OPNsense on **System | Settings | General**. Suppose your OPNsense system is receiving DNS server configuration through a DHCP configuration on a **wide-area network** (**WAN**) interface, for example. In that case, you can find which DNSs are in use by doing a DNS lookup on the **Interfaces | Diagnostics | DNS Lookup** page.

• **Gateway**: By default, OPNsense will send the IP address configured in this interface as the gateway to hosts. To not send any gateway address, just type `none` in the textbox. To change to another IP, fill it in with this option.

> **Important Note**
>
> While using a virtual IP address as a gateway, as we will see in *Chapter 17, Firewall High Availability*, you must set the gateway to the virtual IP address to avoid hosts losing connectivity if the master node fails.

• **Domain name**: The domain name provided to network hosts can be set in this option. If left blank, OPNsense will set its configured domain name as default. For example, if your host has a hostname of `pc` and the domain is `example.com`, this would result in a **fully qualified domain name** (**FQDN**) of `pc.example.com`.

• **Domain search list**: If your network uses more than one domain, you can set them here separated by a semicolon. Hosts will try to resolve short names using these domains first rather than querying the DNS server first.

• **Default lease time (seconds)**: Set the time a DHCP lease will last before expiring when the host doesn't request the lease time. The default is 7,200 seconds (2 hours).

• **Maximum lease time (seconds)**: Specify the maximum time a DHCP lease can last before expiring when the host asks for the expiration time. If you leave it blank, OPNsense will set it to the default: 86,400 seconds (1 day).

• **Response delay (seconds)**: This will set a time delay expressed in seconds before the DHCP server will respond to a client trying to acquire an IP address. The default value is 0 seconds, which means without delay.

• **Interface MTU**: This will define the **maximum transmission unit** (**MTU**) size when the client requests it.

- **Failover peer IP**: While using a **high availability** (**HA**) setup, you must set the network interface address (real IP) on each HA node; on the primary node, set the secondary IP, and vice versa.

- **Failover split**: Using HA capabilities, you can set load-balancing clients between the primary and secondary nodes. The possible range is from 0 (no clients) to 256 (100% of clients) on the primary node. The default value is 128 (50% of clients on the primary node).

- **Static ARP**: If you check **Enable Static ARP entries**, only hosts added to the **DHCP Static Mappings for this interface** section will communicate with OPNsense.

> **Important Note**
>
> Proceed with caution with this option, as it will persist configuration even when the DHCP server service is stopped or disabled.

- **Time format change**: Checking the **Change DHCP display lease time from UTC to local time** option will change the default time zone displayed on the DHCP **Leases** page (**Services | DHCPv4 | Leases**) to the local time zone.

- **Dynamic DNS**: This option will register the DHCP client names in an external DNS server. To enable it, check the **Enable registration of DHCP client names in DNS** option. You have to fill in each option related to the dynamic DNS option as required (depending on your dynamic DNS service/server).

- **MAC Address Control**: It is possible to allow only specific hosts to use their **media access control** (**MAC**) address in the **allow** textbox or deny some hosts by adding them to the **deny** textbox. On both, you need to fill in comma-separated addresses. It is possible to specify just the **organizationally UI** (**OUI**) part of the MAC address. In this way, it will allow permitted hosts of a specific(s) vendor(s).

- **NTP servers**: Specify time servers (**Network Time Protocol**, or **NTP**) that will be set on hosts using the DHCP server.

- **TFTP server**: If some hosts on the network need a **Trivial File Transfer Protocol** (**TFTP**) server address, you can set their IP addresses in this option. Some devices use this combined with the **Bootstrap Protocol** (**BOOTP**) for the boot process.

- **LDAP URI**: Some systems may require the **Lightweight Directory Access Protocol** (**LDAP**) server's **Uniform Resource ID** (**URI**) from the DHCP server. This option will set this and send it to clients that ask for it.

- **Enable network booting**: Check the **Enables network booting** option to provide a network-booting service for devices on the network.

- **WPAD**: This tells OPNsense's DHCP server to tell the client where it can find the **Web Proxy Auto-Discovery (WPAD)** protocol file; so, the **Uniform Resource Locator (URL)** will be set to something like this: `http://fw.example.com/wpad.dat`. Please note that you also need a firewall rule and/or a reverse proxy in place to allow that access. Check the **Enable Web Proxy Auto Discovery (WPAD)** option to provide it through the DHCP service. It is also necessary to configure the WPAD file on the web proxy (**Services | Web Proxy | Administration**) to make it work.

- **Enable OMAPI**: **Object Management Application Programming Interface (OMAPI)** is an **application programming interface (API)** that allows automation between the DHCP server and other specialized systems. If using this integration, fill in the required options to enable it by checking the **Enables OMAPI** checkbox and filling in all the necessary following options.

- **Additional Options**: If some other option you will need isn't implemented on the web **user interface (UI)**, you can specify it by adding the DHCP code and the required parameters in this option. You can check all the available options on this page: `https://www.iana.org/assignments/bootp-dhcp-parameters/bootp-dhcp-parameters.xhtml`.

We can add a static IP address based on the host's MAC address in the **DHCP Static Mappings for this interface** section. To do that, click on the + button. Adding a static lease will open a new page where you can fill in specific configurations for the host, as the following screenshot shows:

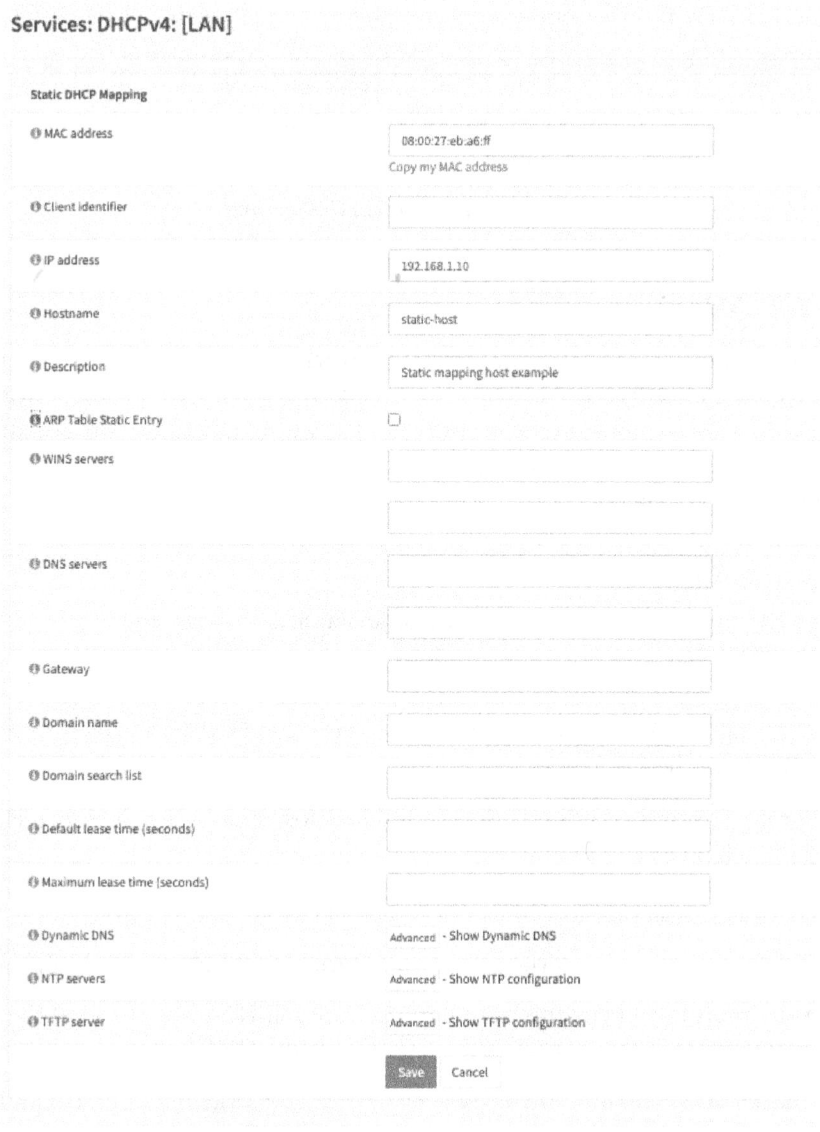

Services: DHCPv4: [LAN]

Static DHCP Mapping

ⓘ MAC address	08:00:27:eb:a6:ff
	Copy my MAC address
ⓘ Client identifier	
ⓘ IP address	192.168.1.10
ⓘ Hostname	static-host
ⓘ Description	Static mapping host example
ⓘ ARP Table Static Entry	☐
ⓘ WINS servers	
ⓘ DNS servers	
ⓘ Gateway	
ⓘ Domain name	
ⓘ Domain search list	
ⓘ Default lease time (seconds)	
ⓘ Maximum lease time (seconds)	
ⓘ Dynamic DNS	Advanced - Show Dynamic DNS
ⓘ NTP servers	Advanced - Show NTP configuration
ⓘ TFTP server	Advanced - Show TFTP configuration

Save   Cancel

Figure 11.2 – DHCP static mapping entry example

After filling in the desired options, click on the **Save** button to add the static mapping new entry.

As we can see, the DHCP server options are compelling and allow us to have different pools with many configurations for each one. Based on our experience at CloudFence managing dozens of OPNsense systems, I can tell you that OPNsense is usually the preferred DHCP server in the network, even when a Windows server with these capabilities is present in the same network.

Now, let's explore the DHCP relay option present on OPNsense.

DHCP relay

While using a remote DHCP server (on another network segment), the DHCP relay service will need to be configured. This is required because the DHCP protocol works using broadcasting requisitions on the network, and as we already know, the broadcast domain is limited to the same network in which the hosts reside. OPNsense supports both DHCPv4 and DHCPv6 relaying.

> **Important Note**
> The DHCP relay service can only be used on an interface that has the DHCP server disabled.

To start configuring the DHCP relay service, go to **Services | DHCPv4 | Relay** or **Services | DHCPv6 | Relay** for the IPv6 protocol.

For both IP protocol versions, v4 and v6, the configuration options are the same. Here are descriptions of each one:

- **Enable**: Check this option to enable the DHCP relay service.
- **Interface(s)**: Select each interface to be configured as the relaying service.
- **Append circuit ID**: Check this option to add a circuit ID—which means the OPNsense interface number—and the agent ID to the DHCP request.
- **Destination servers**: Fill in the DHCP server's IP address(es) (comma-separated) that requests are being relayed to.

As we can see, the DHCP relay service has only a few options to configure compared to the DHCP server. While running as a DHCP relay, OPNsense will forward configurations from the DHCP server it is relaying requests to.

In the following section, we will explore the diagnostics tools available in OPNsense to help troubleshoot DHCP-related issues.

Diagnostics

This section will explore some of the common issues related to the DHCP service and which tools OPNsense has to help us solve them.

Let's start by checking the online hosts on the DHCP **Leases** page. On this page, we can check the leases attributed by the DHCP server. To access the DHCPv4 leases page, go to **Services | DHCPv4 | Leases**, as illustrated in the following screenshot:

Services: DHCPv4: Leases (6)

Interface	IP address	MAC address	Hostname	Description	Start	End	Status	Lease type	
BridgeLAN	192.168.1.4	08:00:27:1f:76:05 *PCS Systemtechnik GmbH*	OPNsense		2021/10/10 13:30:34 UTC	2021/10/10 15:30:34 UTC	..il	active	+
BridgeLAN	192.168.1.10	0c:4d:e9: *Apple, Inc.*	cirrus	Mac JC			..il	static	
BridgeLAN	192.168.1.11	00:0c:29:77:b7:19 *VMware, Inc.*	OPNsense-VM	OPNsense VMWare			⊘	static	
BridgeLAN	192.168.1.13	00:1c:42:98:74:2a *Parallels, Inc.*	VMWIN10JCC				⊘	static	
BridgeLAN	192.168.1.14	00:0c:29:cb:24:b1 *VMware, Inc.*	kali	Kali VM			⊘	static	
	192.168.2.11	74:e6:e2: *Dell Inc.*	WIN10-JCHOME		2021/10/10 13:24:31 UTC	2021/10/10 15:24:31 UTC	..il	static	

Show all configured leases

Figure 11.3 – DHCPv4 leases page

As the preceding screenshot shows, on the DHCP **Leases** page, we can check each lease with the following information: **Interface**, **MAC address**, **Hostname**, **Description** (if it is added in the static mapping), the **Start** and **End** time of the lease, **Status**, and **Lease type**. If you need to check expired leases, click on the **Show all configured leases** button at the bottom of the page. There is also a way to add a static mapping on this page. To do so, click on the + button that is placed at the end of each dynamic lease line.

The **Leases** page is a good starting point to solve DHCP issues. Once there, you can check which hosts are alive and get some additional information about them. Common issues you can check on this page are noted here:

- **Check host's connectivity**: Checking whether some host is grabbing an IP address from the DHCP server is a good starting point to check its connectivity.

- **DHCP pool maximum limit**: Sometimes, it is possible to check whether the number of hosts on the network is larger than the available DHCP pool limit by cross-checking the network subnet size and the pool's configured range.

- **Inconsistent information**: It isn't a rare occurrence while troubleshooting local network connectivity issues to get inconsistent information from the user while checking on the **Leases** page—for example, a user is telling you that their IP address is leased to another device. By comparing the device vendor (displayed on the **MAC address** column) and the MAC address, you can quickly check this.

Another valuable resource while solving issues related to the DHCP server is the **Services | DHCPv4 | Log File** page. On that page, you can check all information generated by the dhcpd daemon (which is the DHCP service). If something goes wrong with this service (such as an unexpected service stop), you will find relevant information on it.

As we explored, the DHCP service is quite simple to troubleshoot, and you may not have any major troubles with it on OPNsense. The webGUI has enough resources to help you find issues and solve them.

Summary

In this chapter, we started to explore services available on OPNsense that extend beyond a simple network firewall. We learned how to enable and configure a DHCP server and relay and check for issues related to them on the webGUI. You can now configure your OPNsense system as a DHCP server on a local network using all the powerful resources it supports. In the next chapter, we will explore DNS services that can be integrated with DHCP, making OPNsense an even more robust solution.

12
DNS Services

This chapter will cover **Domain Name System (DNS)** resolvers, what are the available options in OPNsense core, and the features available in each one. We will also take a brief look at dynamic DNS, explore some DNS plugins available, and see how to troubleshoot common issues with DNS resolving. By the end of this chapter, you will be able to configure and manage DNS services in OPNsense.

Specifically, we'll be looking at the following topics:

- Core DNS services
- DNS plugins
- **Dynamic DNS (DDNS)**
- Troubleshooting

Technical requirements

For this chapter, you'll need good knowledge of the **Domain Name System (DNS)**, an understanding of how it works, and an awareness of common tools used to test DNS. A running OPNsense system to test configuration is optional but advised.

Core DNS services

The DNS protocol is the base of our modern internet. It makes it easy for users to find websites and other services available on the internet without remembering every website's **Internet Protocol** (**IP**) address. To resolve website names to IP addresses, we have two classes of DNS services—servers that host domain entries, and recursive resolvers that translate domain names to IP addresses. OPNsense can become a DNS server and host domain entries. It isn't recommended to transform a firewall into a DNS server for security best practices, so we'll explore only the DNS resolver feature in this chapter.

Default DNS resolvers on OPNsense

OPNsense has two DNS resolver services installed by default—*Dnsmasq* and *Unbound*. The last one became the default service since the 17.7 OPNsense version. We'll discuss both services one by one.

Unbound

As the maintainers define it, "*Unbound is a validating, recursive, caching DNS resolver. It is designed to be fast and lean and incorporates modern features based on open standards.*" It runs on most **Berkeley Software Distribution** (**BSD**) variants (including macOS), Linux, and Windows. It is a robust DNS resolver option, and it is a fellow countryman of OPNsense. To name just a few attributes, it supports both **DNS-over-TLS** (**DoT**) and **DNS-over-HTTPS** (**DoH**), had a rigorous audit in 2019, and so on. To check more about Unbound, you can visit its official site at the following address: `https://www.nlnetlabs.nl/projects/unbound/about/`.

Once Unbound is the default DNS resolver option, it is enabled by default. To start configuring it, go to **Services | Unbound DNS | General**, as illustrated in the following screenshot:

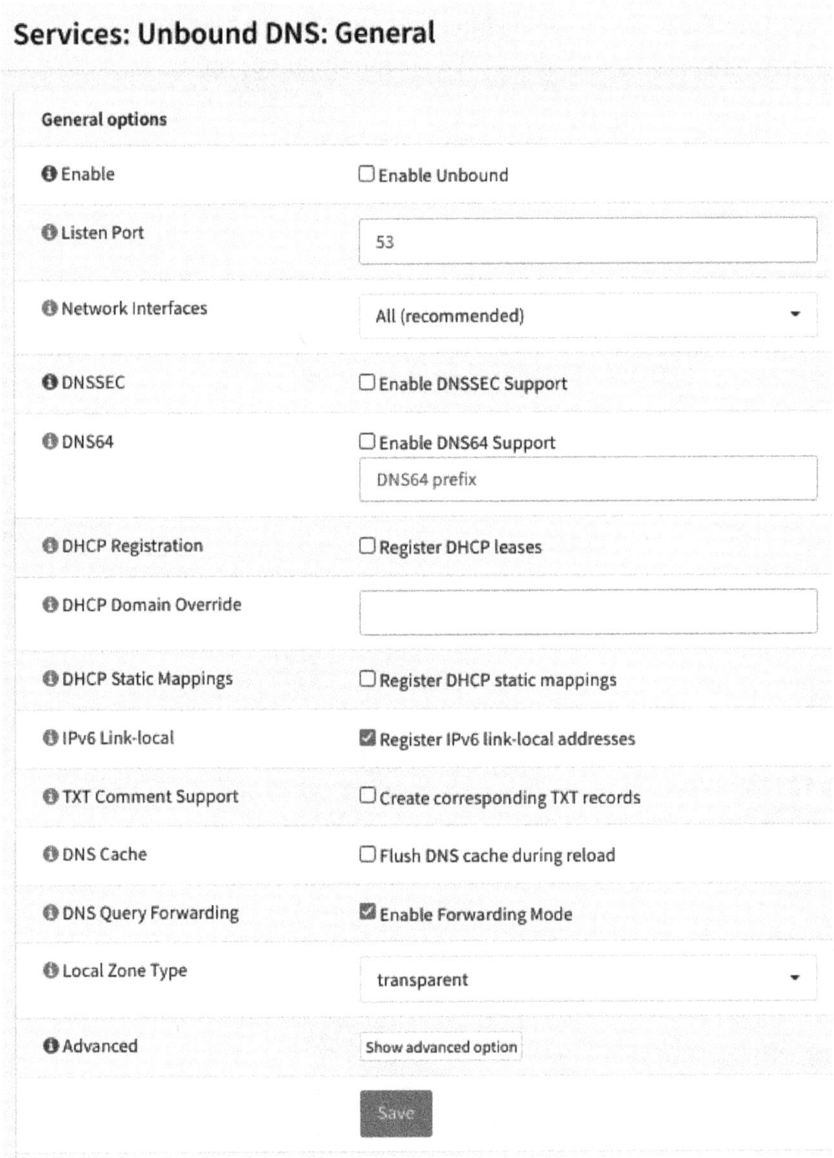

Services: Unbound DNS: General

General options

❶ Enable	☐ Enable Unbound
❶ Listen Port	53
❶ Network Interfaces	All (recommended) ▾
❶ DNSSEC	☐ Enable DNSSEC Support
❶ DNS64	☐ Enable DNS64 Support
	DNS64 prefix
❶ DHCP Registration	☐ Register DHCP leases
❶ DHCP Domain Override	
❶ DHCP Static Mappings	☐ Register DHCP static mappings
❶ IPv6 Link-local	☑ Register IPv6 link-local addresses
❶ TXT Comment Support	☐ Create corresponding TXT records
❶ DNS Cache	☐ Flush DNS cache during reload
❶ DNS Query Forwarding	☑ Enable Forwarding Mode
❶ Local Zone Type	transparent ▾
❶ Advanced	Show advanced option

Save

Figure 12.1 – Unbound DNS general settings page

On the general settings page, the following options will be available:

- **Enable | Enable Unbound**: Check this option to enable the Unbound DNS as a resolver.

- **Listen Port**: Set the port that will respond to DNS queries. By default, port 53 is used.

- **Network Interfaces**: Select which interfaces the Unbound daemon will listen to client DNS requests. By default, the **All (recommended)** option is selected, which means that it will listen on all configured network interfaces.

- **DNSSEC**: The **DNS Security Extensions** (**DNSSEC**) enable digital signatures to authenticate DNS requests. If your DNS servers provide this functionality, it is recommended to enable it.

On the **command-line interface** (**CLI**), you can test DNSSEC by using the `drill` command, as follows:

```
root@bluebox:~ # drill -D fabian-franz.eu
;; ->>HEADER<<- opcode: QUERY, rcode: NOERROR, id: 29094
;; flags: qr rd ra ad ; QUERY: 1, ANSWER: 2, AUTHORITY: 0,
ADDITIONAL: 0
;; QUESTION SECTION:
;; fabian-franz.eu.        IN      A

;; ANSWER SECTION:
fabian-franz.eu.           3600    IN      A       78.47.250.154
fabian-franz.eu.           3600    IN      RRSIG   A 13 2
3600 20220324000000 20220303000000 19848 fabian-franz.eu.
xKyLm+X58jQzrlZgpj8t82Qvi8vwLoC3SxeZ36wabl+XErMC8KJpzEm/
L28XvhbWR8whgIl+hPTaT/oyFG4ZJw==

;; AUTHORITY SECTION:

;; ADDITIONAL SECTION:

;; Query time: 156 msec
;; EDNS: version 0; flags: do ; udp: 1232
;; SERVER: 127.0.0.1
;; WHEN: Sat Mar 12 17:26:08 2022
;; MSG SIZE  rcvd: 171
```

As we can see in the `drill` command output, the highlighted lines show the *DNSSEC* **Resource Record Set** (**RRset**) *signature* and the ad *flag*, which means that the response was validated by a DNSSEC enabled resolver.

> **Important Note**
>
> To test online if your preferred DNS server supports DNSSEC, you can visit this website and run an online test: `https://dnssec.vs.uni-due.de/`.

- **DNS64**: Checking this option will just translate **address** (**A**) records to AAAA records.

- **DHCP Registration**: Enable this option to register DHCP leases on Unbound. This way, registered hostnames can be resolved by other clients.

- **DHCP Domain Override**: Specify the default domain name to be used in the DHCP registration process.

- **DHCP Static Mappings**: Check this option to register DHCP static mappings as DNS entries on the Unbound service.

- **IPv6 Link-local**: To register each configured **IP version 6** (**IPv6**) link-local network interface address, check this option.

- **TXT Comment Support**: While using DHCP server static mappings, every host description created will add a **text** (**TXT**)-type DNS record.

- **DNS Cache**: The DNS cache helps to speed up domain resolution time by creating a temporary database. To flush the DNS cache every time the Unbound service reloads, check this option.

- **DNS Query Forwarding**: To resolve DNS queries, a service such as Unbound can use designated servers to forward queries and get proper replies. It will use servers configured in the **System | General | Networking** section to resolve DNS names. To learn more about DNS root servers, you can visit this link: `https://www.iana.org/domains/root/servers`.

- **Local Zone Type**: Local zones are DNS records that don't exist on the authoritative servers. Unbound can host local zones that behave in different manners depending on the configured option. As it is an extensive list of options, I recommend you to check the following link for a detailed and updated description of each available option: `https://nlnetlabs.nl/documentation/unbound/unbound.conf/#local-zone`.

- **Advanced**: Click on **Show advanced option** to expand the following options:

  - **Outgoing Network Interfaces**: Select the available network interfaces to send DNS queries. It is usual to keep **All** selected here.

  - **WPAD Records**: To enable the **Web Proxy Auto-Discovery (WPAD)** protocol to be configured by DNS records (**Canonical Name (CNAME)** type), check this option. WPAD is a feature that can help to configure a web proxy on the client's browsers. Remember that it will be necessary to configure it on a web proxy too. We'll explore web proxies in a dedicated chapter later in this book.

> Important Note
>
> A generated WPAD file will be made available using `http://<OPNsense_ IP_address>/wpad.dat` (using the `80` port). If you intend to use WPAD, keep the web **user interface (UI)** listening on the **HyperText Transfer Protocol (HTTP)** port (`80`).

There are other Unbound pages available for configuration. We will talk about each of these briefly, as follows:

- **Overrides**: On this page, it will be possible to create host and domain overrides. An example of its usage is to create inexistent records that will override responses from the authoritative servers, not even consulting the original server. We can create a `www.example.com` host override pointing to a local server, or even a domain override such as `example.com`.

- **Advanced**: On this page, we can set the advanced Unbound configuration parameters. I recommend taking a look at Unbound's official documentation at `https://nlnetlabs.nl/documentation/unbound/` before setting parameters.

- **Access Lists**: Access lists define rules that may be applied in DNS queries by the network source address. We can, for example, deny queries of a specific network by creating an access list, like so:

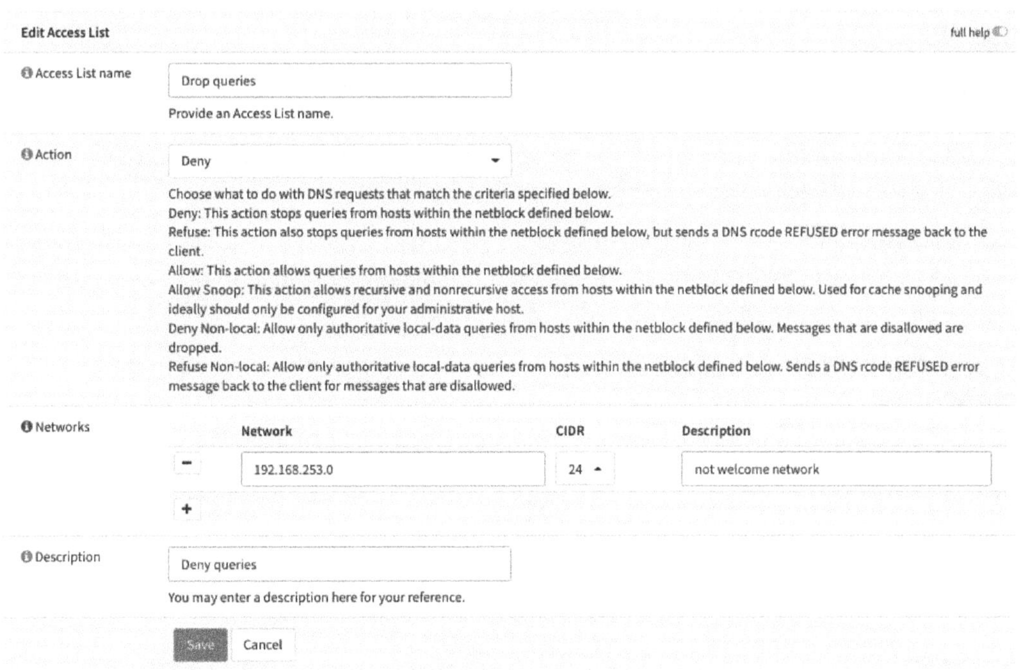

Figure 12.2 – Services | Unbound DNS | Access Lists: New deny rule example

The **Access Lists** page automatically creates rules allowing queries to be sent to local networks. *Additional created rules will override these default rules*:

- **Blocklist**: On this page, it will be possible to add predefined DNS blocklists that'll turn the Unbound service into a DNS filter. This feature is helpful to control access of users to bad domains (domains known as a source of malware) or just block porn or Facebook. You can also define whitelisted domains, private domains, and so on.

- **DNS over TLS**: On this page, we can configure the forwarding of DNS queries to servers using **Transport Layer Security** (**TLS**). To make it work, you'll need a TLS-compatible DNS server to forward requests.

- **Statistics**: On this page, we'll see a list of daemon statistics by process thread.

- **Log File:** To check for Unbound notices or error messages, take a look at this page. It reflects the contents of the `/var/log/resolver.log` file.

Unbound supports an extensible list of features and allows customization to satisfy demands from small to large networks. Next, we'll explore the other core DNS forwarder—Dnsmasq.

Dnsmasq

Dnsmasq is a DNS forwarder that is lightweight and easy to use and maintain. Its official website defines it as "*suitable for resource-constrained routers and firewalls*". OPNsense keeps it for compatibility once it isn't the default DNS resolver. It was the default DNS resolver/forwarder back when OPNsense was forked from pfSense.

Before configuring it, we first need to disable the default DNS forwarder, Unbound. Go to **Services | Unbound DNS | General**, *uncheck* the **Enable** option, click on the **Save** button, and then click on **Apply changes** to disable it.

Go to the **Services | Dnsmasq DNS | Settings** page to configure it, as shown in the following screenshot:

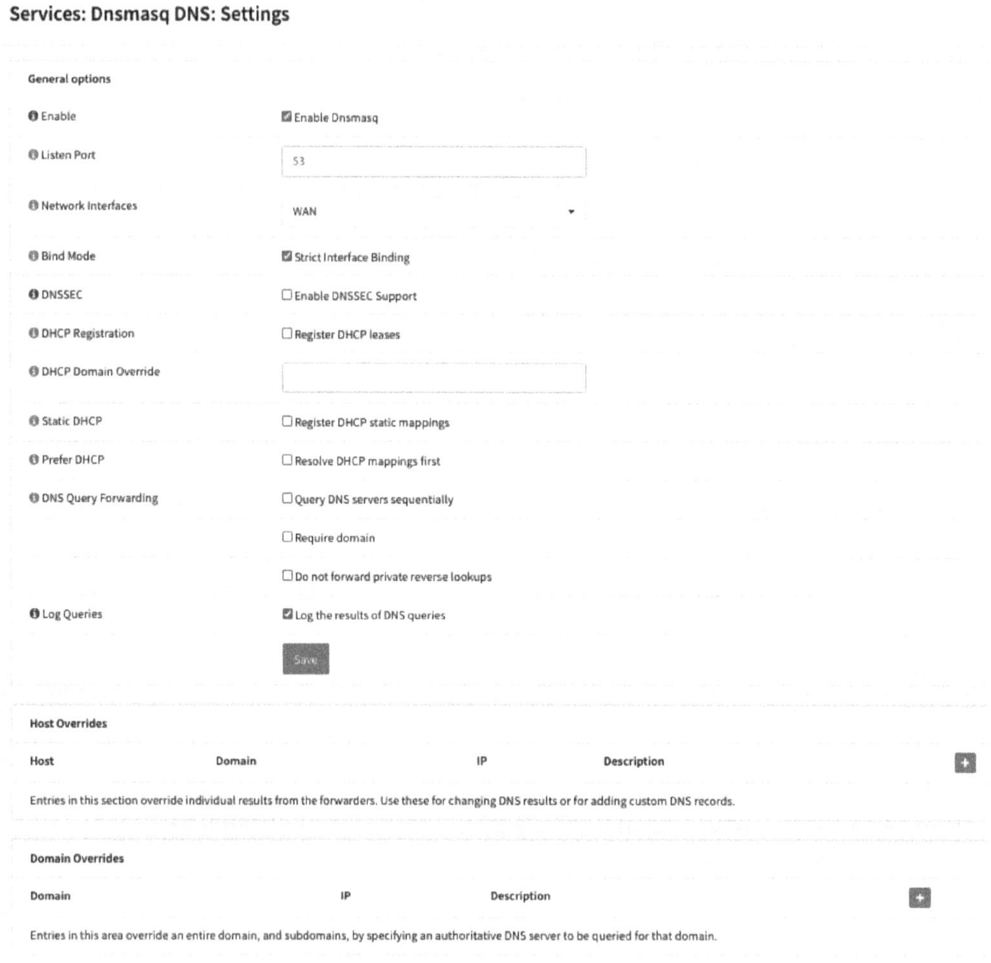

Figure 12.3 – Dnsmasq settings page

We will find the following settings on this page:

- **Enable | Enable Dnsmasq**: Check this option to enable the service.

- **Listen Port**: The *default port is* 53. If you need to change this to another port, specify it in this textbox.

- **Network Interfaces**: Select which network interfaces the service will listen on for DNS queries. The default configuration is to listen on all available network interfaces.

- **Bind Mode | Strict Interface Binding**: Instead of listening to each selected interface, checking this option will only listen for DNS queries on the network interface-configured address. Note that it will not work on IPv6 addresses.

- **DNSSEC**: DNSSEC provides secure extensions for the DNS protocol. To enable Dnsmasq support for it, check this option. Note that the DNS servers responding to OPNsense must support it.

- **DHCP Registration**: While using the DHCP server, every host that leases an address will be registered on Dnsmasq with a record. Thus, the host's name will be available for DNS resolution.

- **DHCP Domain Override**: Check the **Register DHCP static mappings** option to override the default domain configured on OPNsense (verify this on **System | Settings | General | Domain** option) and specify a different domain for DHCP hostname registration.

- **Static DHCP**: As with the **DHCP Registration** option, checking the **Register DHCP static mappings** option will also register static DHCP host entries on the Dnsmasq service for hostname resolution.

- **Prefer DHCP | Resolve DHCP mapping first**: As Dnsmasq allows us to set host and domain overrides, if we prefer to resolve DHCP entry names before manually created entries, we must check this option.

- **DNS Query Forwarding**:

  - **Query DNS servers sequentially**: Check this option to sort the DNS servers' order as they are configured in **System | Settings | General**.

  - **Require domain**: Only queries with the fully qualified domain (host and domain) will be resolved if this option is checked.

  - **Do not forward private reverse lookups**: To avoid sending **pointer record** (**PTR**) queries (reverse name lookups) of **RFC 1918** addresses to external name servers, check this option.

> **Important Note**
>
> To test reverse lookup (PTR), you can use the `dig` command on Linux or macOS, as in the following example:
>
> `dig -x 1.1.1.1 or dig PTR 1.1.1.1`
>
> On a Windows prompt, run `nslookup` and press *Enter* to obtain the following output:
>
> `> set type=PTR`
>
> `> 1.1.1.1`

- **Log Queries | Log the results of DNS queries**: By default, Dnsmasq doesn't log client queries. If you need to register each host DNS query, check this option. It might be helpful for hosts trying to access auditing, for example.

- **Hosts and Domain Overrides**: As we explored with the Unbound service earlier, Dnsmasq also allows host and domain overrides to reply to clients.

- As usual, to finish the Dnsmasq configuration, we must click on the **Save** button.

- **No Hosts Lookup | Don't read the hostnames in /etc/hosts**: Checking this option will avoid resolving hostnames added locally to the `/etc/hosts` file.

- **Log File page** (**Services | Dnsmasq DNS | Log File**): For Dnsmasq daemon messages, errors, or even host queries (with the **Log Queries** option checked), you can refer to this page. This page will show the contents of the `/var/log/dnsmasq.log` file.

Dnsmasq is an alternative to Unbound but with more limited features. We may not need to use it in common scenarios, but as a robust firewall system, OPNsense at its core offers more options for DNS service resolution. In the following section, we will explore some other available DNS plugins that can be used as an alternative to core DNS services.

DNS plugins

One of the features that make OPNsense an extremely flexible system is the plugin additions. To extend DNS core features, we can install some plugins developed by the OPNsense community. Let's take a look at some of these available DNS plugins, as follows:

- **BIND** (`os-bind`): The **Internet Systems Consortium (ISC) Berkeley Internet Domain (BIND)** is one of the most used DNS servers on the modern internet. It is a complete DNS server that can host several domain zones. On OPNsense, its implementation is focused on forwarder/resolver and blocklist capabilities. To install it, add the `os-bind` plugin.

- **DNSCrypt Proxy** (`os-dnscrypt-proxy`): DNSCrypt version 2 is a powerful DNS proxy that supports many features such as DoH and anonymous queries. The OPNsense implementation also allows DNS filtering (blocklists), and it can be a good alternative to dnsmasq or even Unbound in special cases. To install it, add the `os-dnscrypt-proxy` plugin.

After installation, both plugins can be found on the **Services** menu.

Additional plugins aren't always necessary, but they can be a handful in some scenarios where the core features don't supply specific needs. It is recommended to test additional plugins in a lab environment before using them in production.

The following section will explore DDNS features to help gain external administrative access to OPNsense or services such as **virtual private networks** (**VPNs**) or even **network address translation** (**NAT**) rules.

DDNS

Another feature that can be added by using an additional plugin is DDNS. The DDNS feature is most used on home or small business installations where a static or public IP address isn't available. Some services that may use DDNS are OpenVPN, **IP Security** (**IPsec**), NAT rules, or external management access to OPNsense. To add this feature, we need to add the `os-dyndns` plugin. After being installed, the plugin configuration page is on the **Services | Dynamic DNS** menu. To start using it, we'll need an account on a compatible predefined service. The plugin supports several services, so finding a suitable service for your needs won't be challenging.

Troubleshooting

As a fundamental protocol for the internet, the DNS protocol must be working smoothly in our OPNsense system to avoid network issues. When users complain about internet connectivity issues, inexperienced analysts probably start the troubleshooting process by looking at the user's web browser or even on the web proxy or in the firewall ruleset. However, a simple DNS test can sometimes solve the problem and also save a lot of time.

In this section, I'll focus on the *Unbound* service because it is the default DNS service, so it'll be applicable for most cases, but some configurations can be applied to other DNS services.

A common issue related to DNS services is that hosts can't resolve names.

Common issues related to hosts that can't resolve DNS names include the following:

- **While using a DHCP server on OPNsense**: The IP address set on the DHCP server isn't listening for DNS queries. Check the DNS service configuration to check if it is listening for queries on the configured IP address. Check **Listen Port**, **Network Interfaces**, and **Bind Mode** on the *Unbound* configuration page.

- **Service isn't running**: Sometimes, the DNS service is not running. This can happen for a number of reasons, such as the following:

  - **Concurrent DNS services running**: If you try to configure two or more DNS services at the same time (such as Dnsmasq and Unbound, for example), they will try to listen on the default port (when configured on the default value, 53). This will generate an error and the service won't start up.

  - **Configuration errors**: Sometimes, atypical configurations can crash the service, and it won't start up. It is rare to see the latest OPNsense versions because the development team is doing a great job fixing bugs. Still, it's not beyond the realm of possibility that the webGUI configuration doesn't handle some user input configuration properly, resulting in an error.

- **DNS forwarding is not enabled**: To resolve names, the Unbound service must have the **DNS Query Forwarding** option checked.

- Check if **Transmission Control Protocol (TCP)/User Datagram Protocol (UDP)** port 53 is allowed in the firewall rules.

We can always check the log files to look for messages that will help us find out what could be happening. For complete troubleshooting, I advise you to always check both on OPNsense and the client hosts. Some tools such as dig and nslookup will be helpful to uncover DNS resolution issues.

The **DNS Lookup** page is handy too in the webGUI to make tests. To access it, go to **Interfaces | Diagnostics | DNS Lookup**. Pay attention to each server's response and query time. Notice that the name resolution is not always done by the same server (following the configured order), so a single server with a higher query time or different response is enough to give you a big headache in your network.

Making a DNS lookup using the CLI

When I started using OPNsense, I had difficulty finding a tool to resolve DNS names on the CLI. I was pretty familiar with `dig` and `nslookup`, but neither tool is available on the CLI, so one day, I figured this out: drilling could be faster than digging, right? So, if you need to perform DNS tests on the CLI, try the `drill` command.

Summary

In this chapter, we have explored the DNS services available on OPNsense. Now, you can configure both core services—Unbound and Dnsmasq—or an added one by installing plugins such as `dnscrypt-proxy` or BIND. We also learned how DDNS could be enabled and configured to enable external access to OPNsense's services and management. In the next chapter, we will stay on the services track and learn how to integrate them.

13
Web Proxy

A web proxy is one of the top features of a modern firewall solution. With it, you can extend the control capabilities of your system to another level. We will learn about web proxy deployment methods, why to use a web proxy, the core features, the basic configuration, and how to customize it. We will explore the available web filtering options using the core OPNsense capabilities and learn how to block undesired content. Last but not least, we will see some common troubleshooting scenarios and how to read web proxy logs to solve them.

The chapter covers the following main topics:

- Web proxy fundamentals
- Basic configuration
- Web filtering
- Reading logs and troubleshooting

Technical requirements

A clear understanding of the **HyperText Transfer Protocol /HTTPS Secure (HTTP/HTTPS)** protocol is fundamental to understanding this chapter. Knowledge of additional concepts such as **man-in-the-middle (MITM)** attacks (HTTPS interception), **Domain Name System (DNS)**, browser configuration, and knowing how a **Microsoft Active Directory (AD)** and a **Lightweight Directory Access Protocol/LDAP Secure (LDAP/LDAPS)** protocol works are essential to thoroughly understand the web proxy topics discussed in this chapter. A running OPNsense system with a client host connected to it will ease a lot of testing and labs.

Web proxy fundamentals

Back in the 2000s, when the internet bubble burst and the **World Wide Web (WWW)** arose, many companies were adopting it to keep their business connected in a new era. As a young network administrator, I had the mission of keeping users working, even with many distractions popping up on their screens while they worked online. I was using just Linux-based gateways (`iptables`-based firewalls), and they were not enough to carry on. So, I searched online and found an HTTP forwarding proxy—an HTTP *firewall* that could handle HTTP(S) requests from users, cache website contents, and apply some control through **access control lists (ACLs)**. It was the perfect tool for the job! The name of this tool? The *Squid Web Proxy*.

So, this is my personal history about web proxies. Since then, they have remained a fundamental tool in ensuring users' security and control online. But let's dive deep into web proxies and explore how they work and how we can use one on OPNsense to keep networks safe.

We can deploy a web proxy using two types of implementation methods, as detailed next.

The explicit method

The explicit method requires configurations both on the web proxy server and on every host in the network. Using this method, the browser *knows* that it is talking to a web proxy server, thus all communication between both regularly applies the features a web proxy supports without any protocol modification or interception.

An explicit proxy supports authentication, and the overall behavior using the explicit method is better once everything is configured and predicted for both sides (host and server). The cost of this method is that all hosts in the network must be configured to work with the web proxy, and in some cases, this could mean several hours of work. It is possible to automatically deploy the proxy configurations to hosts while using solutions such as Microsoft AD (using Group Policy) or the **Web Proxy Auto-Discovery (WPAD)** protocol, saving precious time. Still, we must consider that if some host has failed to apply this configuration, more effort will be demanded to troubleshoot the issue. Even using the explicit method, much software doesn't support HTTP(S) proxies, so evaluate this before deploying a web proxy in a network environment.

An explicit proxy example topology is shown in the following diagram:

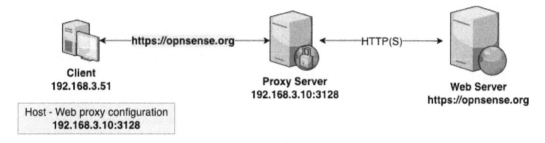

Figure 13.1 – Explicit proxy example topology

As we can see in the preceding diagram, in the explicit method, the host sends the HTTP request to the web proxy that will handle it and then forwards it to the proper web server after all the configured controls are applied. If the web proxy denies the user's requests, it will send a block page back to its web browser.

Using the explicit method has some advantages, mainly while using user authentication or basic HTTPS protocol control. A few advantages of using the explicit method are outlined here:

- User authentication is supported by the explicit method, which can help get better control of hosts/users based on authentication instead of using source **Internet Protocol (IP)** addresses.

- No workaround is necessary—once the browser and operating system *know* that it is a web proxy dealing with HTTP(S) requests, it doesn't require any workaround on the OPNsense system/web proxy to get it running.

- A more straightforward approach—to get this method working, we can just configure the web browsers to use a proxy.

Common deployment methods of web proxy configuration are Microsoft AD's Group Policy or WPAD, which can use **Dynamic Host Configuration Protocol** (**DHCP**) or DNS to send special instructions to users' web browsers.

OPNsense supports WPAD usage in its core features, both in the DHCP server and the Unbound DNS server. Depending on your OPNsense configuration, you may need to install additional plugins such as nginx and HAProxy to support the deployment of the wpad.dat file (for example, `https://<OPNsense_IP_address>/wpad.dat`).

For small-to-medium networks, without resources such as Microsoft AD, another method could be considered to implement the web proxy. Let's look at a good option for such cases.

The transparent method

This method intercepts HTTP(S) requests from the client to the web server and does not require changes in client configurations, which can be an advantage. The proxy server acts as the web server to the clients and also acts as the client to the web server by not modifying the HTTPS requests; therefore, the client and the web server don't know about the web proxy.

It doesn't support user authentication and demands some additional steps to intercept HTTPS traffic. This method is common in small networks or when changing host configuration is a problem or limitation. It requires additional **network address translation** (**NAT**) rules to work, and if **Secure Sockets Layer/Transport Layer Security** (**SSL/TLS**) traffic inspection is necessary, it will be needed to configure a certificate on each host in the network.

Here is an example topology of a transparent method web proxy implementation:

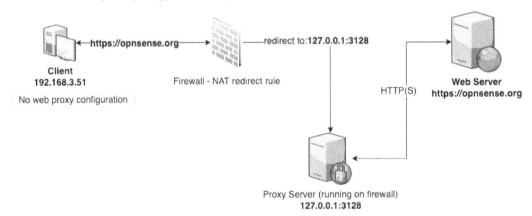

Figure 13.2 – Transparent proxy example topology

As we can see in the preceding diagram, neither the client nor the web server knows about the web proxy. The trick is done by a NAT rule that redirects HTTP(S) traffic to the web proxy server. Sometimes, this method is tricky to configure and get running. This also occurs while using the explicit method.

So, which method may be chosen? There isn't a correct answer here, but a method that will better fit the network is demanded. This will depend on the network requirements and limitations, such as unmanaged hosts in a guest network.

Why use a web proxy?

As a network firewall, should OPNsense with only packet filtering be enough to control users' access to websites? It depends: if we cut off HTTP(S) protocols of users' access, they won't access websites, but the internet will become quite useless without it. So, as we already know, most layer-4-based firewalls can handle IP addresses and **Transmission Control Protocol/User Datagram Protocol** (**TCP/UDP**) ports on OPNsense, even with hostnames and domains. So, if we want to block YouTube access, we must block YouTube's destination IP address on an inbound firewall **local area network** (**LAN**) rule—for example—and the problem is solved, right? Not so simple! A single public IP address can host several websites and domains, so you also block all associated websites by blocking just the IP address. And what about using hostnames and domains in firewall aliases? This also might be tricky, as a single host or domain can resolve to several different IP addresses. So, something working on layer 7 might be necessary, and here, a web proxy is a good solution! It will treat every single domain and **Uniform Resource Locator** (**URL**) as unique, allowing accurate web browsing control.

OPNsense web proxy core features

OPNsense has web proxy capabilities in its core features. The core features are capable of serving as a decent HTTP(S) proxy with basic filtering features. The service used under the hood is the Squid caching proxy, which means that it has one of the better open source solutions to get this job done. Some of the web proxy features available on OPNsense that we can name are listed here:

- Authentication (internal and external).
- Antivirus support—for example, through the **Internet Content Adaptation Protocol** (**ICAP**). ICAP can also be used by other solutions that modify JavaScript and **HyperText Markup Language** (**HTML**) for advertisement blocking.
- **Access Control List** (**ACL**).
- Blacklists (can be based on categories).

- OPNsense supports **proxy auto-configuration** (**PAC**) and WPAD.

- Transparent mode with HTTPS (TLS) inspection.

- Custom error pages.

Let's move on to the practical part and see how to configure a web proxy on OPNsense.

Basic configuration

We'll start the basic web proxy configuration by heading the browser to the **Services | Web Proxy | Administration** page. The default tab is **General Proxy Settings**, as illustrated in the following screenshot:

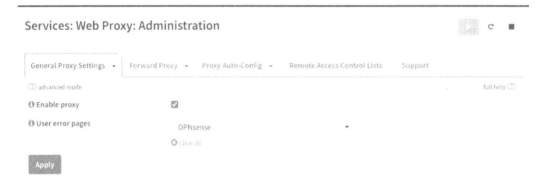

Figure 13.3 – Web Proxy | Administration page

As the web proxy has several configuration pages with tons of options, it unfortunately isn't possible to describe every configuration option, as done in the other chapters. Otherwise, this chapter would have hundreds of pages, and the editors would be crazy with me. Still, I'll try to cover the most relevant options (based on my experience) to explore the main web proxy functionalities.

Let's start by describing the main configuration options, as follows:

- **Enable proxy**: Check this option to start the web proxy service.

- **User error pages**: This is a recent feature, and it is a cool one! Before this feature was implemented, Squid error pages were only the default ones, which are not the fanciest, in my humble opinion. So, in this option, you can choose from **OPNsense**, **Squid**, or **Custom** web proxy error pages that will be shown to users when some web proxy error is detected.

Before moving on, let's take a look at the custom error pages page and how to configure it.

Custom error pages

To start customizing error pages, we must select **Custom** in the **User error pages** option on **General Proxy Settings** and then click on the **Error Pages** tab.

The following screenshot shows the numbered steps to take for customizing error pages:

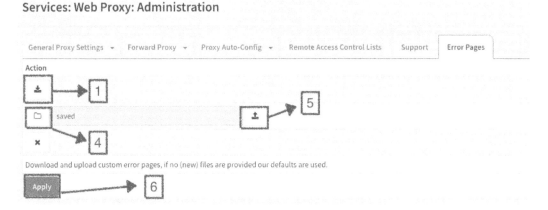

Figure 13.4 – Creating a custom error page template

In the following steps, we'll change how the web proxy's error pages look:

1. Click on the **Download** button to save the proxy ZIP file, as illustrated in the following screenshot:

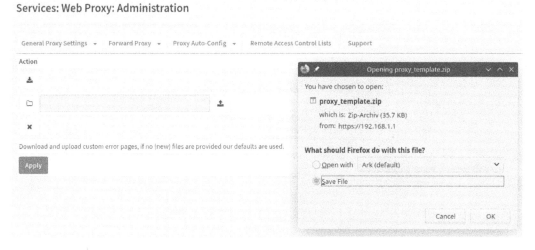

Figure 13.5 – Saving the proxy ZIP file

2.　Uncompress the ZIP file and edit the **Cascading Style Sheets** (**CSS**) file, as illustrated in the following screenshot:

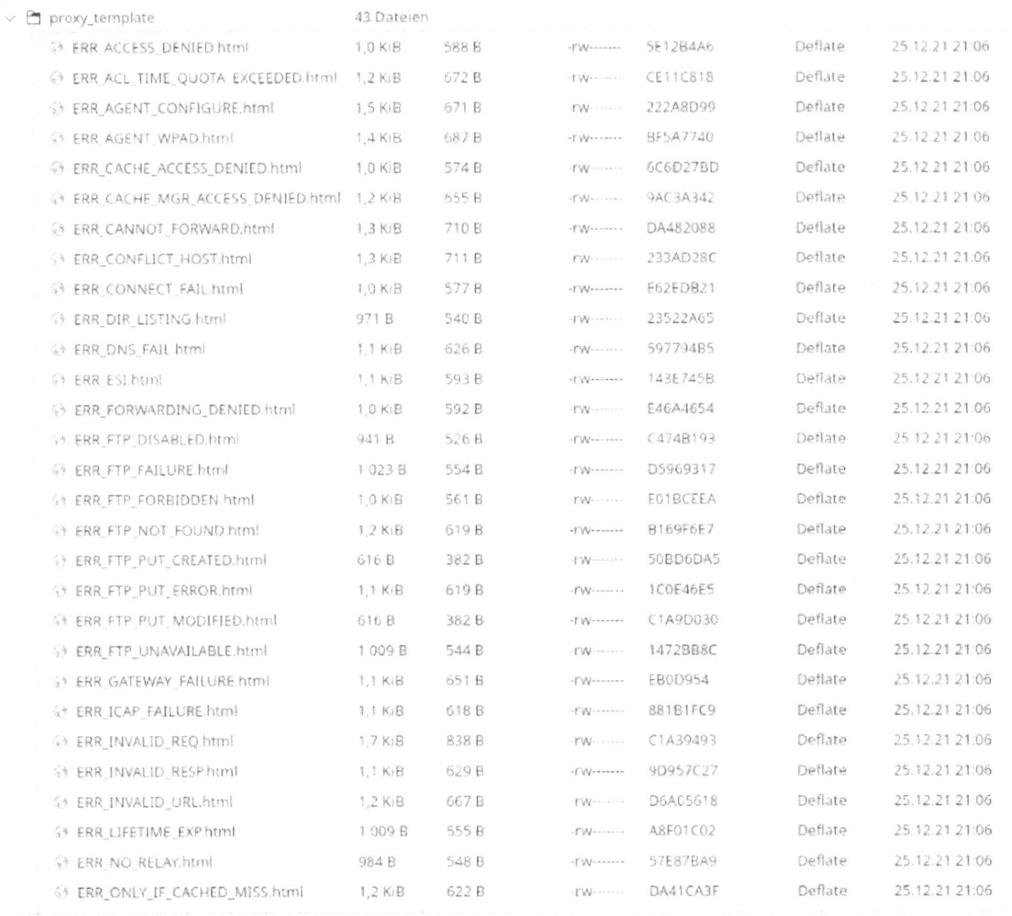

proxy_template	43 Dateien					
ERR_ACCESS_DENIED.html	1,0 KiB	588 B	-rw-------	5E12B4A6	Deflate	25.12.21 21:06
ERR_ACL_TIME_QUOTA_EXCEEDED.html	1,2 KiB	672 B	-rw-------	CE11C818	Deflate	25.12.21 21:06
ERR_AGENT_CONFIGURE.html	1,5 KiB	671 B	-rw-------	222A8D99	Deflate	25.12.21 21:06
ERR_AGENT_WPAD.html	1,4 KiB	687 B	-rw-------	BF5A7740	Deflate	25.12.21 21:06
ERR_CACHE_ACCESS_DENIED.html	1,0 KiB	574 B	-rw-------	6C6D27BD	Deflate	25.12.21 21:06
ERR_CACHE_MGR_ACCESS_DENIED.html	1,2 KiB	655 B	-rw-------	9AC3A342	Deflate	25.12.21 21:06
ERR_CANNOT_FORWARD.html	1,3 KiB	710 B	-rw-------	DA482088	Deflate	25.12.21 21:06
ERR_CONFLICT_HOST.html	1,3 KiB	711 B	-rw-------	233AD28C	Deflate	25.12.21 21:06
ERR_CONNECT_FAIL.html	1,0 KiB	577 B	-rw-------	E62EDB21	Deflate	25.12.21 21:06
ERR_DIR_LISTING.html	971 B	540 B	-rw-------	23522A65	Deflate	25.12.21 21:06
ERR_DNS_FAIL.html	1,1 KiB	626 B	-rw-------	597704B5	Deflate	25.12.21 21:06
ERR_ESI.html	1,1 KiB	593 B	-rw-------	143E745B	Deflate	25.12.21 21:06
ERR_FORWARDING_DENIED.html	1,0 KiB	592 B	-rw-------	E46A4654	Deflate	25.12.21 21:06
ERR_FTP_DISABLED.html	941 B	526 B	-rw-------	C474B193	Deflate	25.12.21 21:06
ERR_FTP_FAILURE.html	1 023 B	554 B	-rw-------	D5969317	Deflate	25.12.21 21:06
ERR_FTP_FORBIDDEN.html	1,0 KiB	561 B	-rw-------	E01BCEEA	Deflate	25.12.21 21:06
ERR_FTP_NOT_FOUND.html	1,2 KiB	619 B	-rw-------	B169F6E7	Deflate	25.12.21 21:06
ERR_FTP_PUT_CREATED.html	616 B	382 B	-rw-------	50BD6DA5	Deflate	25.12.21 21:06
ERR_FTP_PUT_ERROR.html	1,1 KiB	619 B	-rw-------	1C0E46E5	Deflate	25.12.21 21:06
ERR_FTP_PUT_MODIFIED.html	616 B	382 B	-rw-------	C1A9D030	Deflate	25.12.21 21:06
ERR_FTP_UNAVAILABLE.html	1 009 B	544 B	-rw-------	1472BB8C	Deflate	25.12.21 21:06
ERR_GATEWAY_FAILURE.html	1,1 KiB	651 B	-rw-------	EB0D954	Deflate	25.12.21 21:06
ERR_ICAP_FAILURE.html	1,1 KiB	618 B	-rw-------	881B1FC9	Deflate	25.12.21 21:06
ERR_INVALID_REQ.html	1,7 KiB	838 B	-rw-------	C1A39493	Deflate	25.12.21 21:06
ERR_INVALID_RESP.html	1,1 KiB	629 B	-rw-------	9D957C27	Deflate	25.12.21 21:06
ERR_INVALID_URL.html	1,2 KiB	667 B	-rw-------	D6A05618	Deflate	25.12.21 21:06
ERR_LIFETIME_EXP.html	1 009 B	555 B	-rw-------	A8F01C02	Deflate	25.12.21 21:06
ERR_NO_RELAY.html	984 B	548 B	-rw-------	57E87BA9	Deflate	25.12.21 21:06
ERR_ONLY_IF_CACHED_MISS.html	1,2 KiB	622 B	-rw-------	DA41CA3F	Deflate	25.12.21 21:06

Figure 13.6 – Editing the CSS file

3. Click on any HTML file inside the decompressed folder (named `proxy_template`) to check the original file layout. You should see something like this:

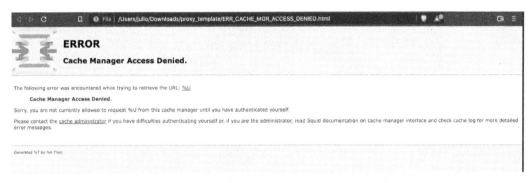

Figure 13.7 – Default template error page example

4. Inside the `proxy_template` directory, edit (with your favorite text editor) the `errorpage.css` file.

5. To change the template's image, find the following line of code:

```
background: url('data:image
/svg+xml;base64,PD94bWwgdmVyc2lvbj0iMS4wIiBlbmNvZGluZz0i
VVRGLTgiIHN0YW5kYWxvbmU9Im5vIj8+CjwhRE9DVFlQRSBzdmcgUFVC
TElDICItLy9XM0MvL0RURCBTVkcgMS4xLy9FTiIgImh0dHA6Ly93d3cud
zMub3JnL0dyYXBoaWNzL1NWRy8xLjEvRFRREL3N2ZzExLmR0ZCI
```
...

6. The image is encoded using the Base64 algorithm. To change it, we need to convert the new image to Base64 format. An easy way to do this is by using the following website: `https://www.base64-image.de/`. On this website, you can drag and drop the new image and it will convert the image for you like a charm following the steps to convert it, as illustrated in the following screenshot:

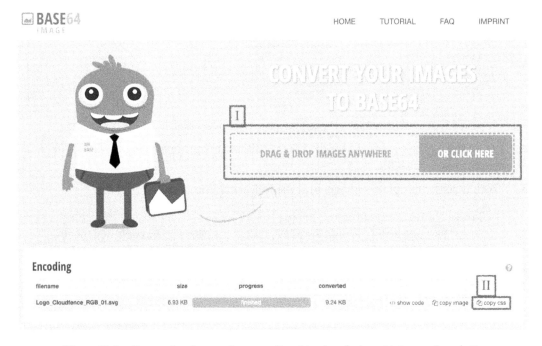

Figure 13.8 – Converting the new image to Base64 using the base64_image.de website

7. Drag and drop any new image as you wish (I have used a **Scalable Vector Graphics (SVG)** file).

8. Click on the **copy css** button. You should then see the process running, as follows:

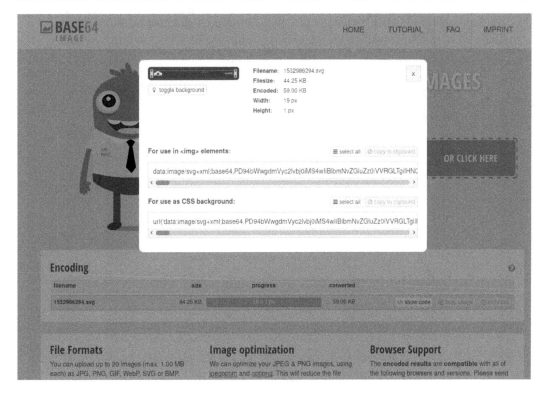

Figure 13.9 – The copy css process running

9. Go back to the CSS file, and paste the copied content substituting the entire url(...) old block, keeping just the new one copied from the website. A template CSS file with the image changed can be found at the following address: https://git.io/JM8Tj.

10. Save the modified CSS file and zip the proxy_template directory again.

11. Click on any HTML file inside the `proxy_template` directory decompressed folder to check the modified template version file layout. Check out my version example version here:

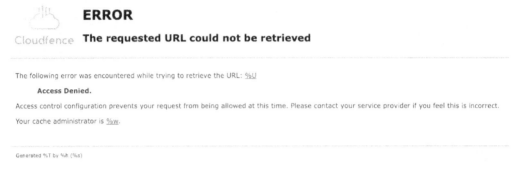

Figure 13.10 – Custom user error page example

12. Click on the **load file** button.

13. Click on the **upload** button.

14. Click on the **Apply** button to change the custom template.

OK! We're ready to go! Now, web proxy users will see a fancy customized error page! Let's go back to web proxy configuration.

Configuring a web proxy with the explicit method

We'll use the explicit method in the following configuration. Get the web proxy enabled and running through these steps:

1. Double-check if the **Enable proxy** field is *checked* on the **General Proxy Settings** option and click the **Apply** button to start the service. You should see a running icon, as shown in the following screenshot:

Figure 13.11 – Web proxy service running

2. If the service does get up and running for you, it may be necessary to check if the proper network interface is configured. Let's move to the next page's tab, **Forward Proxy**, and check this out! This tab will define which network interfaces the web proxy service will listen to if it will work as a transparent proxy and intercept SSL/TLS traffic. Let's start checking this page's options, as follows:

I.   **Proxy interfaces**: This option will select each network interface that the web proxy service will listen to. If you had some issue starting the web proxy service in the last step, check this option. Note that this option doesn't allow setting the default **wide-area network** (**WAN**) interface. For this configuration, **LAN** might be selected.

II. **Proxy port**: This will define which TCP port the web proxy service will listen to for HTTP(S) requests. The default is 3128 (Squid's default). We can leave the default port configured.

> **Important Note**
>
> As we are configuring the explicit method, we'll not touch the following options related to transparent mode/SSL interception in this section.

III. **Enable Transparent HTTP proxy**: This option will configure the web proxy service to work with the transparent proxy method. Nowadays, without SSL inspection, this option is quite useless as most internet traffic is based on HTTPS (TCP/443 port), therefore it isn't recommended to use it without the **Enable SSL inspection** option. It will require adding a NAT rule (port forwarding) to work. To ease this rule addition, the OPNsense team added a shortcut in this option description. Click on the **i** icon to show the option's description.

IV. **Enable SSL inspection**: This option will enable the SSL inspection feature on the web proxy. It will also require an additional NAT rule (as with the transparent HTTP proxy option, it also can be added in the same way). While enabled, the Squid daemon will act as a MITM attack, so sensitive data could be extracted from the user's HTTPS traffic. Be aware of the implications of doing this! The SSL inspection will require all hosts using the web proxy to have an OPNsense-configured **Certificate Authority** (**CA**) installed and trusted by the host operating system. Without a Microsoft AD server, enabling it could become a Herculean job with many hosts on the network.

V. **Log SNI information only**: **Server Name Indication (SNI)** is a TLS extension and acts as a **unique identifier (UID)** for hostnames. Enabling this option will configure the web proxy not to decode (decrypt) the HTTPS traffic and just log it. In this way, we can just see accessed domains on the access log, not entire URLs like when using HTTP- or HTTPS-decrypted traffic.

VI. **SSL Proxy port**: The port the web proxy uses to listen to HTTPS traffic. Note that this must only be redirected by the added NAT rule and not configured on the user's web proxy hosts.

VII. **CA to use**: While intercepting HTTPS traffic, we need to generate a CA. We can easily do this by clicking on this option description in the **CA Manager** link. Select a CA that might be used for HTTPS traffic decrypting. Remember to add this CA as trusted on each host using the web proxy.

VIII. **SSL no bump sites**: While using HTTPS inspection, it might be desirable not to inspect all users' encrypted traffic. Good examples of websites that shouldn't be inspected are online banking sites and ones containing users' personal information.

> **Important Note**
>
> Some regulations, such as the **European Union's (EU's) General Data Protection Regulation (GDPR)** or the Brazilian **General Personal Data Protection Law (LGPD)**, control personal information usage. Before implementing this feature in your company, get advice from the legal department about local laws.

To finish the configuration, just click on the **Apply** button.

Testing the web proxy

After configuring the web proxy with the explicit method, it is time to see it in action.

We'll need a desktop with a web browser running to do this. If you followed this book's steps and created an Ubuntu **virtual machine (VM)**, it is probably running without any **graphical user interface (GUI)**. You can skip *Steps 1-3* if you already have a GUI installed on your Ubuntu VM. Otherwise, let's install the MATE GUI on our Ubuntu host, as follows:

1. First, update your Ubuntu host by running the following commands:

```
sudo apt-get update && sudo apt-get upgrade
```

2. In your Ubuntu host, run the following command to install the MATE core desktop:

    ```
    sudo tasksel install ubuntu-mate-core
    ```

3. To get the GUI up and running, run the following command:

    ```
    sudo service display_manager start
    ```

We can also reboot your VM to test. The GUI might be up and running automatically after the reboot.

After getting the Ubuntu host running on the GUI, you might see something like this:

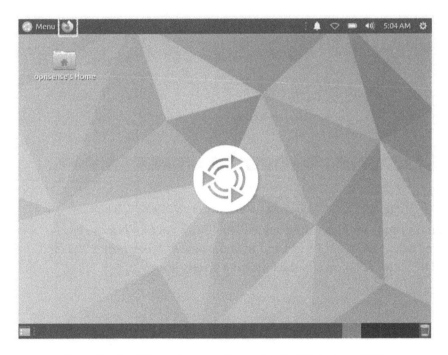

Figure 13.12 – Ubuntu host running on the MATE desktop (GUI)

4. Click on the Firefox web browser icon to open it (as highlighted in the preceding screenshot).

5. With Firefox open and running, click on the **Settings** menu and search for `proxy`, as illustrated in the following screenshot:

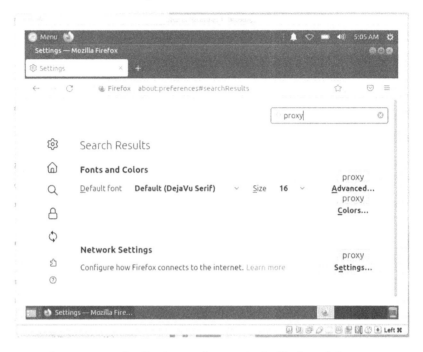

Figure 13.13 – Configuring a web proxy in the Firefox web browser

With the proxy settings dialog opened, fill it with your OPNsense *LAN IP address* (in my case, `192.168.56.1`) and with the default proxy port `3128`, and click on the **OK** button, as shown in the following screenshot:

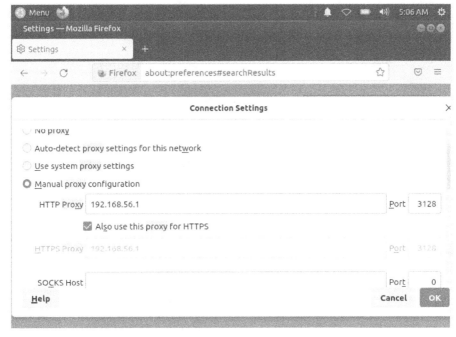

Figure 13.14 – Firefox proxy settings dialog

6. Try to open any website using the Firefox web browser, as depicted in the following screenshot:

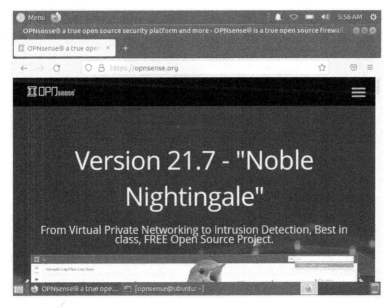

Figure 13.15 – OPNsense website in Firefox using the new proxy settings

7. Now, let's check the web proxy access logs in OPNsense's web GUI by going to **Services | Web Proxy | Access Log**, as illustrated in the following screenshot:

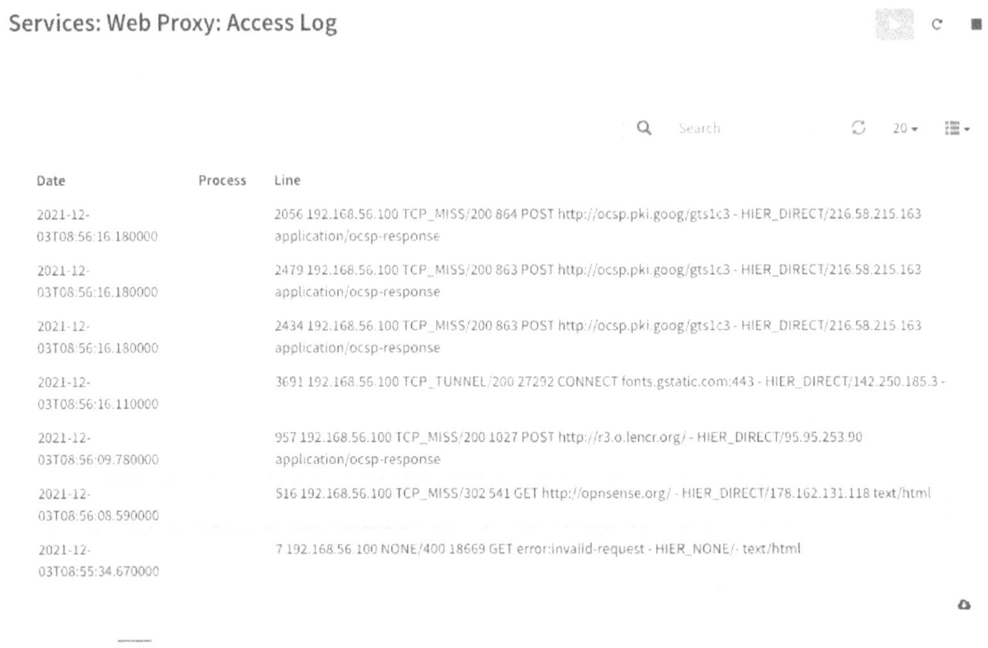

Services: Web Proxy: Access Log

Date	Process	Line
2021-12-03T08:56:16.180000		2056 192.168.56.100 TCP_MISS/200 864 POST http://ocsp.pki.goog/gts1c3 - HIER_DIRECT/216.58.215.163 application/ocsp-response
2021-12-03T08:56:16.180000		2479 192.168.56.100 TCP_MISS/200 863 POST http://ocsp.pki.goog/gts1c3 - HIER_DIRECT/216.58.215.163 application/ocsp-response
2021-12-03T08:56:16.180000		2434 192.168.56.100 TCP_MISS/200 863 POST http://ocsp.pki.goog/gts1c3 - HIER_DIRECT/216.58.215.163 application/ocsp-response
2021-12-03T08:56:16.110000		3691 192.168.56.100 TCP_TUNNEL/200 27292 CONNECT fonts.gstatic.com:443 - HIER_DIRECT/142.250.185.3 -
2021-12-03T08:56:09.780000		957 192.168.56.100 TCP_MISS/200 1027 POST http://r3.o.lencr.org/ - HIER_DIRECT/95.95.253.90 application/ocsp-response
2021-12-03T08:56:08.590000		516 192.168.56.100 TCP_MISS/302 541 GET http://opnsense.org/ - HIER_DIRECT/178.162.131.118 text/html
2021-12-03T08:55:34.670000		7 192.168.56.100 NONE/400 18669 GET error:invalid-request - HIER_NONE/- text/html

Figure 13.16 – Web proxy access log detail

As we can see in the preceding screenshot, the web proxy is logging Ubuntu's host requests. If you can see that in your lab, congratulations!! You just have configured a web proxy on OPNsense!

Let's now take a brief look at the transparent mode configuration.

Transparent web proxy configuration

We'll use the already configured option in the explicit method to configure the transparent one. The necessary changes are shown in the following screenshot:

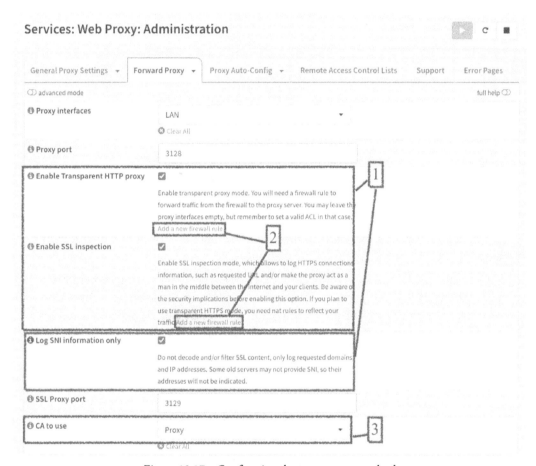

Figure 13.17 – Configuring the transparent method

Follow these next steps (numbered in the preceding screenshot):

1. Check the **Enable Transparent HTTP proxy** and **Enable SSL inspection** options. Also, check the **Log SNI information only** option—this will only log HTTPS requests without breaking the cryptography (not inspecting the ciphered content).

2. Click on the **Add a new firewall rule** links (in the **Enable Transparent HTTP proxy** and **Enable SSL inspection sections**) to add NAT rules for both HTTP/HTTPS ports. This will add a new NAT rule required by the transparent method, as illustrated in the following screenshot:

Figure 13.18 – NAT rules required for HTTP(S) redirection traffic using the transparent method

3. It will be required to select a CA in the **CA to use** option. This option also contains a link to create a new CA in the information/description.

4. On the lab host (Ubuntu, in our example), remove the proxy settings previously configured (set it to **No proxy configuration**) and try to access some website (for example, `https://wikipedia.org`). On the **Access Log** page, you should see requests from the host, as illustrated in the following screenshot:

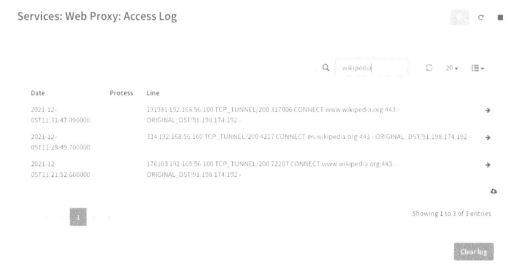

Figure 13.19 – Access Log page with HTTP redirected requests

With that, we have finished our transparent method configuration using OPNsense. Before moving on to web filtering capabilities, let's briefly look at other relevant web proxy configurations and some considerations about each one.

Additional web proxy configurations

The additional pages to configure a web proxy on OPNsense are described in the following sections.

User authentication

Go to the configuration page via **Services | Web Proxy | Administration - Forward Proxy - Authentication Settings**.

A web proxy supports all authentication backend options configured on **System | Access | Servers** to be enabled for authenticating users using the web proxy. The limitation is that **single sign-on** (**SSO**) can't work without plugin additions while using a Microsoft AD service.

Local cache

To enable this option, navigate to the configuration page via **Services | Web Proxy | Administration - General Proxy Settings - Local Cache Settings**.

In the modern internet, due to fast internet connections and dynamic content, the local cache isn't something that enhances performance as it used to do in the past. In larger networks, the local cache could help while enabling the **Enable Linux Package Cache** and **Enable Windows Update Cache** options, which would both create a local cache of operating system updates (for Linux for major distributions such as Debian, Red Hat, Arch, and so on) and save some time while updating the hosts' operating systems. It can also be used to traffic rate limitations to avoid excessive bandwidth consumption.

Web proxy traffic management

To enable this option, navigate to the configuration page via **Services | Web Proxy | Administration | General Proxy Settings | Traffic Management Settings**.

Web proxy traffic management can help control users' bandwidth usage while browsing the internet. As some networks may have other protocols' concurrent traffic, it is good to restrict how much of the web proxy bandwidth is used.

General settings

The general configuration page can be found here: **Services | Web Proxy | Administration | General Proxy Settings**.

Depending on the network host's requirements, some specific web proxy configurations may be changed from the default. There are many reasons to do that, but while troubleshooting a web proxy issue, I recommend you always check this page to search for something that helps in reviewing the configurations, as doing so may help a lot!

Parent proxy settings

The configuration page can be found here: **Services | Web Proxy | Administration | General Proxy Settings | Parent Proxy Settings**.

Larger network environments may have several web proxies deployed in different places. An example is a branch office that needs to connect to the main office for internet browsing. In that case, you will probably need to configure the main office as the parent proxy for the branch office.

FTP proxy settings

The general configuration page can be found here: **Services | Web Proxy | Forward Proxy | FTP Proxy Settings**.

First of all: try to avoid **File Transfer Protocol** (**FTP**) usage! It is an insecure protocol and causes many headaches for network administrators, who prefer **Secure FTP** (**SFTP**) instead, but if this isn't possible, you can configure FTP clients to use a web proxy, and in this configuration page, it will be possible to enable FTP support in the web proxy. Using this feature will tunnel FTP through HTTP. However, it would be best to consider using the os-ftp-proxy plugin as a preferred solution to perform FTP communication using a transparent proxy.

ICAP settings

The general configuration page can be found here: **Services | Web Proxy | Forward Proxy | ICAP Proxy Settings**.

The **Internet Content Adaptation Protocol (ICAP)** is a protocol that allows sending HTTP content to external solutions such as antivirus servers, to filter content. In the web proxy, you can set it to scan files that have been downloaded by users to check whether they might be affected by malware. It is possible to achieve an antivirus feature combining the C-ICAP and ClamAV plugins. OPNsense's documentation has a dedicated guide on how to do that, found at the following link: `https://docs.opnsense.org/manual/how-tos/proxyicapantivirusinternal.html#setup-anti-virus-protection-using-opnsense-plugins`.

SNMP settings

The general configuration page can be found here: **Services | Web Proxy | Forward Proxy | SNMP Agent Settings**.

Squid supports sending data using the **Simple Network Management Protocol (SNMP)** to monitor performance. Many good tools can monitor and produce beautiful graphs based on SNMP data. A good example is Zabbix: `https://www.zabbix.com/`.

Additional configuration pages are mentioned in the next section. You can also always count on OPNsense's official documentation to explore any feature in depth. Let's move on and explore web filtering!

Web filtering

As the modern internet works on top of protocols such as HTTP and DNS, web filtering is one of network administrators' most desired features in their firewalls to keep network users under some control. On OPNsense, this feature doesn't have a dedicated functionality as a core feature or an official plugin; instead, the web proxy achieves web filtering by configuring it with some additional features the OPNsense team has added since the project started. As the web proxy functionality on OPNsense is Squid-based, let's take a closer look at the Squid web proxy project to better understand this filtering matter.

The Squid project website definition is reproduced here: "*Squid is a caching proxy for the web supporting HTTP, HTTPS, FTP, and more. It reduces bandwidth and improves response times by caching and reusing frequently-requested web pages. Squid has extensive access controls and makes a great server accelerator.*" (`http://www.squid-cache.org/`)

As we can see, the project's primary goal isn't web filtering, so it is expected to have some known limitations when used as a web filtering solution. To overcome some of these known limits, some auxiliary projects such as SquidGuard, DansGuardian, and others were created to improve Squid's web filtering features. Neither of the aforementioned projects is officially maintained anymore, but I mentioned them because they were the most used tools to turn Squid into an accurate web filtering tool for a long time. Many users migrated from pfSense to OPNsense, and we (at CloudFence), as one of them, missed some features that were only provided by using a tool such as SquidGuard (that was supported in pfSense), for example. Back in 2018, when we were trying to migrate customers from pfSense to OPNsense, we needed to keep the same web proxy features that pfSense has, and we stumbled upon the web filtering feature. Back then, OPNsense had minimal web proxy features to get this job done, so we decided to write a piece of code that became the Web Filter plugin later. If you want to know more about this plugin, please take a look at the following link: `https://wiki.cloudfence.com.br`.

Since the beginning, the OPNsense project has improved the core web proxy features to support web filtering. Other web proxy plugins improve the core features, but I'll focus only on the core features to keep things concise in this book.

The following filtering features can be achieved using the core web proxy features:

- **Blacklisting**: It is possible to download a Squid-compatible blacklist and install it on OPNsense to block sites based on categories.

- **ACLs**: In **Forward Proxy | Access Control List**, it is possible to configure some fine-tuned filters such as *clicking on the* **advanced** *mode button to show all following options*:

  - The defining of allowed subnets, and unrestricted and banned IP addresses.

  - **Whitelist** destinations such as URLs, domains, or even file extensions (download control) using **regular expressions** (**regexes**).

  - **Blacklist** destinations using the same criteria described under whitelist destinations.

- **Block browser/user-agents** is a handy feature, especially if you need to apply some compliance regulation or policy that forbids some web browsers in the network. Another example is to block some known bad user agents used by malware or other harmful applications.

- **Block specific MIME type reply**: Combined with a good regex defined in the blacklist option, this can block some content from being accessed, such as video streaming, music, and others. Consider that you can always control web traffic content better using a solution such as the Zenarmor plugin that improves OPNsense, turning it into a **next-generation firewall** (**NGFW**) solution with more granular and precise web traffic control.

> **Important Note**
>
> If **Multipurpose Internet Mail Extensions** (**MIME**) types don't sound familiar to you, then I recommend you take a look at the following web page:
>
> `https://developer.mozilla.org/en-US/docs/Web/HTTP/Basics_of_HTTP/MIME_types`

- **Google GSuite restricted**: This option will deny access to Google G Suite (Google Drive, Gmail, and others), with only access to accounts that an organization manages. To enable this option, you need to type the DNS domain of the organization in the textbox.

- **YouTube Filter**: This option will enable YouTube native filtering by setting predefined levels: **Strict** or **Moderate**. To get more information about this Google service feature, take a look at the following web page: `https://support.google.com/a/answer/6212415?hl=en`.

> **Important Note**
>
> Both Google *G Suite* and *YouTube Filter* features require HTTPS inspection enabled to work, and they were ported from CloudFence's web filter plugin—proudly.

- **HTTP TCP port control**: The **Allowed destination TCP port** and **Allowed SSL ports** options will configure which web proxy will allow HTTP/HTTPS ports. Some web servers have different ports configured (for example, `8081`, `8443`, and so on) and this must be configured in these options to work with a web proxy.

After all these pages talking about web filtering, you might be wondering: *Is it really possible to filter web content?* Let's do a lab and find out!

Web filtering practice

To follow the steps in this lab, you must have a running OPNsense system with a host connected on its LAN with a web proxy configured on it. Here are the steps to filter web content:

1. To blacklist a website, go to the **Forward Proxy | Access Control List** menu and click on the **Blacklist** option, and then enter facebook.com, for example. You can see an illustration of this in the following screenshot:

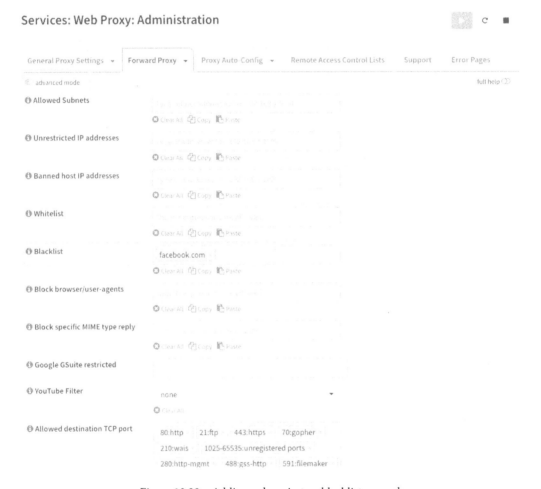

Figure 13.20 – Adding a domain to a blacklist example

2. Try to open the `facebook.com` domain on the host. You should then receive a message like this:

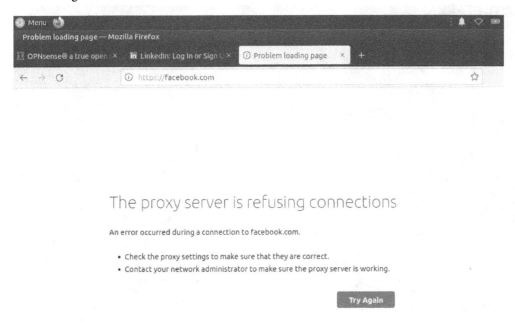

Figure 13.21 – Blocked Facebook website

As the website enforces the HTTPS protocol, the web proxy error page will not be displayed. To display error pages using HTTPS websites, it would be necessary to configure SSL/TLS interception, which requires installing the CA generated by OPNsense in each host using the web proxy.

3. To confirm that the website was blocked by the web proxy, go to the **Services | Web Proxy | Access Log** page. You should then see output like this:

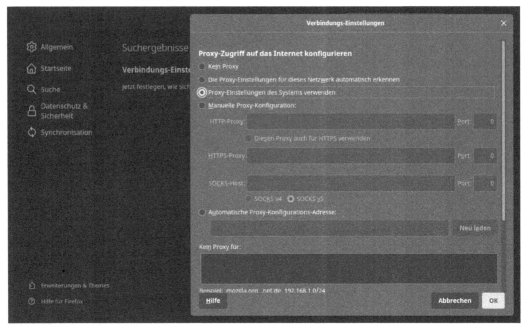

Figure 13.22 – Facebook website request blocked in the Access Log page

It's confirmed—the web proxy is blocking the blacklisted domain! Congratulations!!

Web filtering – final thoughts

It is possible to implement a decent web filtering level using OPNsense, but this will require some additional steps. To mention some, I'll describe a cookbook followed by some reference guides from the official documentation to achieve some web filtering goals. You can view these in the following table:

Goal	Guide	Reference
Automatic proxy deployment	Setup WPAD / PAC	`https://docs.opnsense.org/manual/how-tos/pac.html#`
Antivirus protection using a web proxy	Setup Anti Virus Protection (using OPNsense plugins)	`https://docs.opnsense.org/manual/proxy.html#setup-icap-anti-virus-malware-engine`
Transparent proxy with SSL inspection	Setup Transparent Proxy	`https://docs.opnsense.org/manual/how-tos/proxytransparent.html`
Add a blocklist (category-based)	Setup Web Filtering	`https://docs.opnsense.org/manual/how-tos/proxywebfilter.html`

Unfortunately, due to the extensive web proxy content, it isn't possible to include all the preceding topics in this chapter, but I tried to cover the more theoretical part of a web proxy implementation here to serve as a base while setting up these other possible deployments.

Reading logs and troubleshooting

Web proxy troubleshooting strongly relies on log reading, so it is fundamental to know how to read logs. Despite the web GUI having a logs page, I will focus on the **command-line interface (CLI)** in this section. Why? Because web proxy logs are really dynamic and most times will demand filters (using the `grep` command, for example) and will help you to see things *happening in real time*.

Before touching the keyboard to log on to the CLI, I recommend you this must-read Squid documentation:

`https://wiki.squid-cache.org/SquidFaq/SquidLogs`

Pay special attention to the *Squid result codes* and *HTTP status codes* sections. Without knowing what these codes mean, reading the logs would be like reading a foreign language that you don't speak. Maybe you'll be able to understand part of the information, but not all of it.

Log files

The web proxy log files are located in the /var/log/squid/ directory and are described in more detail here:

- access.log: This log file contains all host request information.
- cache.log: Squid's daemon-related messages are stored in this file.
- store.log: This file contains information about cached objects. It may not be relevant while not using a local cache.

Here are some examples while troubleshooting web proxy issues:

Looking for host requests for troubleshooting: When a user complains a website isn't loading or has some web proxy error, try looking for a clue in the access.log file.

Here are some example commands:

- *Filtering by source host IP or username:*

```
tail -f /var/log/squid/access.log | grep 192.168.1.10
tail -f /var/log/squid/access.log | grep santos.dumont
```

- *Filtering by destination:*

```
tail -f /var/log/squid/access.log | grep deciso.com
```

- *Filtering by result or HTTP status code:*

```
tail -f /var/log/squid/access.log | grep DENIED
  (blocked requests)
tail -f /var/log/squid/access.log | grep TCP_TUNNEL/200
(HTTPS requests)
```

The tail -f command outputs the file continuously (-f parameter). In this way, it will ease finding requests in real time. Of course, you can combine several grep filters to have more powerful results.

Web proxy service issues

As a full-featured configuration service, a web proxy can sometimes present issues. It is rare but not impossible to face a problem that results in the Squid daemon not getting up and running. A good way to check what is going on is to run the following steps:

1. Check the Squid configuration, as follows:

```
squid -k check
```

2. Try to reload the Squid service with the cache.log file opened using the tail command, as follows:

```
tail -f /var/log/cache.log
```

3. Sometimes, the web GUI can't stop or start the web proxy service. Try to reload the Squid daemon manually and look for error messages by running the following command:

```
squid -k reconfigure
```

> **Author's Note**
>
> You will probably never find these steps in the official documentation (even in the future). Still, please consider that my comments are (heavily) based on years of personal experience and our CloudFence support team discussions, so before suggesting something to you here, I had long discussions with them. They have thousands of *OPNsense flying hours*, managing hundreds of firewalls in different environments, and I consider their expertise invaluable. Thanks, @ cloudfenceteam S2.

Last but not least, you can try to use the **Reset** button on the **Support** tab in the web GUI (**Services | Web Proxy | Administration**) to try solving some service-/configuration-related issues.

Summary

I have tried to cover the most relevant content related to a web proxy service in this chapter. The topic of web proxies could easily produce an entire book due to its complexity. Still, our goal in this book is to complement the existing documentation and share a little bit of experience of years working with network security and OPNsense. Now, you might be able to install and configure a web proxy service for small-to-medium networks using OPNsense. We have explored the basics, deployment methods, basic configuration and customization, enabling web filtering, and troubleshooting common issues. In the next chapter, we'll move on to another great OPNsense feature to control user access using web browsers: the Captive Portal. See you on the following pages!

14

Captive Portal

In this chapter, we will learn about Captive Portal concepts, how to configure and use the Captive Portal with OPNsense, the most common issues while configuring a Captive Portal, and how to solve them. By the end of this chapter, you will be able to set up a guest network to authenticate and control guest hosts using a Captive Portal and web proxy with OPNsense. Specifically, we'll be looking at the following topics:

- Captive Portal concepts
- Setting up a guest network
- Web proxy integration
- Common issues

Technical requirements

To follow along with this chapter, you need to have a clear understanding of the **HyperText Transfer Protocol/HTTP Secure (HTTP/HTTPS)**, **Transport Layer Security (TLS)**, and **Domain Name System (DNS)** protocols. It is also advisable to have an OPNsense system with a host connected to its **local area network (LAN)**. My suggestion is to use the previously configured lab scenario (*Chapter 9, Multi-WAN – Failover and Load Balancing*) with OPNsense and Ubuntu **virtual machines (VMs)**.

Captive Portal concepts

Everyone with a smartphone or laptop has tried to access a public wireless network at least once and got an authentication page before proceeding with internet access—haven't you?

This method used to redirect a user trying to access the internet to an authentication page is a **Captive Portal**, which OPNsense has as a core feature. Captive Portals are usually used on guest networks Captive Portals are usually used on guest networks, such as airports and hotels' wireless networks hotspots, pay-as-you-go Wi-Fi guest networks, or even on wired networks in business centers.

OPNsense Captive Portal implementation

On OPNsense, the method used to redirect users is an HTTP redirect. Once the user tries to access a web page, a **network address translation (NAT)** rule will redirect it to a web service that runs the Captive Portal authentication page.

The following authentication methods are available through the OPNsense backend fabric: **LDAP**, **RADIUS**, **Local users**, and **Vouchers**. We can also configure a Captive Portal with no authentication (not recommended) or with a combination of the available methods (configured on the **System | Access | Servers** page).

Setting up a guest network

To start configuring a Captive Portal, follow the next steps:

1. Go to **Services | Captive Portal | Administration**, as illustrated in the following screenshot:

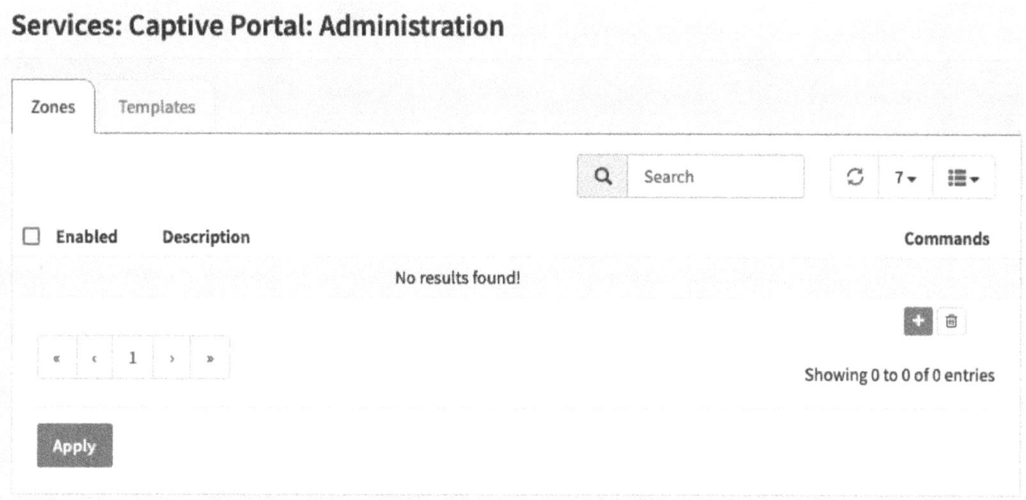

Services: Captive Portal: Administration

Zones	Templates

<table>
<tr><td></td><td></td><td>Q Search</td><td>C 7▾ ▤▾</td></tr>
<tr><td>☐ Enabled</td><td>Description</td><td></td><td>Commands</td></tr>
<tr><td colspan="4" align="center">No results found!</td></tr>
<tr><td></td><td></td><td></td><td>➕ 🗑</td></tr>
<tr><td>« ‹ 1 › »</td><td></td><td></td><td>Showing 0 to 0 of 0 entries</td></tr>
</table>

Apply

Figure 14.1 – Captive Portal configuration page

As we can see in the preceding screenshot, the configuration page has two tabs: **Zones** and **Templates**. These tabs are described in more detail here:

I. **Zones**: The Captive Portal is divided into different zones, each with its own configuration. Working this way, OPNsense can manage different guest networks. We can add multiple zones in OPNsense's Captive Portal feature.

II. **Templates**: Each template can use a different form of authentication or show a splash page to agree with the internet usage terms.

2.   Let's start the configuration by creating a new zone. Clicking on the + sign will take you to the following screen:

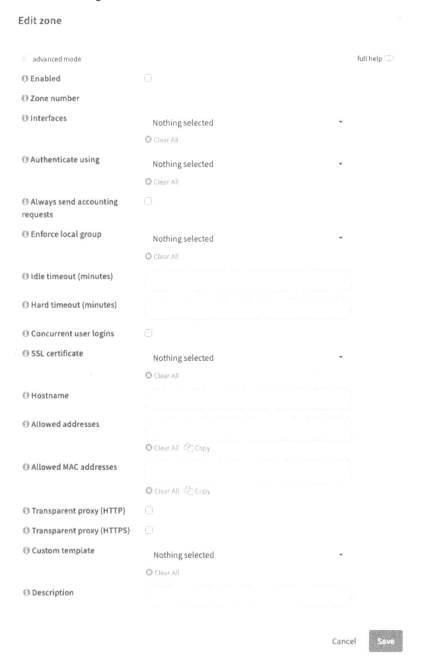

Figure 14.2 – Edit zone page with advanced mode on

The following options will be available to configure the **Edit zone** page:

I.   **Enabled**: Check this option to enable the zone.

II.   **Zone number**: This is a label and can't be configured. The first created zone starts with 0 (zero).

III. **Interfaces**: Select on which network interfaces the Captive Portal should be enabled.

IV.   **Authenticate using**: Select a configured authentication backend. If nothing is selected, the authentication will be disabled.

V.   **Always send accounting requests**: Check this option to enable session accounting. The **Remote Authentication Dial-In User Service** (**RADIUS**) authentication protocol has an accounting process that starts as soon as the user has access granted.

VI.   **Enforce local group**: This option will restrict users to a local group created on OPNsense.

VII. **Idle timeout (minutes)**: This option will set a timeout for user inactivity. After the timeout, the user will need to authenticate again. The default value is 0 (zero), which disables this timeout.

VIII. **Hard timeout (minutes)**: This option will set a timeout to force users to disconnect even with activity. The default value is 0 (zero), which disables this timeout.

IX.   **Concurrent user logins**: If this option is checked, then a user can log in to more than one machine at once, such as a laptop and a mobile phone. For security reasons, it isn't recommended to enable this option to avoid users sharing their credentials.

X.   **SSL certificate**: When an HTTP request is redirected, it will be served by an HTTP-only authentication portal. To enable an HTTPS authentication portal, set a **Secure Sockets Layer** (**SSL**) certificate in this option.

XI.   **Hostname**: By default, HTTP redirects will use the OPNsense **Internet Protocol** (**IP**) address. To use a hostname instead, set it in this option. Remember that hosts in the network must be able to resolve this hostname. An easy way is to set the OPNsense server as the default DNS server and create a local hostname entry. While using an SSL certificate, the hostname and the **Subject Alternative Name** (**SAN**; or **Common Name** (**CN**), as labeled in the web **user interface** (**UI**)) must match.

XII. **Allowed addresses**: List IP addresses including subnets that will not need authentication. This is like a bypass option by IP address.

XIII. **Allowed MAC address** (needs **advanced mode** marked): Once OPNsense is serving the Captive Portal in a local network, it will be able to get the localhost's **media access control** (**MAC**) address. This option will bypass authentication for the MAC addresses specified in it. It is a little bit more secure than using an IP address, but remember that a person with bad intent who uses a *MAC spoofing technique* can bypass this kind of control.

> **Important Note**
>
> **MAC spoofing**: It is possible to change the MAC address despite the MAC address being hardcoded in the network interface, thus spoofing it.

XIV. **Transparent proxy (HTTP)** and **Transparent proxy (HTTPS)**: Both options will forward HTTP(S) traffic to the web proxy using the transparent proxy method. These options are helpful while filtering or just logging user access.

XV. **Custom template**: While using a custom web authentication page template, we can select a template here. We'll explore custom templates in the next section.

XVI. **Description**: Put a description of this zone that will make sense to other OPNsense administrators.

3. To create a new zone, click on the **Save** button. This will add a new zone to the **Services** | **Captive Portal** | **Administration** page, as illustrated in the following screenshot:

Services: Captive Portal: Administration

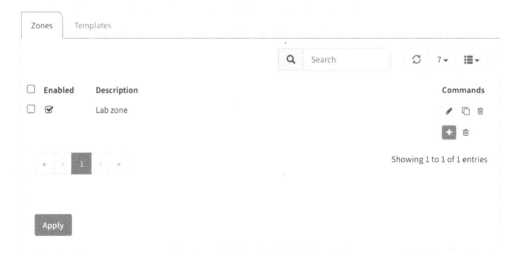

Figure 14.3 – A newly added Captive Portal zone

To apply the new configuration, you must click on—guess what? Yes! The **Apply** button!

Testing the configuration

We'll need a connected host on OPNsense's LAN to test the Captive Portal configuration. Using the Ubuntu host, I'll assume the same lab scenario from the previous chapters. Follow the next steps to test the configuration:

1. Disable any NAT redirect for HTTP/HTTPS if your OPNsense system has any.

2. Create the following firewall rules in the LAN interface:

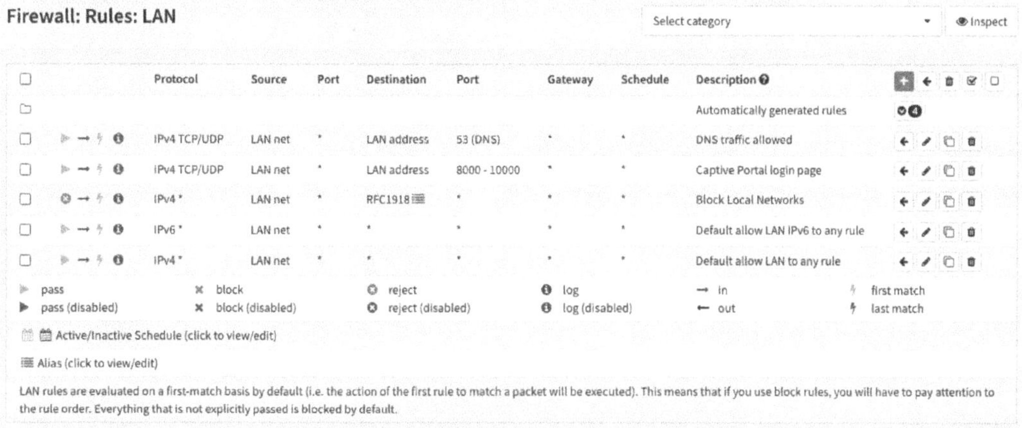

Firewall: Rules: LAN

		Protocol	Source	Port	Destination	Port	Gateway	Schedule	Description ❓	
☐									Automatically generated rules	
☐	▶ → ⅓ ❶	IPv4 TCP/UDP	LAN net	*	LAN address	53 (DNS)	*	*	DNS traffic allowed	
☐	▶ → ⅓ ❶	IPv4 TCP/UDP	LAN net	*	LAN address	8000 - 10000	*	*	Captive Portal login page	
☐	⊘ → ❶	IPv4 *	LAN net	*	RFC1918 ▦	*	*	*	Block Local Networks	
☐	▶ → ⅓ ❶	IPv6 *	LAN net	*	*	*	*	*	Default allow LAN IPv6 to any rule	
☐	▶ → ⅓ ❶	IPv4 *	LAN net	*	*	*	*	*	Default allow LAN to any rule	

| ▶ | pass | ✖ | block | | ⊘ | reject | | ❶ | log | | → | in | ⅓ | first match |
| ▶ | pass (disabled) | ✖ | block (disabled) | | ⊘ | reject (disabled) | | ❶ | log (disabled) | | ← | out | ⅘ | last match |

Active/Inactive Schedule (click to view/edit)

Alias (click to view/edit)

LAN rules are evaluated on a first-match basis by default (i.e. the action of the first rule to match a packet will be executed). This means that if you use block rules, you will have to pay attention to the rule order. Everything that is not explicitly passed is blocked by default.

Figure 14.4 – Captive Portal LAN's firewall rules

3. The first rule will allow DNS traffic to the local DNS service running on OPNsense (in this lab, we'll assume that is Unbound).

4. The second rule will allow traffic that the HTTP redirect rule produces (automatically created by the Captive Portal service) to the authentication page.

5. The third rule is *rejecting* any traffic to local networks. To ensure that nothing will evade this rule, I recommend you create an *alias* with **RFC 1918** networks in its content: `10.0.0.0/8`, `172.16.0.0/12`, and `192.168.0.0/16`.

6. The two final rules allow any traffic outgoing to the internet (only) once the third rule assures that no traffic escapes to the addresses of the local network. Go to the host and open a new web browser. Notice that we are considering just the **IP version 4 (IPv4)** network in this chapter. If you are using IPv6, you must set rules for both IP protocol versions.

> **Important Note**
> We assume the LAN network is our guest network in our lab to avoid additional steps in creating a GuestNet-only network, but in a real environment, a dedicated guest network interface is recommended to be added.

7. Try to open an *HTTP-only* address in the Captive Portal authentication page, as illustrated in the following screenshot:

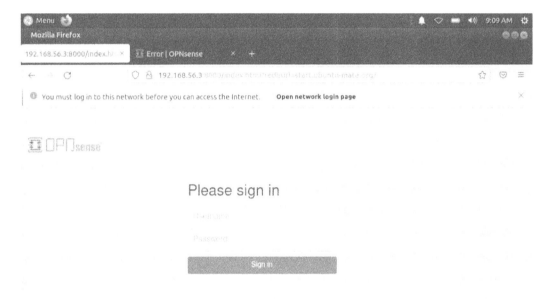

Figure 14.5 – Captive Portal authentication page while trying to open the http://start.ubuntu-mate.org URL

8. If you try opening an HTTPS URL, the browser will show an error page, as displayed in the following screenshot:

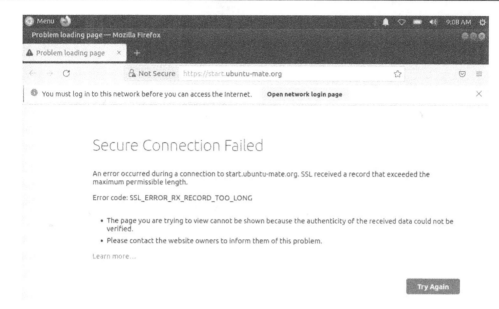

Figure 14.6 – Error page while trying to access an HTTPS URL

9. This issue is caused for the same reason that error pages don't work on the transparent proxy method; the presented certificate will not match the expected by the browser, so this SSL error occurs. We will see how to solve this issue in the *Common issues* section. It can quickly be solved by changing `https` to `http` in the URL bar.

10. To continue, fill in the username and password fields with previously created user credentials. If you didn't create any additional users, go to **System** | **Access** | **Users** (the OPNsense webGUI) and create a new one.

11. After authenticating, the browser will be redirected to the typed URL—in our example, this is `https://start.ubuntu-mate.org`.

Important Note

Note that the HTTP to HTTPS redirection in this last step was done by the remote web server, not by OPNsense, once we had typed the URL using HTTP-only.

---

12. To confirm the existing session, go back to the OPNsense webGUI and go to the **Services | Captive Portal | Sessions** page, as illustrated in the following screenshot:

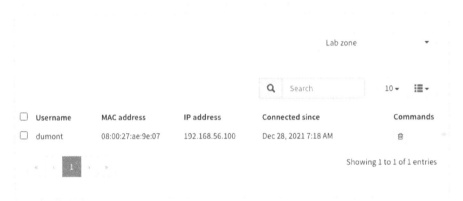

Figure 14.7 – Active Captive Portal session

As shown in the preceding screenshot, the user (dumont in my lab) is logged in and has an active session. We can check the user's **IP address** and **MAC address** values and the *connection start time* (found in the **Connected since** column). To terminate a user session, we can click on the trashcan icon.

With this last step, we have finished configuring our first Captive Portal zone on OPNsense! Congratulations!

Using voucher authentication

Let's suppose that our guest network is configured in an airport, where it would be impractical to create a new user for each new guest who wants to use it. Some techniques to handle this better are outlined here:

> **Important Note**
> Vouchers are stored in a local SQLite database that is not backed up while the OPNsense configuration is exported.

- Social login (based on social media authentication). This method is not supported by OPNsense.

- Social **identifier** (**ID**), which can create a problem with regulations such as the **General Data Protection Regulation** (**GDPR**).

- Vouchers.

This last one is an option that OPNsense supports natively, and we'll see how to configure it in the following steps:

1. Go to the **System | Access | Servers** page and click on the + button to add a new option.

2. On the new configuration page, select the **Voucher** option and a **Type** value.

3. Provide a description such as Captive Portal, for example.

4. Fill the **Username** and **Password length** fields with the number 4. Notice that users will need to type *eight digits in total*: four for the username and four for the password. For security best practices, you should consider an eight-character username and an eight-character password.

5. Click on the **Save** button to finish adding the new voucher authentication backend.

6. Go back to the captive portal configuration page via **Services | Captive Portal | Administration** and edit the current zone.

7. Change the **Authenticate using** option to the new voucher backend we created and click on the **Save** and **Apply** buttons.

8. Now, we must create vouchers. To do that, go to **Services | Captive Portal | Vouchers** and click on the **Create vouchers** button.

9. A window will be shown; change the **Number of vouchers** option to 10, for example, and click on the **Generate** button, as illustrated in the following screenshot:

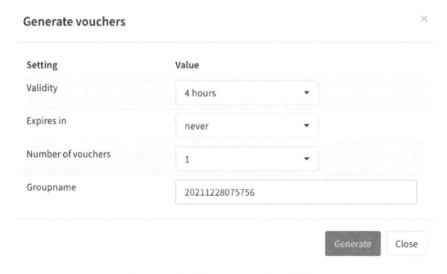

Figure 14.8 – Generate vouchers dialog

10. A browser dialog will be shown to download a **comma-separated values** (CSV) file for the voucher, as illustrated in the following screenshot. Save it on your computer:

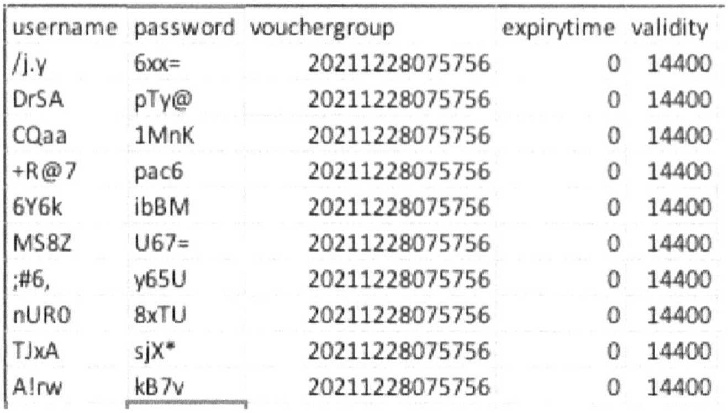

username	password	vouchergroup	expirytime	validity
/j.y	6xx=	20211228075756	0	14400
DrSA	pTy@	20211228075756	0	14400
CQaa	1MnK	20211228075756	0	14400
+R@7	pac6	20211228075756	0	14400
6Y6k	ibBM	20211228075756	0	14400
MS8Z	U67=	20211228075756	0	14400
;#6,	y65U	20211228075756	0	14400
nUR0	8xTU	20211228075756	0	14400
TJxA	sjX*	20211228075756	0	14400
A!rw	kB7v	20211228075756	0	14400

Figure 14.9 – CSV file content

11. We can also check for vouchers on the **Services | Captive Portal | Vouchers** page, as illustrated in the following screenshot:

Services: Captive Portal: Vouchers

Captive Portal   ▾   20211228075756   ▾   🗑

Q Search   10 ▾   ▥ ▾

	Voucher	Valid from	Valid to	Expires at	State
☐	/j.y	Dec 28, 2021 7:52 AM	Dec 28, 2021 11:52 AM		unused
☐	DrSA	Dec 28, 2021 7:52 AM	Dec 28, 2021 11:52 AM		unused
☐	CQaa	Dec 28, 2021 7:52 AM	Dec 28, 2021 11:52 AM		unused
☐	+R@7	Dec 28, 2021 7:52 AM	Dec 28, 2021 11:52 AM		unused
☐	6YGk	Dec 28, 2021 7:52 AM	Dec 28, 2021 11:52 AM		unused
☐	MS8Z	Dec 28, 2021 7:52 AM	Dec 28, 2021 11:52 AM		unused
☐	;#6,	Dec 28, 2021 7:52 AM	Dec 28, 2021 11:52 AM		unused
☐	nUR0	Dec 28, 2021 7:52 AM	Dec 28, 2021 11:52 AM		unused
☐	TJxA	Dec 28, 2021 7:52 AM	Dec 28, 2021 11:52 AM		unused
☐	A!rw	Dec 28, 2021 7:52 AM	Dec 28, 2021 11:52 AM		unused

Showing 1 to 10 of 10 entries

[1]

Expire selected vouchers 🗑   Drop expired vouchers 🗑   Create vouchers ◆

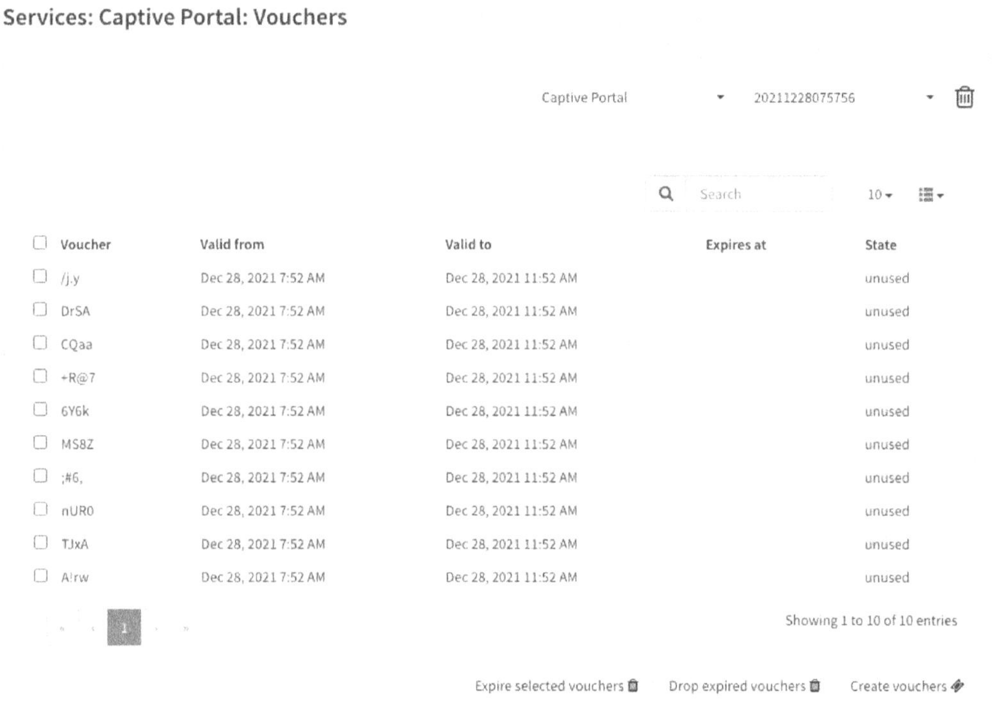

Figure 14.10 – New generated vouchers listed

12. Now that we have changed the *authentication of vouchers*, go back to **Services |
Captive Portal | Sessions** and click on the trashcan icon to drop the previous active
session.

13. Try opening an *HTTP-only* URL on the host and wait for the Captive Portal
authentication page. To avoid web cache issues, use a private-mode window in your
browser.

14. Pick a voucher *username* and *password* from the previously downloaded CSV file
when the authentication page is shown, and authenticate using them.

15. After successfully authenticating, check the active session on **Services | Captive
Portal | Sessions**. You should see the current voucher in use in the **Username**
column, as illustrated in the following screenshot:

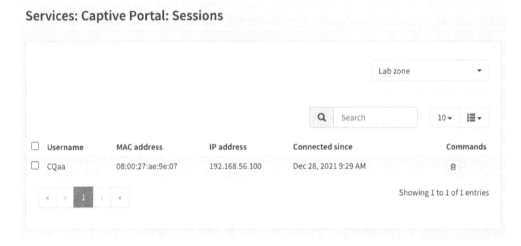

Figure 14.11 – Active session using voucher authentication

As we can see, voucher-based authentication will work very similarly to local database user
authentication, but with the advantage that we can export a list of vouchers and create an
administrative procedure to share each voucher with guests. Next, we'll take the necessary
steps to enable web filtering on the guest network to integrate it with the web proxy.

Web proxy integration

To control what guests can access on the internet, we can apply some web proxy controls.
As we explored in the last chapter, the native web proxy features on OPNsense provide
reasonable control of internet content. It is a good idea to control what guests access on
the internet. Depending on each country's laws, it may not be allowed to inspect users'
web browsing traffic, so check whether you can do this first!

In this section, we'll take the necessary steps to enable web filtering on the GuestNet, integrating it with the web proxy, as follows:

> **Important Note**
>
> Before we start working through the following steps, make sure that the web proxy is up and running. If you aren't sure how to do that, please refer to *Chapter 20, API – Application Programming Interface*, for the steps to configure the web proxy with the transparent method.
>
> Remember to disable or remove the NAT port forwarding rules created in the previous chapter to redirect HTTP(S) traffic to the web proxy. Captive Portal uses the `ipfw` packet filter, and it might conflict with the port forwarding NAT rules created on the webGUI.

1. Go to **Services | Captive Portal | Administration** and edit the current configured zone.
2. Check the **Transparent proxy (HTTP)** and the **Transparent proxy (HTTPS)** options and apply the configuration.
3. On **Services | Web Proxy | Administration - General Forward Settings** submenu, check that the **Enable Transparent HTTP proxy**, **Enable SSL inspection**, and **Log SNI information only** options are marked, and apply the configuration.
4. Try browsing on the host and check if requests are being logged by the web proxy (**Services | Web Proxy | Access Log**).

You should see HTTP(S) requests from your host logged by the web proxy. It is done! Our Captive Portal and web proxy integration are working! Now, we can choose which filtering rules we want to apply in the GuestNet.

> **Important Note**
>
> I recommend you revise the web proxy configuration steps in *Chapter 20, API – Application Programming Interface*, if you have any issues following the preceding steps.

Let's now look at some common issues while using the Captive Portal on OPNsense.

Common issues

As we have explored so far, the OPNsense Captive Portal implementation works pretty well, but things in life are not always a bed of roses. We sometimes face issues while using the Captive Portal, such as HTTPS URLs' redirection to the Captive Portal authentication page, as I mentioned previously in this chapter.

I will mention some common issues we need to handle daily with our CloudFence support team. Of course, you might find dozens of issues on the official forum, but it will not be possible to cover each one here as we have a limited number of pages for this chapter.

HTTPS page redirection while using the Captive Portal

We have two different approaches to try to address this issue, as outlined next.

Using DNS

This is an effective way for most modern mobile devices to access specific DNS entries while detecting a Captive Portal before going out to the internet. It might also work with most modern desktop browsers too. This technique consists of adding domain overrides in the local DNS server (Unbound, for example) to resolve the IP configured in the OPNsense GuestNet network interface. In this way, a device will find the authentication portal faster. Each device operating system/browser has its own domain list, also called **walled garden domains**. If you google it, you might find lists for each type of device.

Using DHCP

A special option can be configured on the DHCP server to help devices identify a Captive Portal authentication page. This option must be set as DHCP option 114 (for IPv4). An example using OPNsense would look like this:

Figure 14.12 – Services | DHCPv4 | LAN – Additional Options configuration example

Naturally, the HTTPS protocol's protection mechanisms will often result in errors on devices that don't recognize these mechanisms for identifying a Captive Portal correctly, but it is a protection mechanism this protocol was designed for.

Users failing authentication often while using vouchers

Try reducing vouchers' complexity. This will penalize the security level but might help in this case. To generate simpler vouchers, you can set the **Use simple passwords** field (less secure) on **System | Access Servers** (editing an existing voucher's backend). Notice that if existing vouchers are created, you will need to recreate them to apply this new configuration.

Users complaining about vouchers expiring without using them

First of all, explain to users the basics: once a voucher is activated, the time will count even if the user doesn't access the internet. Another issue is that users sometimes have difficulties understanding the authentication page. You can customize a template (on **Services | Captive Portal | Administration**) and upload it on the Captive Portal. A real-life example: as we (at CloudFence) support many Portuguese-speaking companies, it is common to translate the default template and customize it to customers' needs. The official documentation has a well-written guide on creating custom templates and adding some other features that we didn't cover in this chapter. I strongly recommend you read it: `https://docs.opnsense.org/manual/how-tos/guestnet.html`.

As we explored, the Captive Portal is a really useful OPNsense feature, and there are many other possibilities to authenticate and control guest internet access. You can try combining other OPNsense features as well.

Summary

As we explored, OPNsense has a pretty decent Captive Portal implementation, and it can work in combination with other features such as web proxy but is not limited to it. We also took a brief look at common issues and how to overcome them. Now, you can set up an entire GuestNet using all these and other possibilities. As we can observe, the more we learn about OPNsense, the more we realize that it is a flexible and powerful security solution. In the next chapter, we'll dive into outstanding **intrusion detection systems (IDS)** and **intrusion protection systems (IPS)** features.

15
Network Intrusion (Detection and Prevention) Systems

In this chapter, we will explore the **Intrusion Detection System** (**IDS**) and **Intrusion Prevention System** (**IPS**) concepts to understand their functionality. This will help us implement a good network perimeter defense using them. We will explore how OPNsense employs Suricata and combines it with Netmap to implement an outstanding IDS and IPS open source solution. By the end of this chapter, you will know how to use an IDS/IPS solution to monitor and block traffic using OPNsense.

In this chapter, we will cover the following topics:

- IDS and IPS definition
- Suricata and Netmap
- Rulesets
- Configuration
- SSL fingerprint
- Troubleshooting

Technical requirements

Good TCP/IP networking knowledge will be enough for you to understand the concepts in this chapter. To follow the configuration steps, you will need a working version of OPNsense with a host connected to it, along with an active internet connection.

IDS and IPS definition

The rise of different types of attacks on the internet has pushed firewall solutions to increase their defense mechanisms. A layer 4-only approach became inefficient against more sophisticated attacks, such as techniques that are used to exploit a known vulnerability, requiring a new approach to detect and block the latest threats. Let's look at a practical example.

Suppose that a layer 4-only firewall allows LAN to internet-only connections to well-known internet protocols such as DNS and HTTP, and connections that are used for email communication, such as POP3, SMTP (submission), and IMAP. So, to bypass the firewall, the attacker could install malware that uses the same ports as the HTTP protocol, such as `80` and `443`, so that the malware can transmit data using an allowed port without any problem. Now, suppose that this kind of technique became a trend, and all attackers started bypassing firewalls in this way! Cybersecurity professionals would get in trouble, right? Not so fast – thanks to skilled security researchers, the **IDS** concept was created, and some excellent tools have risen since then. An IDS can detect attacks using allowed ports, alerting the security teams about possible harmful traffic.

The IDSs that are used in network borders such as firewalls are known as **Network IDSs (NIDSs)**. There are **host-based IDS (HIDS)** tools that are installed on each host of the network, also known as endpoints.

> **Important Note**
> An outstanding HIDS open source project is **Open Source HIDS SECurity (OSSEC)**: `https://www.ossec.net/`. This project was founded by Daniel B. Cid, who is a Brazilian tech entrepreneur – which I am proud to mention, by the way.

Now, let's look at the next level of protection! What if a NIDS could automatically block attacks based on signatures? It would be heaven for any security team! So, let's introduce the concept of **IPSs**. Most NIDSs have blocking capabilities, turning them into NIPSs or **Network Intrusion Detection and Prevention Systems (NIDPSs)**. Let's dig a little deeper and explore how all *this magic* happens.

Suricata and Netmap

The OPNsense IDPS implementation is based on the Suricata project, a truly open source I(DP)S that's supported by the **Open Information Security Foundation** (**OISF**). Suricata is an excellent open source NIDS solution with superb support for signatures from companies such as Proofpoint, for example. On OPNsense, Suricata has Netmap support, which means fewer CPU resource requirements to detect threats, which results in good performance. The Netmap framework is driver-dependent, and it is essential to check whether the network device that's being used supports it before activating a feature that uses Netmap. In this chapter, we will do that while enabling IPS mode.

> **Note**
> We discussed how to implement Netmap in OPNsense in *Chapter 2, Installing OPNsense*. There, you can find the Netmap devices that are supported on FreeBSD.

The OPNsense project steps closer to the OISF, so you can always expect a better Suricata implementation on it. The Suricata and Netmap combination means good performance for I(DP)S inspection. One of the most challenging things to implement on a busy network is a NIPS because, with high traffic, packet inspection in real time is a CPU-consuming task. Without an exemplary implementation, like what OPNsense has, using a kernel-based version of Netmap and Suricata, sometimes even with powerful hardware, isn't enough to enable the IPS on larger networks.

Another significant advantage OPNsense has, compared to other NIPS open source projects, is that it only blocks the traffic that matches the NIPS signature. Some solutions block the source/destination IP for a while, which can cause a lot of issues in a false positive case.

Its support for Proofpoint signatures is another vast improvement that the OPNsense project achieved. Having better signature rulesets means fewer false positives and better accuracy while blocking threats.

Some of the key features of the Suricata project are as follows (extracted from the project's official website – `https://suricata.io/features/all-features/`):

- **NIDS** engine
- **NIPS** engine
- **Network Security Monitoring** (**NSM**) engine
- Offline analysis of PCAP files

- Traffic recording using a PCAP logger
- Unix socket mode for automated PCAP file processing

As you may have noticed, the OPNsense project is always pursuing state-of-the-art open source security. This is also true for I(DP)S features.

Next, let's check out the available signatures and rulesets.

Rulesets

The I(DP)S rulesets are a group of rules that you can enable to detect certain types of traffic – for example, a signature that's been designed to prevent attacks on web servers. In OPNsense, it is possible to enable different rulesets simultaneously. By default, the available rulesets are as follows:

- `Abuse.ch`: These are rulesets that are provided by the `Abuse.ch` project. They focus on blacklists based on an IP address's reputation.
- **Proofpoint's Emerging Threat Open** (**ET Open**): This is the community version of the Proofpoint ruleset. It's more limited than the ET Pro version.
- **OPNsense Application Detection**: This is OPNsense's project ruleset. It contains rules for controlling web applications such as YouTube, Netflix, Dropbox, and others.

Whatever ruleset you decide to use, you must download and install these rulesets before enabling them.

Some additional rulesets that are available as plugins are as follows:

- **Proofpoint ET Pro**: To use this plugin, you need a valid subscription. Compared to ET Open, it has more than double the amount of signatures, and it is updated more frequently.
- **Proofpoint ET Pro Telemetry edition**: This plugin is free and provides the same features as the Pro edition but requires you to register. It is free because, when installed, it shares events that have been logged by the I(DP)S with Proofpoint.

- **IDS PT Research**: The attack detection team ruleset is only available for non-commercial usage.

- **IDS Snort VRT**: This requires you to register/subscribe. It is a Talos ruleset that's been developed to the Snort IDS, but on OPNsense, we can use it by combining it with Suricata.

Before we configure an I(DP)S in OPNsense, we need to enable some rulesets. To do that, go to **Services | Intrusion Detection | Administration** and click the **Download** tab via the webGUI:

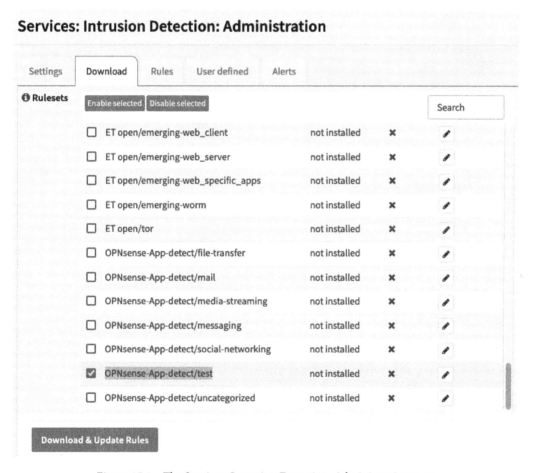

Figure 15.1 – The Services: Intrusion Detection: Administration page

Check the **OPNsense-App-detect/test** rule, click the **Enable selected** button, and click the **Download & Update Rules** button:

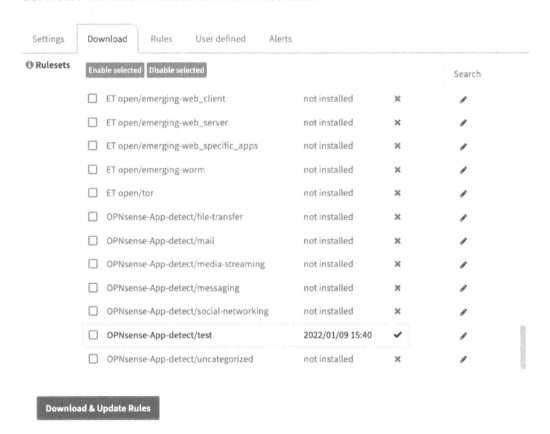

Figure 15.2 – Installing and enabling the OPNsense-App-detect/test ruleset

Once it has been installed, you may see the rule as it's shown in the preceding screenshot. Pay attention to the **Last updated** column. It must show a recent date. If you see **not installed**, this means that the ruleset hasn't been installed yet.

Now that we have at least one ruleset installed and enabled, let's learn how to configure an I(DP)S in OPNsense.

Configuration

To configure an I(DP)S, we need to go to **Services | Intrusion Detection | Administration** and go to the **Settings** tab. This can be seen in the following screenshot:

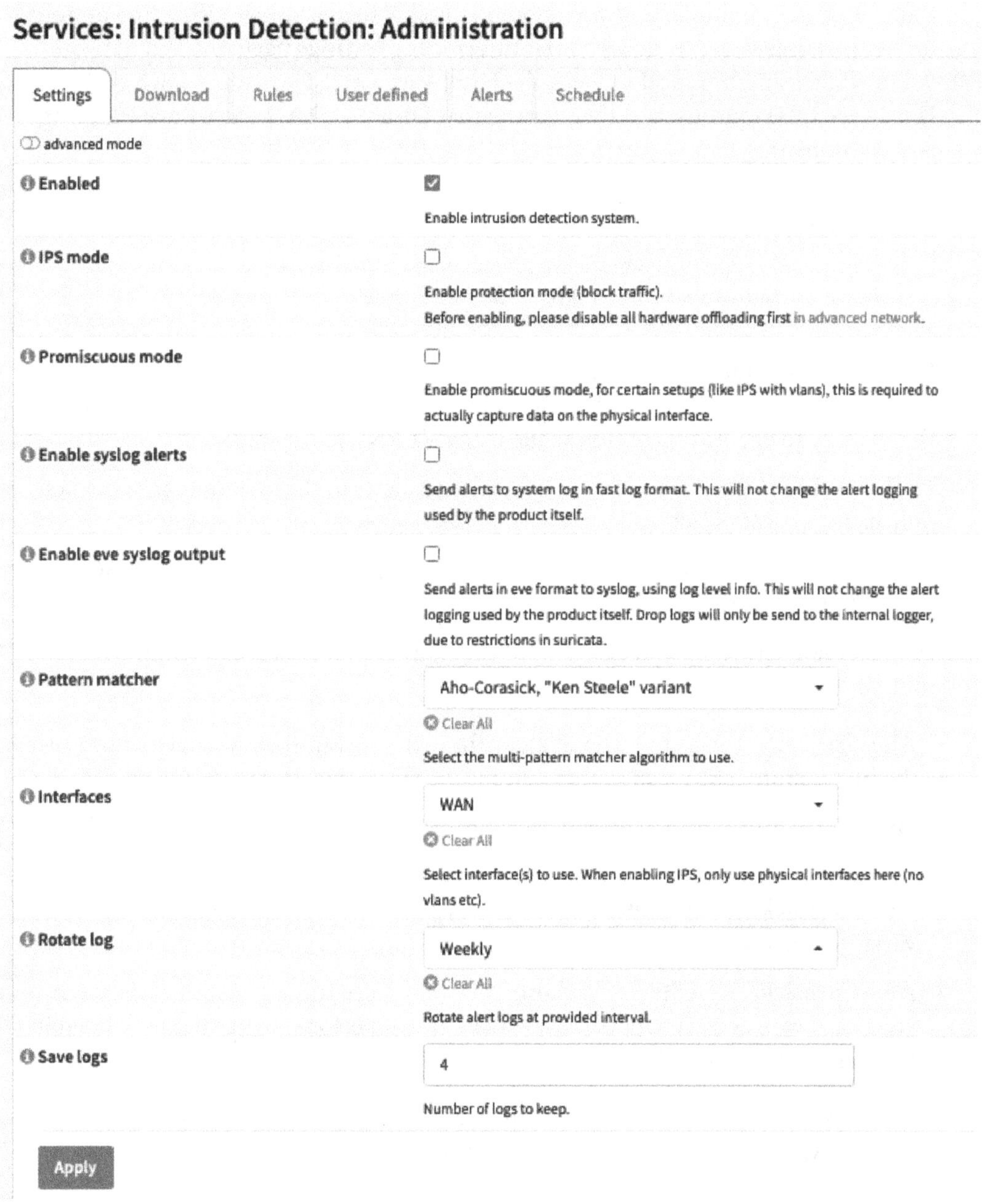

Services: Intrusion Detection: Administration

| Settings | Download | Rules | User defined | Alerts | Schedule |

advanced mode

ⓘ Enabled   ☑

Enable intrusion detection system.

ⓘ IPS mode   ☐

Enable protection mode (block traffic).
Before enabling, please disable all hardware offloading first in advanced network.

ⓘ Promiscuous mode   ☐

Enable promiscuous mode, for certain setups (like IPS with vlans), this is required to
actually capture data on the physical interface.

ⓘ Enable syslog alerts   ☐

Send alerts to system log in fast log format. This will not change the alert logging
used by the product itself.

ⓘ Enable eve syslog output   ☐

Send alerts in eve format to syslog, using log level info. This will not change the alert
logging used by the product itself. Drop logs will only be send to the internal logger,
due to restrictions in suricata.

ⓘ Pattern matcher

| Aho-Corasick, "Ken Steele" variant   ▾ |

✕ Clear All

Select the multi-pattern matcher algorithm to use.

ⓘ Interfaces

| WAN   ▾ |

✕ Clear All

Select interface(s) to use. When enabling IPS, only use physical interfaces here (no
vlans etc).

ⓘ Rotate log

| Weekly   ▴ |

✕ Clear All

Rotate alert logs at provided interval.

ⓘ Save logs

| 4 |

Number of logs to keep.

[Apply]

Figure 15.3 – IDS configuration page

Here, you will see the following options:

- **Enabled**: Check this option to enable the IDS service.

- **IPS mode**: Enabling this option will turn the IDS into an IPS service, blocking traffic instead of only alerting you about it. Disable the hardware offloading options for network interfaces by going to the **Interfaces | Settings** page to avoid network issues while using IPS mode. Check the **Hardware CRC**, **Hardware TSO**, and **Hardware LRO** options, and then ensure that **Disable VLAN Hardware Filtering** is also **disabled**.

> **Important Note**
> The Suricata daemon in IPS mode uses Netmap in OPNsense. It doesn't work well with hardware offloading, so keep it disabled while using IPS mode.

- **Promiscuous mode**: This mode allows all the traffic to be sent to the CPU to be inspected instead of all the traffic only being processed by the network controller. It is necessary to enable this option while using VLANs, for example.

- **Enable syslog alerts**: Check this option to send alerts to the system log, instead of only using Suricata's log.

- **Enable eve syslog output**: Check this option to send alerts using the EVE JSON format to the system log.

- **Pattern matcher**: Here, you can choose which pattern matcher algorithm will be used for traffic inspection:

  - **Aho-Corasick**: This is the default algorithm. This string-searching algorithm isn't the best option in terms of performance.

  - **Hyperscan**: This high-performance algorithm is supported by CPUs with SSE3 extensions.

  - **Aho-Corasick, reduced memory implementation**: You should use this algorithm variation for systems with limited RAM.

  - **Aho-Corasick, "Ken Steele" variant**: You should use this algorithm variation if your hardware doesn't support Hyperscan. It will perform better than the default version of Aho-Corasick.

- **Interfaces**: Here, you can choose which interfaces will have IDS enabled. While using IPS, OPNsense only supports physical interfaces.

> **Important Note**
>
> If you are using Zenarmor Sensei, you can't use the same interface that has been configured on it with IDS/IPS (Suricata).

- **Rotate log**: Here, you can select in which rotation period the logs will be rotated: **Weekly**, **Daily**, or every 4 days (default).

- **Save logs**: Before deleting old log files, you can specify how many logs will be kept on the disk. The default is four logs.

The available advanced options are as follows (you must ensure that **advanced mode** is checked):

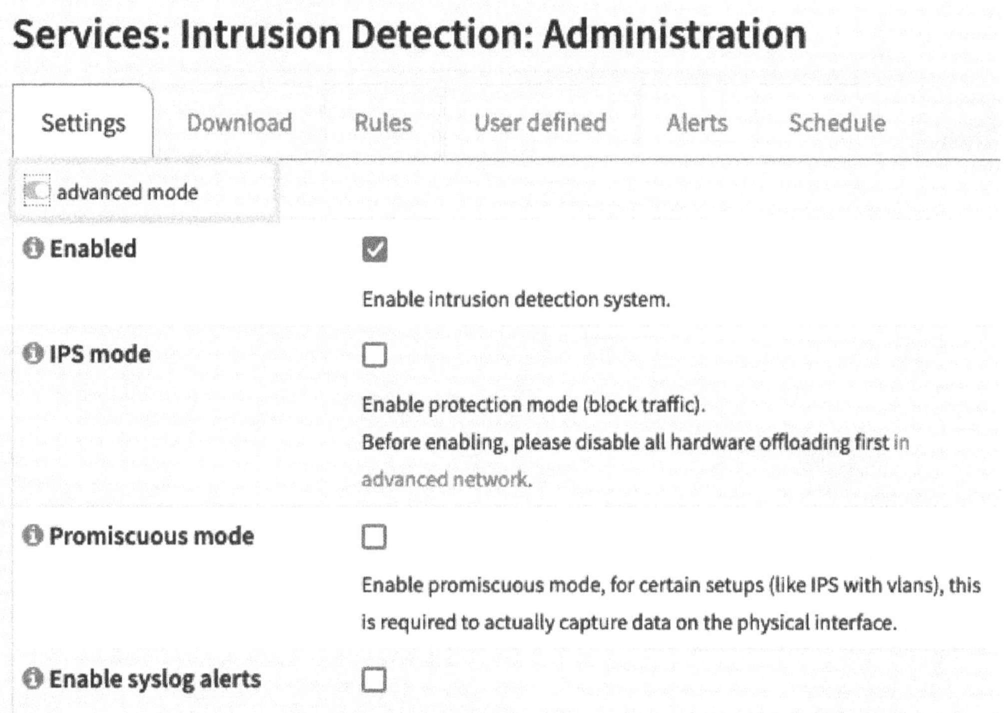

Services: Intrusion Detection: Administration

Settings	Download	Rules	User defined	Alerts	Schedule

advanced mode

ⓘ Enabled ☑

Enable intrusion detection system.

ⓘ IPS mode ☐

Enable protection mode (block traffic).
Before enabling, please disable all hardware offloading first in advanced network.

ⓘ Promiscuous mode ☐

Enable promiscuous mode, for certain setups (like IPS with vlans), this is required to actually capture data on the physical interface.

ⓘ Enable syslog alerts ☐

Figure 15.4 – Checking advanced mode to display the advanced options

Some of these additional options are as follows:

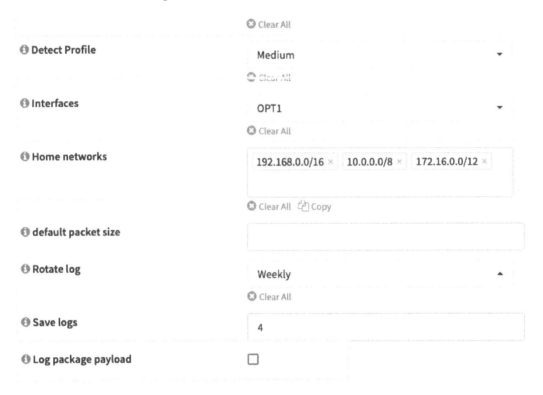

Figure 15.5 – Additional options

Let's look at these options in more detail:

- **Detect Profile**: You can use this option to optimize the signature-matching process. The IDS engine can group the signatures to improve performance and memory usage, inspecting only packets that match the grouped signatures. The default option will set it to **High**, which is the recommended option. For the **Custom** option, you can set your own values. This will allow you to choose from the **ToClient** and **ToServer** options. As a parameter, the **High** profile uses a value of **75** for both **ToClient** and **ToServer**.

- **Home networks**: By default, the *RFC1918 networks* are configured; these are the local networks. You must set it to your existing local networks once the signatures use this to define which traffic is local and which isn't.

- **default packet size**: This specifies the default size of the packets that will be processed by the IDS, which is *1514*. You can set it to a bigger size but this will decrease performance.

- **Log package payload**: Check this option to save the package payload in a log. Be aware that this will increase disk usage.

To finish configuring and setting up the service, click on the **Apply** button. Now, let's test this configuration.

Testing

To test an I(DP)S in OPNsense, you'll need a host connected to OPNsense's LAN. To follow these steps with ease, we'll use the lab that's already been configured using an Ubuntu host, but you can use your own if you wish.

In this exercise, you will detect and block Facebook usage using the **OPNsense-App-detect/social-networking** ruleset. Follow the steps described in the *Rulesets* section to enable this ruleset.

Go to **Services | Intrusion Detection | Administration** via the webGUI and follow these steps:

1. Enable the **LAN interface** option by going to the **Settings** tab and selecting **LAN** under **Interfaces**.

2. On the **Download** tab, enable the **OPNsense-App-detect/social-networking** ruleset by following the steps highlighted in the following screenshot:

Figure 15.6 – Enabling the OPNsense-App-detect/social-networking ruleset

These steps are as follows:

I.   Check the box next to **OPNsense-App-detect/social-networking**.

II.   Click the **Enable selected** button.

III. Click **Download & Update Rules** to download and make these rules active.

3. Go to the **Rules** tab to enable Facebook rules:

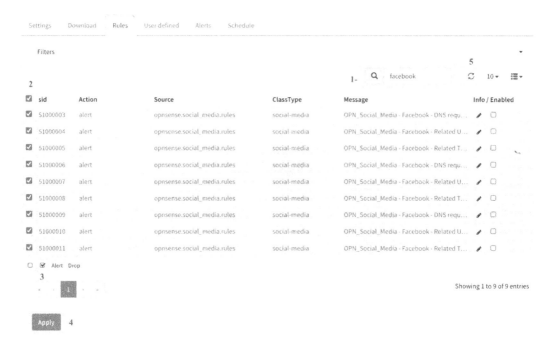

Figure 15.7 – Enabling the OPNsense-App-detect/social-networking ruleset

Let's look at the steps that are highlighted in the preceding screenshot:

I.   First, type `facebook` to find the rules you want to enable.

II.   Next, check all the Facebook rules (**social-media ClassType**).

III. Next, *enable* these rules by checking the box shown in the preceding screenshot.

IV. Now, click the **Apply** button.

V. Finally, click the refresh button to ensure the rules are enabled. They will still be disabled until the list is refreshed.

4. Go to the host's browser (Ubuntu, in our example) and open Facebook.

5. Go back to the webGUI and go to the **Alerts** tab:

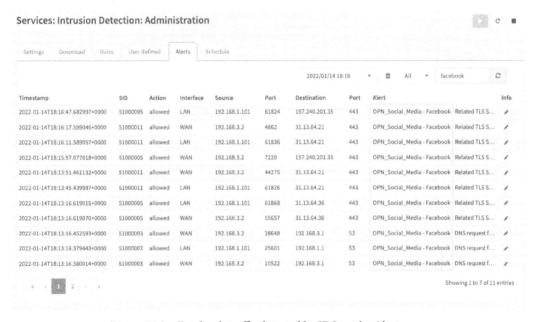

Figure 15.8 – Facebook traffic detected by IDS on the Alerts page

You should see all the alerts related to **OPN_Social_Media - Facebook**.

Great – the IDS is working! Congratulations!

Now, let's make things more interesting: *Let's block Facebook access using the IPS.* Follow these steps:

1. Go back to the **Settings** tab and enable **IPS mode** by checking the box next to it. *Remember to disable all the hardware offloading options* before doing this, as we explored in the *Configuration* section of this chapter:

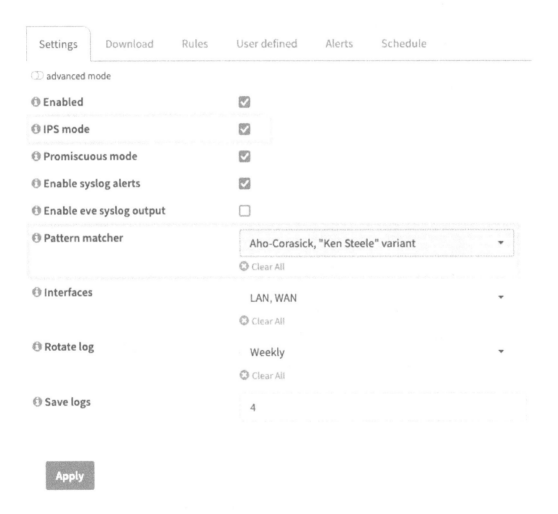

Figure 15.9 – Enabling IPS mode

2. If you use OPNsense as a virtual machine, I suggest setting the **Pattern matcher** option to **Aho-Corasick, "Ken Steele" variant**.

3. Apply these changes by clicking the **Apply** button.

4. Now, let's change the Facebook rules from **Alert** to **Drop**:

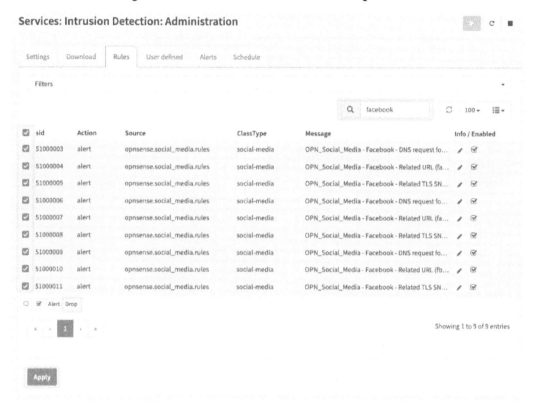

Figure 15.10 – Changing the rules from Alert to Drop

5. In the search box, type Facebook to filter only the Facebook rules.

6. Select all the displayed rules.

7. Click the **Drop** button to change the default action for these rules.

8. Click the **Apply** button to apply these changes.

You should see that the rules have their default action set to **drop**:

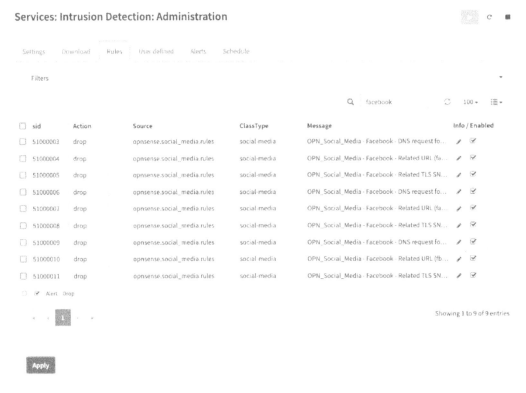

Figure 15.11 – Rules changed to drop

9. Go back to the host's browser and open Facebook again. I suggest that you open a new private window to avoid using cached content.

You should see the following error:

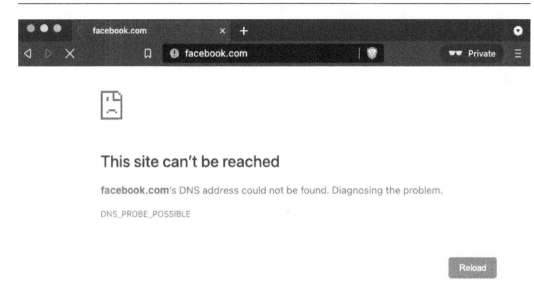

Figure 15.12 – This site can't be reached

10. On the webGUI, go back to the **Alerts** tab. You will see that the alerts have been **blocked**:

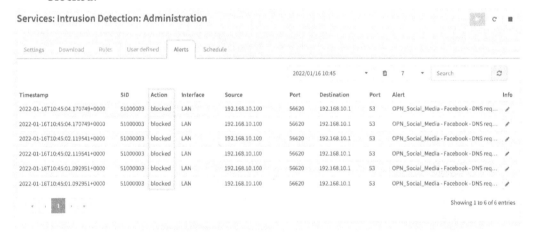

Figure 15.13 – Facebook blocked on the Alerts page

As you can see, an I(DP)S is very efficient and can detect and block content based on traffic. To block Facebook, as we did, using only firewall rules would be difficult, but not impossible, depending on the results you want to achieve.

In this example, we blocked outgoing traffic, but IPS is often used as an effective barrier for dangerous incoming traffic. A good example is attacks based on recent vulnerabilities that can be detected and blocked using an effective ruleset. A recent example is the *Log4j* vulnerability (also known as Log4Shell), which had rules released by Proofpoint quickly.

SSL fingerprint

The Suricata project that's used by OPNsense, known as `a192.168.0.1s`, has an I(DP)S engine that isn't capable of decrypting SSL/TLS traffic without external tools. Still, you can create your own rules based on the SSL/TLS certificates' SHA fingerprints. SHA is an algorithm that checks data integrity, and in OPNsense, we can extract it from a website certificate to match traffic and create custom rules for alerting or even blocking traffic.

> **Note**
> This isn't a stable feature. Test it before implementing it in a production environment. For more accurate SSL/TLS filtering in a production environment, it is advisable to use the Zenarmor plugin instead.

To start, go to **Services | Intrusion Detection | Administration**. Then, go to the **User defined** tab and click + to create a new custom rule. The following screenshot shows the **Rule details** page:

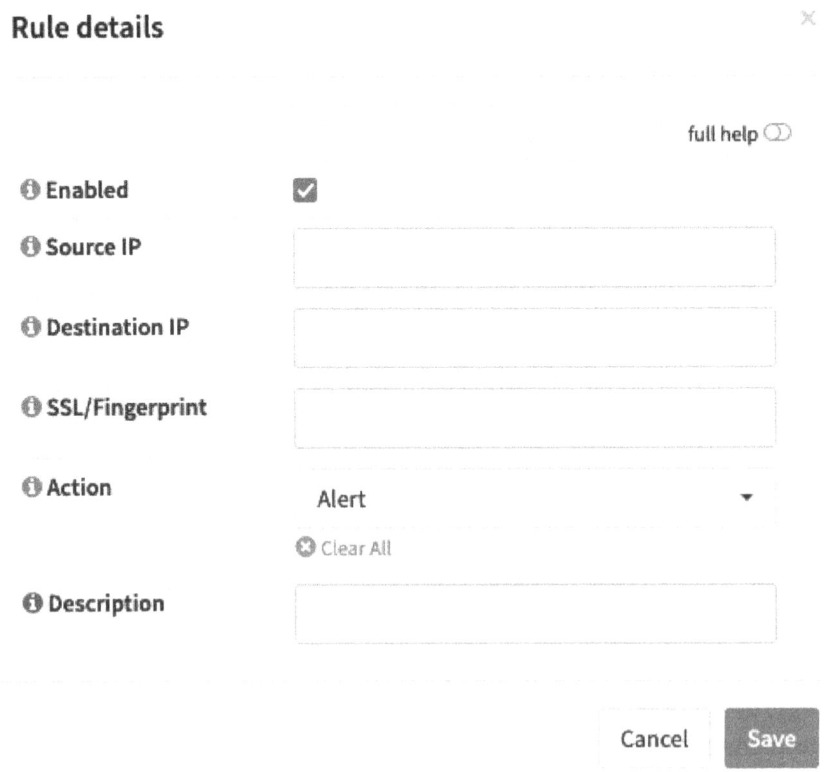

Rule details

full help

- 🛈 **Enabled** ✅
- 🛈 **Source IP**
- 🛈 **Destination IP**
- 🛈 **SSL/Fingerprint**
- 🛈 **Action** Alert
  - ⊗ Clear All
- 🛈 **Description**

Cancel Save

Figure 15.14 – The Rule details page

On the **Rule details** page, you can define the following options:

- **Enabled**: Check this box to enable or disable the rule.

- **Source IP**: The source IP address that matches this rule. For *any*, leave it empty.

- **Destination IP**: The destination IP address that matches this rule. For *any*, leave it empty.

- **SSL/Fingerprint**: The SHA fingerprint of the SSL/TLS certificate. To extract it using Firefox, follow these steps:

    I.   Open the desired website – for example, `https://instagram.com`.

    II. Click on the *padlock* icon located on the address bar and choose the **Connection secure** option:

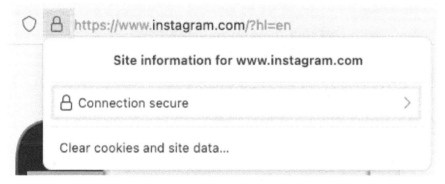

Figure 15.15 – Extracting a website's certificate SSL fingerprint on Firefox

    III. Click on the **More information** option:

Figure 15.16 – Extracting a website's certificate SSL fingerprint on Firefox

    IV. Click on the **View Certificate** option in the displayed dialog.

V.   On the **certificate details** page, in the **Fingerprints** section, copy the **SHA-1** fingerprint:

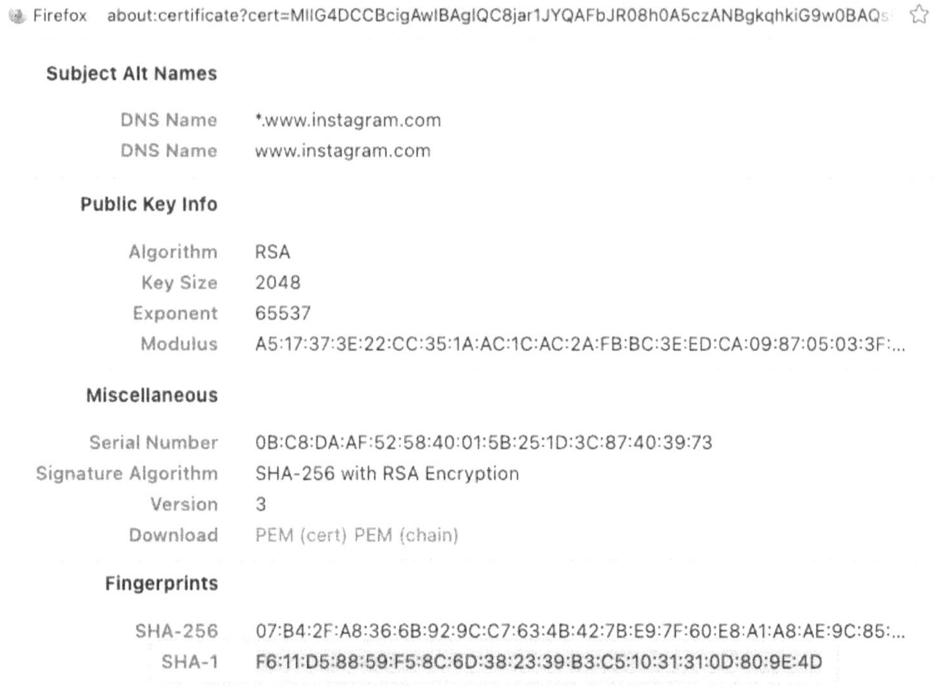

Firefox   about:certificate?cert=MIIG4DCCBcigAwIBAgIQC8jar1JYQAFbJR08h0A5czANBgkqhkiG9w0BAQs   ☆

Subject Alt Names

DNS Name   *.www.instagram.com
DNS Name   www.instagram.com

Public Key Info

Algorithm   RSA
Key Size   2048
Exponent   65537
Modulus   A5:17:37:3E:22:CC:35:1A:AC:1C:AC:2A:FB:BC:3E:ED:CA:09:87:05:03:3F:...

Miscellaneous

Serial Number   0B:C8:DA:AF:52:58:40:01:5B:25:1D:3C:87:40:39:73
Signature Algorithm   SHA-256 with RSA Encryption
Version   3
Download   PEM (cert) PEM (chain)

Fingerprints

SHA-256   07:B4:2F:A8:36:6B:92:9C:C7:63:4B:42:7B:E9:7F:60:E8:A1:A8:AE:9C:85:...
SHA-1   F6:11:D5:88:59:F5:8C:6D:38:23:39:B3:C5:10:31:31:0D:80:9E:4D

Figure 15.17 – Firefox's certificate details page

VI.   Paste it into the **SSL/Fingerprint** box, back on the **Rule details** page.

- **Action**: Select **Alert**, **Drop**, or **Pass**, depending on which action you want for this rule.

- **Description**: Describe what this rule is supposed to do.

- Apply the changes by clicking on the **Apply** button.

Before the **Zenarmor Sensei** plugin, it was considered useful to block websites and web applications using OPNsense. Nowadays, the Zenarmor Sensei plugin is more effective in blocking localhost traffic rather than collecting every website/app certificate you want to block. A high number of certificates have a short time when they're valid, such as those generated using the **Let's Encrypt** project, which means that the fingerprint could change every 60 days. In the next chapter, we'll explore the Zenarmor Sensei plugin, which has **next-generation firewall** capabilities.

Now, it's time to explore the common issues surrounding IDS and IPS and how to solve them.

Troubleshooting

In this section, we will explore some of the common issues you may face while using an I(DP)S in OPNsense and also how to solve each:

- **Poor performance while using IPS**: First things first – plan your IPS deployment with care! IPS mode is a resource-consuming feature, and you will need suitable hardware with a high-clock CPU, multicores, and good network cards to get the IPS working well in a high-traffic network. The recommended pattern matcher is Hyperscan, which works well with supported CPUs (SSE3-capable). A long list of enabled rulesets will demand a lot of RAM if you wish to inspect traffic without issues. A common symptom of this issue is high CPU usage and low bandwidth, especially from WAN, while using IPS mode. If your hardware isn't working reasonably with IPS mode on, it might be good to keep it only in IDS mode, which will consume fewer resources.

- **A lot of false positives**: A good deployment plan will avoid this kind of situation. Only choose those rulesets that will match the traffic of the monitored network. Adjust the rules that are causing the false positives. You can adjust each rule individually by going to the **Services | Intrusion Detection | Policy** page, then the **Rule adjustments** tab. It is possible to edit the enabled rules and change their actions or even disable them.

- **Service isn't up and running**: If the IDS service isn't running, even after you've started/restarted, this could indicate a corrupted ruleset or rule. This is rare but not impossible. To check whether there is something wrong with the ruleset or even with the Suricata daemon, go to the **Services | Intrusion Detection | Log File** page and read the logs carefully.

- **The enabled rulesets are not updating automatically**: We need to enable the **Schedule** job to update the rulesets automatically. Go to the **Services | Intrusion Detection | Administration – Schedule** tab to enable it. A dialog will appear. Enable it to ensure that the job will be created. The default job's configuration is to update daily at 00:00 hours.

- **Loss of connectivity (while IPS is enabled)**: If some connections have been dropped (TCP stream/UDP transmission), it might be a good idea to check IPS alerts. Some connections might be detected as false positives.

Of course, these are not the only issues that you may face while using an I(DP)S in OPNsense, but its community is engaging with it as a fantastic open source project. You may find many topics on this in the official forum talking about the issues other users have faced while using it.

Summary

In this chapter, you learned about IDSs and IPSs and how to configure OPNsense as an IDS to gather traffic alerts for traffic based on rulesets. You also learned about IPS mode, as well as how to enable it and block traffic on the local network. In the *Troubleshooting* section, we explored common issues that you may face while using an I(DP)S in OPNsense and learned how to solve each.

In the next chapter, we'll take traffic inspection to the next level by using the next-generation capabilities that are available via Sunny Valley's Zenarmor plugin.

16

Next-Generation Firewall with Zenarmor

The Sensei plugin broke the commercial-only next-generation firewalls barrier and introduced the open source world to Zenarmor's outstanding features. In this chapter, we will explore the Sensei plugin's features and how to install and apply layer7 control to the network. Finally, you will deploy OPNsense as a next-generation firewall solution that extends OPNsense's capabilities so that they're at the same level as the premium commercial solutions, which filter packets in layer4 to layer7 in a few steps.

In this chapter, we will cover the following topics:

- Layer7 application control with Zenarmor
- Choosing a Zenarmor edition
- Installing and setting up the Zenarmor plugin

Technical requirements

To complete this chapter, you will need to install plugins in OPNsense. Having OPNsense running with a host connected to its LAN interface will help with the steps we'll cover.

Layer7 application control with Zenarmor

As we have explored so far, OPNsense is a stateful firewall with some extra features, such as an **Intrusion Detection System** (**IDS**) and an **Intrusion Prevention System** (**IPS**), that can extend its filtering capabilities. But to compete head-to-head with the well-known commercial firewall solutions from giant cybersecurity tech firms, an open source firewall must have all the capabilities those solutions offer. There was a chasm between commercial and open source network firewalls due to the layer7 filtering feature being present only in the commercial ones. The ability to detect traffic, despite the TCP/IP port number, is a must-have feature these days, especially when malware tries to bypass stateful inspection by mimicking legitimate traffic.

The OPNsense project implemented some application control features, thus creating custom signatures for IPS, which we tested in the previous chapter. Still, its application control wasn't at the same level as it was for the commercial firewall solutions. The game changed in 2018 when Sunny Valley launched the former Sensei plugin for OPNsense, a next-generation firewall plugin that enabled OPNsense to reach the next level of network security. Later, in 2021, Sensei was rebranded **Zenarmor-sensei** for copyrights reasons.

The Zenarmor plugin helped OPNsense become a leading open source next-generation firewall project, enabling a **Transport Layer Security** (**TLS**) protocol inspection and application control that left some commercial vendors behind, eating dust. The plugin's frontend is open source, but its traffic engine inspection is proprietary with closed source code.

If you are a Linux user, you might be yelling: *Iptables has been doing Layer7 filtering for years!* OK! Calm down! I know it, but you must admit, Iptables does not do it the same way as commercial firewalls and Zenarmor does. You should probably become a Linux guru to keep a layer7 firewall based on it working, and a commercial solution does with tons of knowledge and lines of code. The commercial solutions ease this process, which means that a Linux-based firewall will also provide a decent firewall solution. Moreover, some of the most famous commercial firewall solutions run a Linux kernel. The big difference compared to OPNsense with Zenarmor and a Linux kernel with layer7 filtering enabled iptables is the way it is configured and maintained. Sunny Valley has made a significant effort to keep the layer7 filtering updated and working with newer applications. Besides that, they have cloud-enabled filtering that uses DNS, an **artificial intelligence** (**AI**)-enabled solution, and they are a company that makes money on it! Sometimes, it is better to count on a company that's dedicated to supporting and improving an affordable product rather than trying to build it by yourself.

The following features are listed in Sunny Valley's official documentation:

- Application Control Filtering – control traffic based on the type of application
- Cloud Application Control (Web 2.0 Controls)
- Advanced Network Analytics
- All-Ports Full TLS Inspection (for every TCP port, not just HTTPS)
- Cloud Threat Intelligence
- Encrypted Threats Prevention
- Web Filtering and Security
- User-Based Filtering and Reporting
- Active Directory Integration
- Policy-Based Filtering and QoS
- Application/Web Category-Based Traffic Shaping and Prioritization
- Cloud-Based Centralized Management and Reporting

They also support other platforms such as FreeBSD, pfSense, and some Linux distribution flavors, but we will focus on OPNsense in this book.

Choosing a Zenarmor edition

The plugin has a free edition that is pretty functional and works great in most small network environments. Compared to the paid versions, it has some limitations, but most are features that are only required by complex networks that deserve a paid version subscription.

Before comparing the different available versions, let's check the hardware requirements.

Hardware requirements

As IPS does, Zenarmor also uses the Netmap framework to inspect and filter network packets, so don't enable it in the same configured network interfaces, as you would in IPS mode by going to **Services | Intrusion Detection | Administration – Interfaces**.

Reports are something else that Zenarmor has that demands more memory and CPU than a stock OPNsense installation. It is based on Elasticsearch, which requires significant memory and CPU power to process data and transform it into meaningful graphs. The plugin installation does a hardware requirements test to indicate how many devices the installation will support. To have a good user experience with this plugin, consider having at least 4 GB of RAM and a modern two-core CPU.

Now, let's explore each subscription type. This will help you find out which suits your needs the most.

Paid subscriptions

There are different paid subscriptions. Each one was designed with the size and complexity of the network environment it will protect in mind. For example, the available plans are Home, SoHo, and Business. These names easily define which type of network they are suitable for, as described here:

- A Home subscription is intended for non-commercial usage and can protect up to three different networks, for example, using the different filtering policies.

- SoHo is the first step for small and home offices, and it is a version that can protect networks with up to five different filtering policies.

- A Business subscription is complete and provides businesses with the most advanced security features.

The main differences between these versions are based on the number of policies you can create, resource reporting, authentication integration (for example, Active Directory), malware protection, and so on. The product had a fast evolution, and at the time of writing, new features will probably be added to the product. Look at the plans page and pick the best option for you!

> **Important Note**
> To check the available plans, from free to business, go to `https://www.sunnyvalley.io/plans`.

Now, let's learn how to install and configure the Zenarmor plugin.

Installing and setting up the Zenarmor plugin

To install the Zenarmor plugin, we will follow the same steps we used for OPNsense's other plugins:

1. Go to **System | Firmware | Plugins** and click on the + icon on the right-hand side of os-sunnyvalley. This will install SunnyValley's repository, which will allow you to install Zenarmor:

Figure 16.1 – Sunny Valley repository installation

2. Once you've installed Zenarmor, click on the **Status** tab and click the **Check for updates** button to update the newly installed repository:

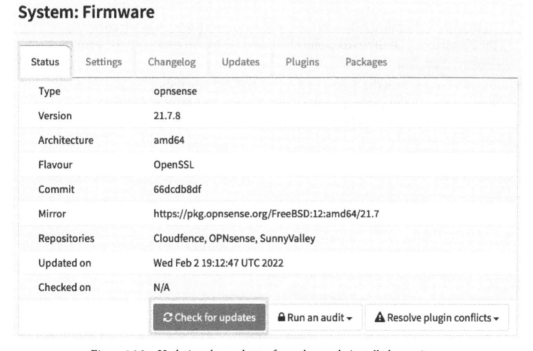

Figure 16.2 – Updating the packages from the newly installed repository

3. Go back to the **Plugins** tab, find the **os-sensei** package, and click on the + icon to install it. The search box can be found at the top left:

System: Firmware

Figure 16.3 – Adding the os-sensei package

> **Important Note**
>
> Remember that the plugin was called Sensei before it was rebranded as Zenarmor, so the package's names will remain as `*-sensei*` for ease of use. The maintainers may update the package's name to `*-zenarmor*` in the future.

4. Once it's been installed, click on the OPNsense logo (in the top-left corner). The **Zenarmor | Dashboard** menu should be visible, so click on it to begin the *configuration wizard* process.

5. You must agree with the **Terms of Service** and **Privacy Policy** documentation to proceed with the configuration wizard. Check the box next to **Check here to indicate that you have read and agree to the Terms of Service and Privacy Policy.** and click **Proceed**:

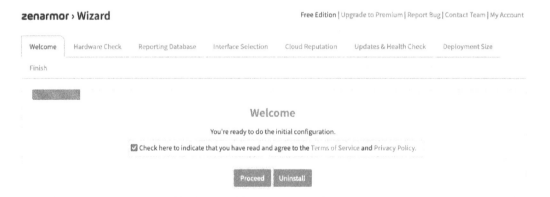

Figure 16.4 – Configuration wizard welcome page

6. The plugin will analyze your hardware performance and will classify it based on its criteria. Click **Next** to continue:

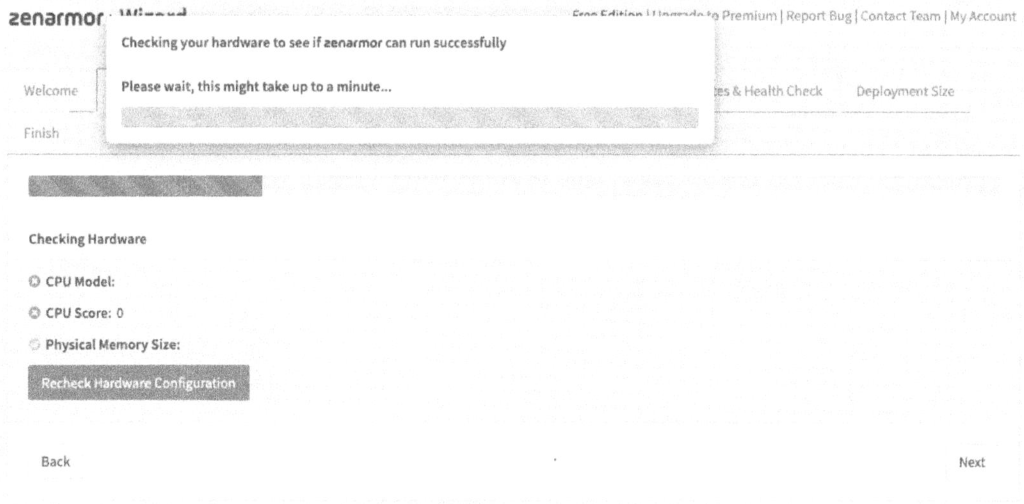

Figure 16.5 – Hardware requirements check page

After the hardware tests, you should see something similar with the following screen:

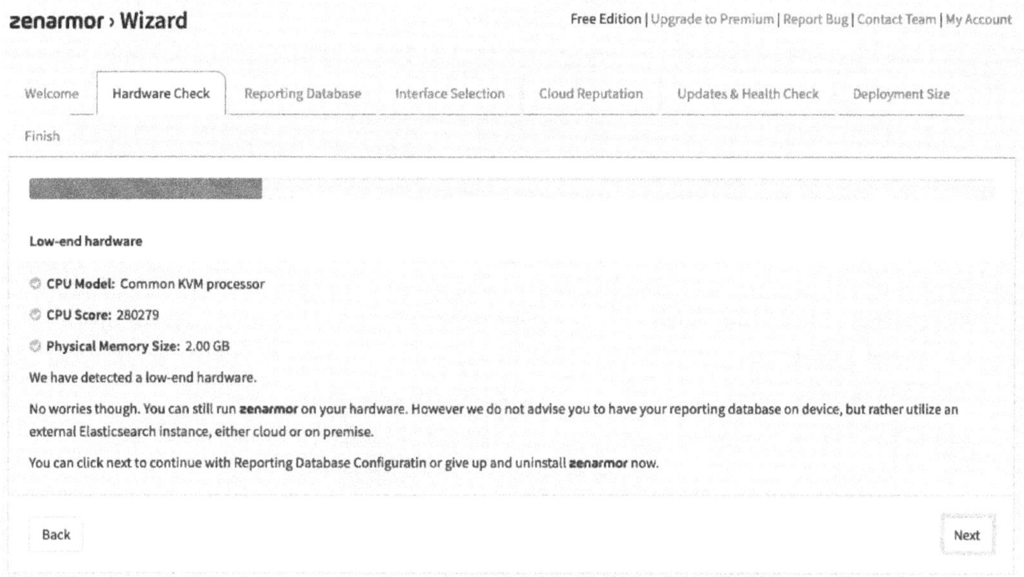

Figure 16.6 – Hardware requirements page once the tests have finished

7. On the next screen, the wizard will ask you about installing the database. We will use a local database for this installation, so we've left the default option selected – that is, **Install a local Mongodb Database**. This database will be used for reporting. It is possible to use a centralized reporting tool with an Elasticsearch database running on it by clicking the **Use a Remote Elasticsearch Database** option. Using a remote database might be helpful if you're installing hardware with limited RAM and CPU. Otherwise, using a local database will demand more CPU and RAM resources from the OPNsense machine. To proceed, click the **Install Database & Proceed** button. Wait for the database installation process to finish and click the **Next** button. *Warning: Don't close the database installation process window! Have a coffee and enjoy while the configuration wizard does its job:*

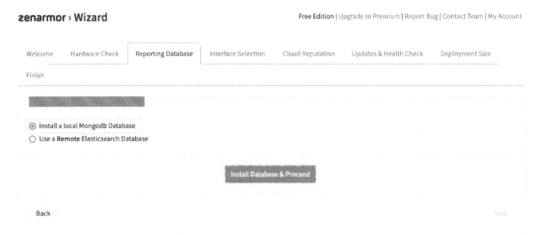

Figure 16.7 – Database installation page

8. On the **Deployment mode** page, leave the option set to **Routed Mode (L3 Mode, Reporting + Blocking) with native netmap driver** if you have a Netmap-capable network card. If you don't have that, you can choose the **Routed Mode (L3 Mode, Reporting + Blocking) with emulated netmap driver** option. The **Passive Mode (Reporting Only)** option will, as its name suggests, only report the traffic without blocking. **Bridge Mode** will work in layer2 as a bridge.

9. In the **Interfaces Selection** section, select the **LAN** interface under **Available Interfaces** and click the >> button to add it to **Protected Interfaces**. Then, click **Next**:

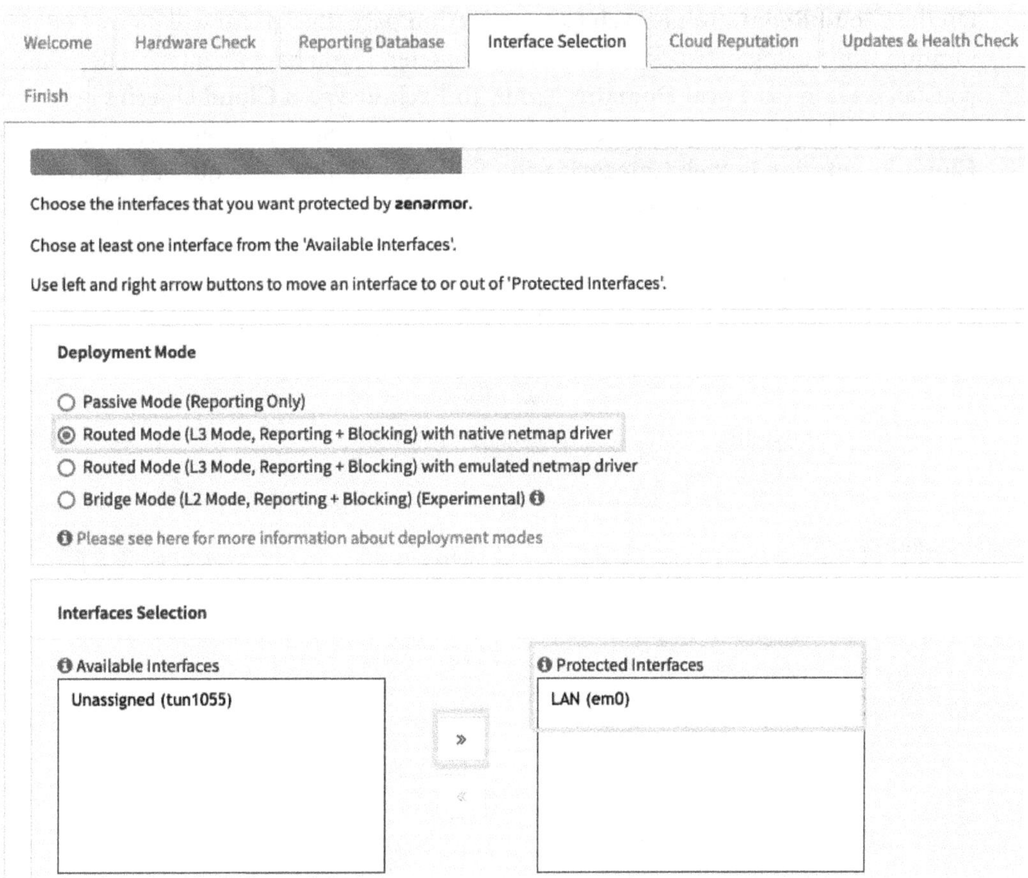

Welcome	Hardware Check	Reporting Database	Interface Selection	Cloud Reputation	Updates & Health Check

Finish

Choose the interfaces that you want protected by **zenarmor**.

Chose at least one interface from the 'Available Interfaces'.

Use left and right arrow buttons to move an interface to or out of 'Protected Interfaces'.

Deployment Mode

○ Passive Mode (Reporting Only)

◉ Routed Mode (L3 Mode, Reporting + Blocking) with native netmap driver

○ Routed Mode (L3 Mode, Reporting + Blocking) with emulated netmap driver

○ Bridge Mode (L2 Mode, Reporting + Blocking) (Experimental) ❶

❶ Please see here for more information about deployment modes

Interfaces Selection

❶ Available Interfaces

Unassigned (tun1055)

»

«

❶ Protected Interfaces

LAN (em0)

Figure 16.8 – Zenarmor mode and interface selection page

Important Note

If the interface you've chosen is being used by IDS, the configuration process
will not move on. As a simple rule of thumb, while using IDS and Zenarmor
together, only keep IDS watching for traffic on WAN interfaces and Zenarmor
on the LAN interfaces.

On the **Cloud Reputation & Web Categorization** page, the wizard will pick a better location while using network latency as a parameter. If you have an internal domain, you can set it in the **Local Domains Name To Exclude From Cloud Queries** box to avoid local domain requests from being forwarded to Sunny Valley's servers. The **Cloud Reputation & Web Categorization** features will filter DNS queries based on the locally configured policies:

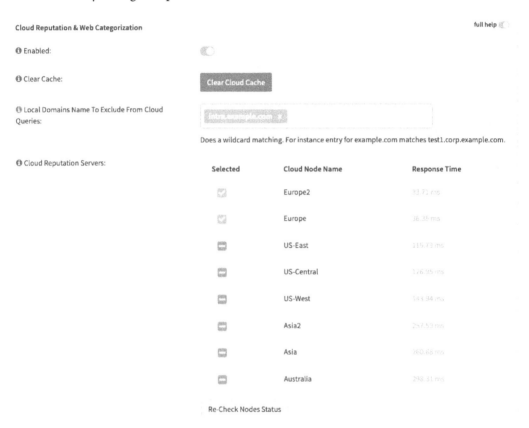

Figure 16.9 – The Cloud Reputation & Web Categorization page

10. You can leave the *default options* as-is for **Updates and Support** and **Health Check** on the next page. Click **Next**.

11. Under **Deployment Size**, the configuration wizard will show the available options based on the hardware test (*Step 3*), so you can choose the best choice for the network you are deploying Zenarmor on:

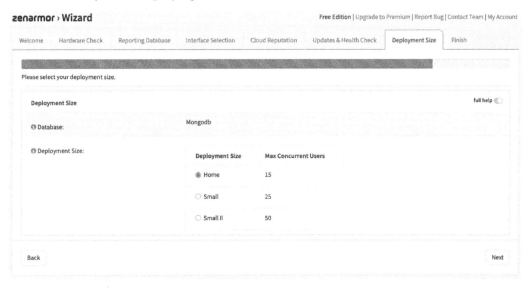

Figure 16.10 – Selecting the deployment size

12. Now, you can sign up for Sunny Valley's newsletter, fill in your email address, and click **Finish** to start using the Zenarmor plugin!

> **Important Note**
>
> If you need to revise the configurations you have set in the configuration wizard, you can go to **Zenarmor | Configuration | General** and set the configurations that your network needs.

To start using the Zenarmor plugin, go to the **Zenarmor** | **Dashboard** menu and check out the graphs for *app categories*, *top local hosts*, *top remote hosts*, and so on. It may take a while for helpful information to appear. Now, it's time to go to OPNsense's LAN-connected host and do some web browsing to generate traffic:

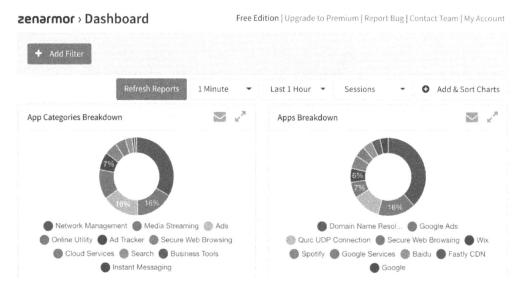

Figure 16.11 – Zenarmor's Dashboard page

On the **Zenarmor** | **Status** page, we can check the engine, rules versions, cloud node status, protected interfaces, and the services' statuses. It is also possible to set the services to start on boot, stop, restart, and so on. Something that might be helpful while you're troubleshooting outgoing traffic blocks is setting the Zenarmor packet engine to bypass mode, which will bypass all the traffic without blocking or reporting it. You can enter bypass mode by clicking the **Enter Bypass Mode** button. *Warning: this mode will not survive a reboot*:

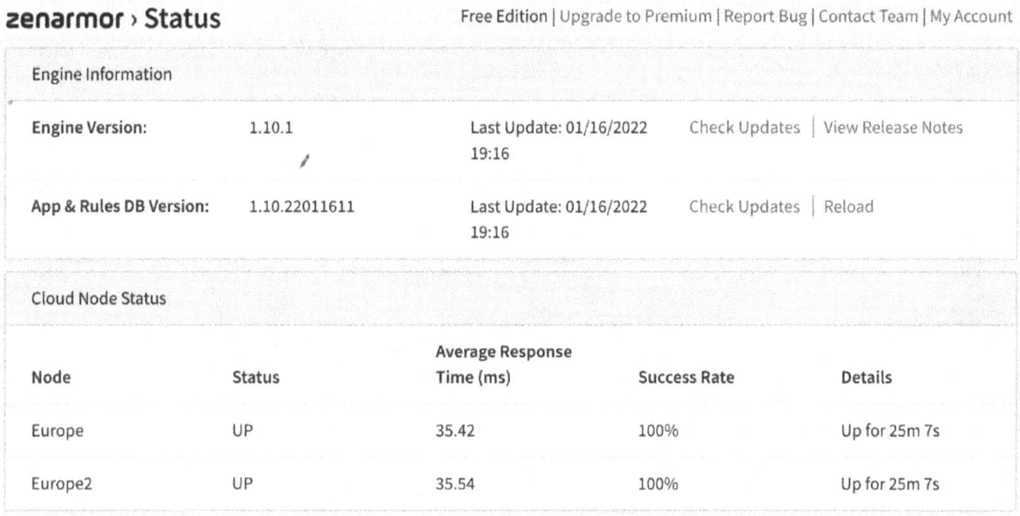
zenarmor › Status
Free Edition | Upgrade to Premium | Report Bug | Contact Team | My Account
Engine Information
Engine Version:
1.10.1
Last Update: 01/16/2022 19:16
Check Updates | View Release Notes
App & Rules DB Version:
1.10.22011611
Last Update: 01/16/2022 19:16
Check Updates | Reload
Cloud Node Status
Node
Status
Average Response Time (ms)
Success Rate
Details
Europe
UP
35.42
100%
Up for 25m 7s
Europe2
UP
35.54
100%
Up for 25m 7s

Figure 16.12 – Zenarmor's Status page

Another thing that's great about Zenarmor is its reporting feature. You can check the available reports by going to **Zenarmor | Reports**. I recommend that you invest some time exploring the available reports and options – it's worth it!

In the **Zernarmor | Policies** menu, it is possible to set up protected network policies. Note that in the free version, only one policy will be available, so if you select two different networks – for example, to protect using Zenarmor – they will use the same rules that were defined in this policy. If you need different policies for different users or networks, I recommend checking out one of the paid subscription options.

Following we will edit the default policy:

1. To start editing a policy, click on the pencil icon button:

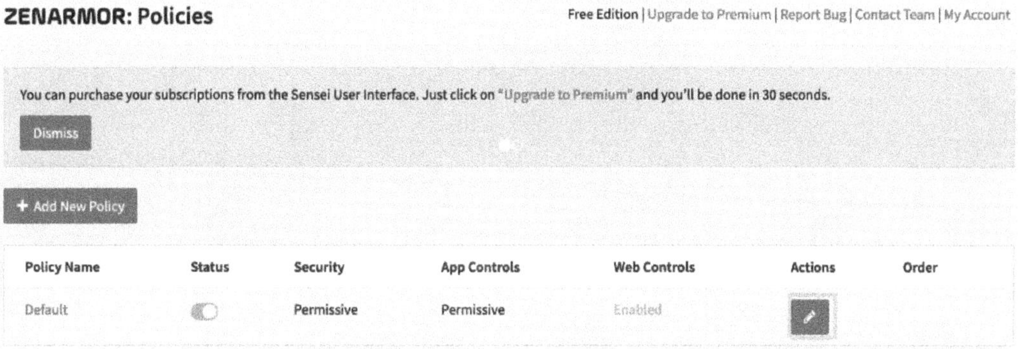
ZENARMOR: Policies
Free Edition | Upgrade to Premium | Report Bug | Contact Team | My Account
You can purchase your subscriptions from the Sensei User Interface. Just click on "Upgrade to Premium" and you'll be done in 30 seconds.
Dismiss
+ Add New Policy
Policy Name
Status
Security
App Controls
Web Controls
Actions
Order
Default
Permissive
Permissive
Enabled

Figure 16.13 – Editing a Zenarmor policy

2. The **Security** tab will appear:

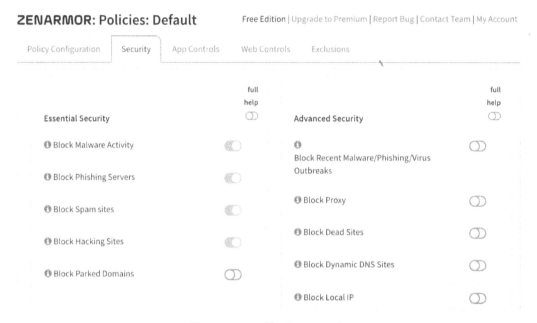

Figure 16.14 – The Security tab

Here, you can select which malware categories will be blocked on the protected networks.

> **Important Note**
> Some options are paid exclusives.

Application control can be configured under the **App Controls** tab. By clicking on it, you can check out all the available applications and select the ones you want to block. By default, all are allowed:

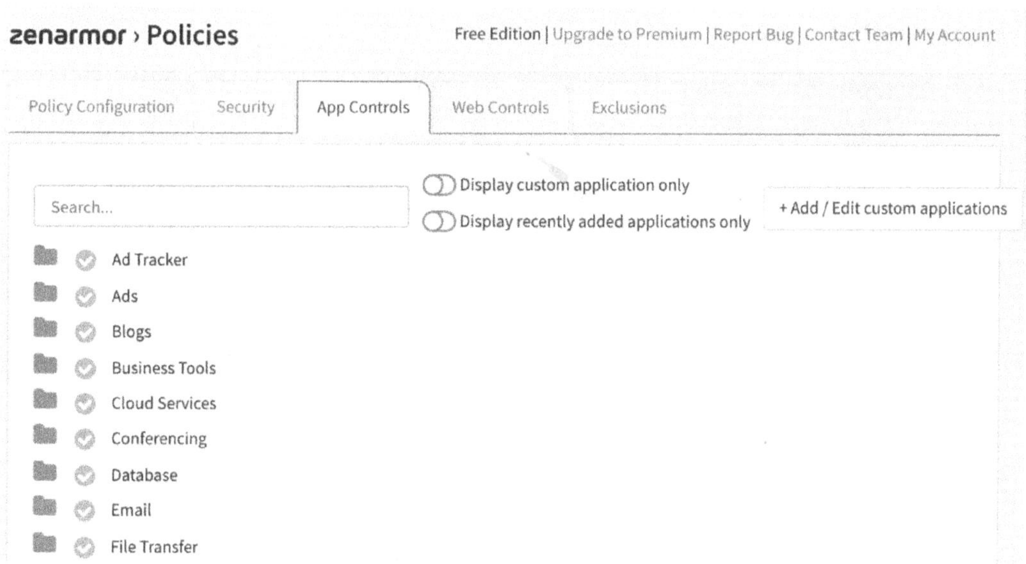

Figure 16.15 – The App Controls page

You can also make the categories block specific web browsing filtering by clicking the **Web Controls** tab. The preset profiles that are available in the free version are **Permissive**, **Moderate Control**, and **High Control**. *To customize these web categories, you will need a paid subscription.*

To allow and block domains for web filtering, go to the **Exclusions** tab and set your own white or blacklists.

> **Important Note**
> The **Policy Configuration** tab is only customizable if you have a paid subscription version.

As we've seen, the Sunny Valley team did an outstanding job integrating Zenarmor with OPNsense webGUI. They created a turning-table product that has led open source firewall projects such as OPNsense to compete with commercial firewall solutions at the same features level.

Summary

In this chapter, you learned how to install the Zenarmor plugin and enable next-generation firewall capabilities on OPNsense. Now, you can deploy OPNsense as a firewall, just like any other commercial solution, with application inspection and control, DNS and web filtering, excellent reports, and the outstanding cloud-enabled threat detection provided by Sunny Valley. In the next chapter, we will discuss the high availability of firewalls.

17
Firewall High Availability

In this chapter, we will learn how to configure high availability by connecting two firewalls to sync configuration, connect states, and preserve network connectivity if something goes wrong with one of our firewalls. By the end of this chapter, you will understand the concepts surrounding high availability and be able to implement them using OPNsense.

In this chapter, we will cover the following topics:

- High availability concepts
- Configuring high availability
- Testing the HA configuration

Technical requirements

You will need two running OPNsense installations on the same network to follow the steps in this chapter. Good knowledge of how to configure OPNsense networking and firewall rules is mandatory.

High availability concepts

Let's introduce this topic with an aviation example. At the beginning of heavier-than-air history, airplanes had just one engine to fly from one location to another. As aviation grew, the demand for long-range flights increased, and new projects that used two or more engines began. Nowadays, it is possible to cross the oceans with a twin-engine plane thanks to reliable engine technology and the **Extended-range Twin-engine Operations Performance Standards** (**ETOPS**). But even with all this technology, two engines are required to keep a long-haul flight within safety standards. Developments similar to the aviation industry also happened in the IT world – redundancy standards/protocols were created to keep the availability of the systems at acceptable levels for the business.

In OPNsense, the **Common Address Redundancy Protocol** (**CARP**) is a protocol that ensures that the network interfaces of two or more firewalls keep operating in case of a hardware failure. As in a twin-engine plane, a firewall cluster must have at least two firewalls working together to keep everything online. Think about a firewall cluster as the airplane and each firewall as the plane's engines. If one fails, you have the other one to keep the network safe and available! The key difference between a twin-engine plane and an OPNsense firewall cluster is that, in the plane, both engines are in use during normal flight conditions, while with a firewall cluster, the second firewall only comes into play if the first one fails. This is called an active-passive cluster. An active-active cluster is when both firewalls are actively working in the network, but OPNsense doesn't support this operation mode.

Active-active and active-passive modes

Active-active mode will load-balance workloads to all cluster nodes, which can appear as an advantage at first glance. But let's suppose that a two-node cluster has an average CPU usage of 60%. What will happen if one of the nodes fails? The second one will probably be overloaded with 100% CPU usage since, in theory, the workload has a 20% deficit compared to the last cluster state (60% + 60% = 120%). So, it isn't a good idea to have a firewall cluster running in this mode. Do you agree?

Active-passive mode takes a more straightforward approach, and it is the OPNsense cluster's default mode. In this mode, considering the two-node cluster, it will have one node as the master and the other one as the backup node, so if the master node fails, the backup will jump in automatically. In OPNsense, the firewall states are synchronized between the nodes using the **pfsync** protocol. This ensures that network connections won't be dropped and recreated in case of a node failure. Users won't even notice that a failure happened in the network.

> **Important Note**
>
> Both CARP and pfsync were forked from OpenBSD to FreeBSD. To learn more about those two protocols, please refer to `https://www.openbsd.org/faq/pf/carp.html`.

To ensure both nodes have the same configuration, OPNsense uses the **Extensible Markup Language Remote Procedure Call** (**XMLRPC**) to transfer the OPNsense configuration from the master node to the backup node.

> **Important Note**
>
> Notice that only the master (primary) node transfers configurations to the backup (secondary) node. The reverse isn't supported.

We will assume that a firewall cluster has two nodes from this point on. Before we look at the configuration, let's dive into some CARP concepts.

CARP – how it works

The master node uses multicasting to frequently advertise in networks that have CARP virtual IPs configured, to let the backup node know that it is alive. Two parameters are configured to set the advertising frequency: `base` and `skew`. `base` specifies how often the advertisements (in seconds) will be multicast to the `224.0.0.18` address (IPv4), while `skew` defines which node is the preferable master. For example, the node with the lesser number will be elected as the master. For instance, considering that 0 is the minimum possible value, this will represent the master node in the cluster.

CARP uses a **virtual host ID** (**VHID**), which identifies the redundancy group to be shared between cluster nodes. Each CARP interface will have the following virtual MAC address format: `00:00:5e:00:<VHID number>`. The VHID uses a password to authenticate the communication between cluster nodes.

Next, we'll learn how to configure CARP and high availability in OPNsense.

The preempt behavior

Even in experienced OPNsense teams, it is common to have a quick discussion about the preempt CARP configuration before defining it as enabled or disabled in OPNsense as a high availability option, as we will explore in the following section. But the key point here is to clarify what it does. When preempt is enabled, the preempt CARP configuration (net.inet.carp.preempt = 1) is enabled; in the case of a single network interface failure, the cluster node will assume a failure. All other CARP configured network interfaces will assume a failure condition, turning the secondary node (backup) into a master on all the networks. Otherwise, while the CARP configuration is disabled (the default in OPNsense), each CARP configured network interface will act individually, and if a failure occurs, just the failed network will have the BACKUP status.

Now that we've learned about the theory behind high availability and how it works in OPNsense, let's learn how to configure it in an OPNsense firewall.

Configuring high availability

First, let's look at the high availability scenario topology shown in the following diagram:

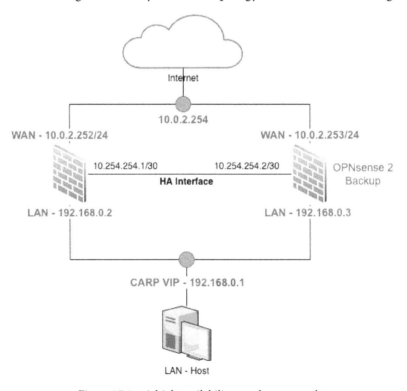

Figure 17.1 – A high availability topology example

In the preceding diagram, we have two OPNsense firewalls connected using a dedicated network interface. This interface, which we will call the **HA Interface**, is the network interface that will keep the firewall states synchronized using the pfsync protocol. We also use it to run XMLRPC to sync **OPNsense 1**'s (master node) configuration to **OPNsense 2** (backup node).

Proposed scenario

You can quickly reproduce this scenario using VirtualBox while configuring the network and the virtual machines using the steps we learned about in the previous chapters of this book (modifying the steps according to your lab environment).

First, configure two OPNsense firewalls, add a network interface (both from OPNsense), and connect them directly using a network interface that we will use for firewall states sync.

You can use a /30 CIDR, which is the address that was suggested in the topology (10.254.254.0/30), with two available IPs for host addressing:

```
ipcalc 10.254.254.0/30
Address:    10.254.254.0          00001010.11111110.11111110.000
000 00
Netmask:    255.255.255.252 = 30  11111111.11111111.11111111.111
111 00
Wildcard:   0.0.0.3               00000000.00000000.00000000.000
000 11
=>
Network:    10.254.254.0/30       00001010.11111110.11111110.000
000 00
HostMin:    10.254.254.1          00001010.11111110.11111110.000
000 01
HostMax:    10.254.254.2          00001010.11111110.11111110.000
000 10
Broadcast:  10.254.254.3          00001010.11111110.11111110.000
000 11
Hosts/Net: 2                      Class A, Private Internet
```

If you need help configuring a new network interface, please go back to *Chapter 3, Configuring an OPNsense Network*, where we explored all the necessary steps for configuring a network interface in OPNsense.

Next, we have the virtual IPs for the networks that OPNsense is serving – LAN and WAN. Each network interface has a virtual IP that's been configured using CARP. Follow these steps to configure CARP virtual IPs on OPNsense 1:

1. Go to **Interfaces | Virtual IPs | Settings** and add a new IP:

Interfaces: Virtual IPs: Settings

Figure 17.2 – CARP virtual IP configuration

Here, you can make the following configurations:

- **Mode**: Select **CARP**.

- **Interface**: Select the network interface you want to configure the CARP virtual IP on (for example, **LAN**).

- **Address**: The virtual IP address.

- **Virtual IP Password**: This password will be used to help CARP authenticate between the cluster nodes. Once XMLRPC syncs the configuration, you don't need to save this password.

- **VHID Group**: Select an available VHID group. To help configure an available VHID, click the **Select an unassigned VHID** button.

- **Advertising Frequency**: The **Base** option, or advertisements' frequency, keeps the default value of **1 second**. For example, in the **Skew** option, you can set **0** for the master node and **100** for the backup nodes.

> **Important Note**
>
> OPNsense automatically creates the CARP virtual IPs in the backup nodes by setting the appropriate skew value (**100** by default).

- **Description**: Type in a description for this virtual IP.

- Click the **Save** button to add the virtual IP.

As a suggested scenario, on the *master* node, create a *CARP virtual IP* for the *LAN* interfaces. You can pick the better IP addresses for your network. This chapter will use the scenario presented previously in the topology diagram.

In our example scenario, the following LAN network address configuration will be used:

- OPNsense-1 (master) interface address: `192.168.0.2/24`

- OPNsense-2 (master) interface address: `192.168.0.3/24`

- CARP virtual IP address: `192.168.0.1/24`

> **Important Note**
>
> Suppose you need help configuring a new virtual IP address, please refer to *Chapter 3, Configuring an OPNsense Network*.

After the configuration, the virtual IP address should look like the following screenshot:

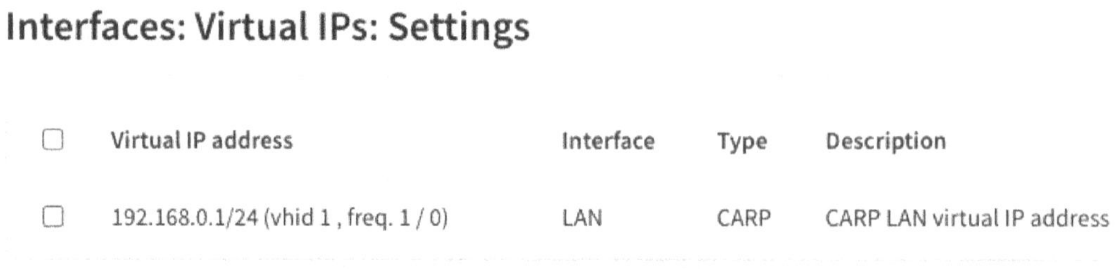

Interfaces: Virtual IPs: Settings

	Virtual IP address	Interface	Type	Description
☐	192.168.0.1/24 (vhid 1 , freq. 1 / 0)	LAN	CARP	CARP LAN virtual IP address

Figure 17.3 – LAN's CARP virtual IP address example

For the WAN configuration, the following settings will be used:

- OPNsense-1 (master) interface address: `10.0.2.252/24`

- OPNsense-2 (master) interface address: `10.0.2.253/24`

- CARP virtual IP address: `10.0.2.1/24`

2. To check the configured virtual IPs, go to **Interfaces | Virtual IPs | Status**:

Interfaces: Virtual IPs: Status

Temporarily Disable CARP	Enter Persistent CARP Maintenance Mode

CARP Interface	Virtual IP	Status
LAN@1	192.168.0.1	▶ MASTER
WAN_A@2	10.0.2.1	▶ MASTER
Current CARP demotion level		0

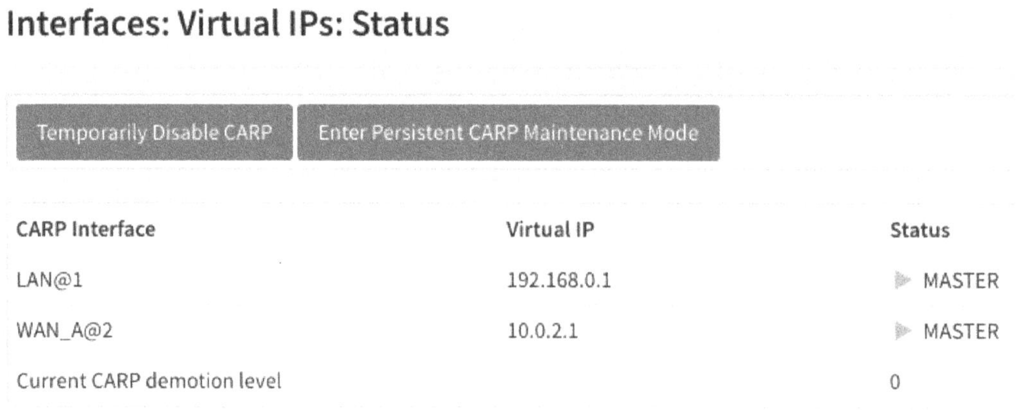

Figure 17.4 – Configured CARP virtual IPs

Notice that the **Status** column for both virtual IP addresses is **MASTER**. You can also check that the *VHID group* is represented after the @ character – for example, **LAN@1**, which means VHID 1.

3. Next, you must add the proper firewall rules to set up the high availability configuration. For the OPNsense (master and backup nodes) firewalls, add the following rules to the HA network interface:

Firewall: Rules: HA

		Protocol	Source	Port	Destination	Port	Gateway	Schedule	Description ❓
☐									
☐	▶ → ⚡ ❶	IPv4 TCP	HA net	*	HA net	443 (HTTPS)	*	*	XMLRPC sync allowed - WebUI
☐	▶ → ⚡ ❶	IPv4 PFSYNC	HA net	*	HA net	*	*	*	pfsync traffic allowed
☐	▶ → ⚡ ❶	IPv4 ICMP	HA net	*	HA net	*	*	*	ICMP allowed - testing purposes

Figure 17.5 – Adding firewall rules

The preceding screenshot shows the following rules that were added while using the *HA network interface IP address* (**HA net**) as **Source** and the **Destination** address:

- The first rule allows traffic in the webGUI port (**433 (HTTPS)**). It is required for XMLRPC to sync the firewall configurations from the master node to the backup node.

- The second rule will allow the firewall states to be synced between the nodes cluster using the *pfsync* protocol.

- The third rule was created just for testing purposes – to run a ping command (*ICMP protocol*) and confirm the communication between the nodes using the HA interface, for example.

4. To confirm that the *HA network interface* is working properly, try running a `ping` command from the *master node*. You can use the webGUI by going to **Interfaces | Diagnostic | Ping** and try to reach the *backup node IP address* or use the CLI to run the `ping` command:

```
# ping -c '3' '10.254.254.2'
PING 10.254.254.2 (10.254.254.2): 56 data bytes
64 bytes from 10.254.254.2: icmp_seq=0 ttl=64 time=0.706 ms
64 bytes from 10.254.254.2: icmp_seq=1 ttl=64 time=0.794 ms
64 bytes from 10.254.254.2: icmp_seq=2 ttl=64 time=0.690 ms
```

```
--- 10.254.254.2 ping statistics ---
3 packets transmitted, 3 packets received, 0.0% packet
loss
round-trip min/avg/max/stddev = 0.690/0.730/0.794/0.046
ms
```

To ensure that the outgoing WAN traffic keeps the same state as the source IP if the master node fails, we need to create an outbound NAT rule. Otherwise, the connections states will use the WAN interface IP address of the active node (the master, in this example), and if this master fails, the backup node won't maintain the same states once the source IP comes from the master node. To solve this issue, the CARP virtual IP must be used as the source IP.

> **Important Note**
>
> Before creating the outbound NAT rule, ensure that NAT outbound mode has been configured as **Hybrid outbound NAT rule generation** on the **Firewall | NAT | Outbound** page.

To address this issue, we need to create an outbound NAT rule:

Manual rules

		Source		Destination	NAT	NAT	Static	
Interface	Source	Port	Destination	Port	Address	Port	Port	Description
☐ ▷ WAN_A	LAN net	*	*	*	10.0.2.1	*	NO	WAN outbound NAT rule

▷ Enabled rule

▶ Disabled rule

Figure 17.6 – NAT outbound rule example

This rule was created by specifying a **Source LAN net** and the *translation address* set to the *WAN CARP virtual IP address*.

Great! We have the HA interface and the basic rules configured.

Now, we have to configure the HA settings to sync the configuration and the firewall states table.

Go to **System | High Availability | Settings** and set as it follows:

- **Synchronize States**: Check this option.

- **Disable preempt**: Leave unchecked.

- **Synchronize Interface**: Select the HA interface.

- **Synchronize Peer IP**: Specify the OPNsense-2 HA interface IP address – for example, `10.254.254.2`.

The next section of this page configures *XMLRPC OPNsense's configurations sync*. Set it as follows:

- **Synchronize Config to IP**: Set it to the *OPNsense-2 HA interface IP address*.

- **Remote System Username**: Set the webGUI username: `root`.

- **Remote System Password**: Set the webGUI username's password.

On the same page, there are settings we must synchronize to the backup node. You can select all the available options to sync all the settings from the master node to the backup node. Click the **Save** button to save all the configurations.

Go to **System | High Availability | Status** to perform the configuration sync and click the **Synchronize config to backup** button:

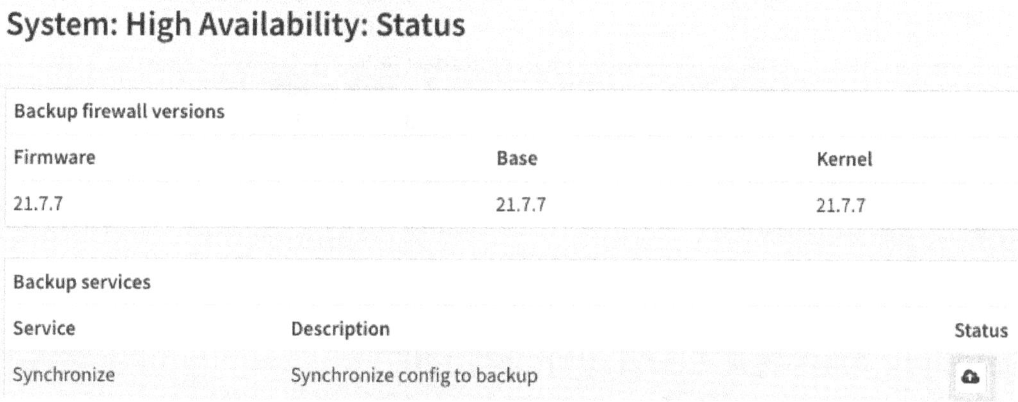

System: High Availability: Status

Backup firewall versions		
Firmware	Base	Kernel
21.7.7	21.7.7	21.7.7

Backup services		
Service	Description	Status
Synchronize	Synchronize config to backup	☁

Figure 17.7 – Synchronize config to backup

After the configuration sync has been completed, you can check the OPNsense-2 (backup) configurations to compare them to OPNsense-1's (master):

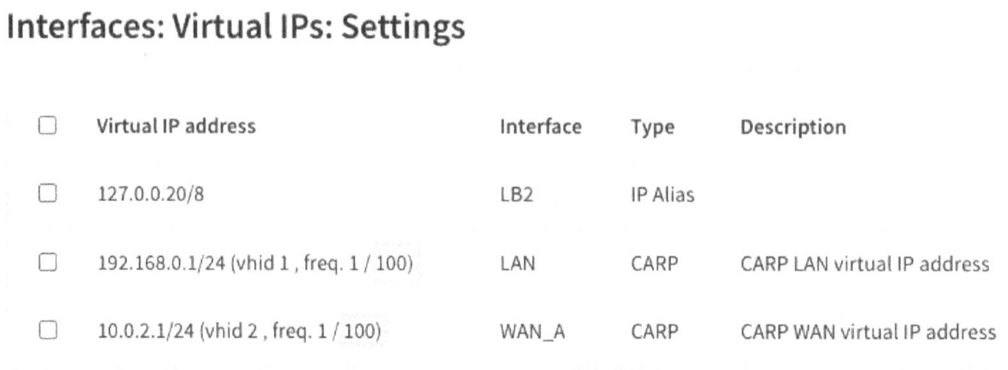

Interfaces: Virtual IPs: Settings

	Virtual IP address	Interface	Type	Description
☐	127.0.0.20/8	LB2	IP Alias	
☐	192.168.0.1/24 (vhid 1 , freq. 1 / 100)	LAN	CARP	CARP LAN virtual IP address
☐	10.0.2.1/24 (vhid 2 , freq. 1 / 100)	WAN_A	CARP	CARP WAN virtual IP address

Figure 17.8 – Virtual IPs created after syncing on the backup node

Notice that the Virtual IPs were created after syncing with *VHID group 100* on OPNsense-2 (backup node).

> **Important Note**
>
> OPNsense requires you to manually synchronize the master to the backup node settings. Some firewalls solutions (as OPNsense used to) automatically sync the settings after any change has been made to the master. Requiring human intervention to sync the changes can prevent some accidental changes from replicating to the backup node, which could lead to a possible crash or unavailability of the cluster. I recommend that you always check the changes on the master node in terms of side effects before replicating to the backup node.

Finally, we can proceed with the high availability testing. We'll learn how to do that in the next section.

Testing the HA configuration

Now that we have both OPNsense firewalls up and running and configured with high availability, let's do some testing!

First, check if the virtual IPs are working accordingly on both OPNsense firewalls, as shown in the following screenshot:

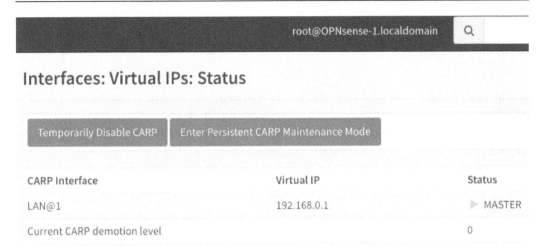

Figure 17.9 – OPNsense-1 (master) running the CARP virtual IP

On both firewalls, you can check the virtual IP's status in the webGUI. Go to **Interfaces |
Virtual IPs | Status**:

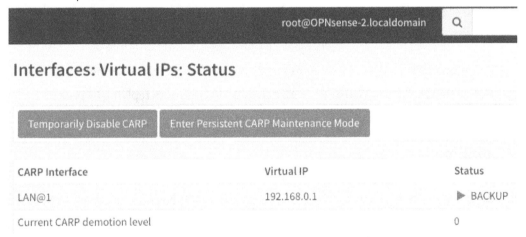

Figure 17.10 – OPNsense-2 (backup) running the CARP virtual IP

If you want to use a DHCP server on the LAN interface, you will need to make some
configuration adjustments, as follows:

- Go to **Services | DHCPv4 | LAN** and set the following settings on the *master node*:

  - Set both the **DNS servers** and **Gateway** fields to the CARP LAN virtual IP; that is,
    `192.168.0.1`.

Remember that every time you change the configuration on the master node, you must
sync to secondary node.

> **Important Note**
> You can skip the DNS server configuration if you want to use external DNS servers.

To avoid both firewalls sending DHCP offers to the network, we must set **Failover peer IP** to the *backup node's* LAN interface address of 192.168.0.3. Once the configurations have been synced, OPNsense automatically fills the **Failover peer IP** field of the backup node with the master node's LAN interface address; that is, 192.168.0.2.

Once you've checked that the CARP virtual IPs are working as expected, you can go to the host that's connected to the LAN and run a ping command. Here, OPNsense automatically fills the **Failover peer IP** field of the backup node with the master node's LAN interface address.

Now, let's perform a functional test. We will run a ping command from the host connected to the OPNsense cluster's LAN and turn off the master node to see what happens with the backup node:

1. Go to the host (Linux) and run the following command:

    ```
    ping 192.168.0.1
    ```

2. Power off OPNsense-1 (master node) or disconnect it from the LAN network.

3. Notice that the CARP virtual IP (192.168.0.1) continues replying to packets to the host.

4. Go to **Interfaces | Virtual IPs | Status** for OPNsense-2 and notice that it became the master node.

5. Take OPNsense-1 back online and wait for the boot process to finish.

6. Repeat *Step 4* and check that the CARP virtual IP on OPNsense-2 changed back to the **backup** status.

Well done – high availability is working!

Caveats

To get a high availability cluster working correctly, you must use an identical hardware (or virtual machine) configuration.

In the webGUI, on the **Interfaces | Virtual IPs | Status** page, it is possible to simulate node failure using both the **Temporarily Disable CARP** and **Enter Persistent CARP Maintenance Mode** buttons. The difference between them is that the former doesn't survive a firewall reboot.

If the primary node fails and you want to keep the secondary as the master node, even after the primary is back online, you can check the **Disable preempt** box on the **System | High Availability | Settings** page.

Marking this option will also make the cluster CARP virtual IPs act like a group. So, if a CARP virtual IP is demoted to backup, all other CARP virtual IPs will also assume the backup status.

To prevent a firewall rule configuration from being synced from the master node to the backup nodes, you can mark the **No XMLRPC Sync** option inside the firewall rule editing page (**Firewall | Rules |** `<Interface Name>`). It is also possible to prevent the firewall states from syncing by checking the **State Type | NO pfsync** box inside the **Advanced** options of each rule.

Troubleshooting

High availability clusters can sometimes present some troubleshooting challenges as a complex configuration.

Let's look at some examples of issues where a high availability configuration can't work correctly.

Same virtual IP status on both cluster nodes

Sometimes, you may find one or more CARP virtual IPs set as master nodes on both firewalls. This can happen when the secondary node can't find the primary. Network connectivity issues often cause this. So, if you are using a physical hardware platform, debugging the switch logs is a good idea. Otherwise, start looking at the virtualization host logs if you are running it on a virtualization platform.

Configuration sync problems

A common issue is when you set the network interfaces differently on the primary and secondary nodes. This kind of misconfiguration can lead to the HA configuration not working correctly. Before configuring the other options, configure the network interfaces on the primary node and export its XML config file to the secondary. The network configuration is not synced between the nodes, so doing so can prevent a network configuration from happening. Other misconfigurations can occur when the **System | High Availability | Settings** page's options are not set correctly.

Summary

In this chapter, you learned how OPNsense implements high availability to guarantee enterprise-grade availability as a firewall solution. We explored how the CARP protocol provides a cluster virtual IP operation and that when combined with pfsync, XMLRPC brings a complete stack for a firewall high availability cluster. Now, you can build solid and highly available firewall clusters using OPNsense that can survive a hardware failure and keep the networks safe and protected. In the next chapter, we will learn how to protect websites and web applications using OPNsense.

18
Website Protection with OPNsense

With the NGINX plugin, OPNsense becomes a full-featured, solid **Web Application Firewall** (**WAF**). It can help you to protect your network and your web servers with the addition of the NGINX plugin. By the end of this chapter, you will be able to use OPNsense as a reverse proxy, WAF, and web server load balancer.

In this chapter, we will explore the following topics:

- Publishing websites to the world
- About the NGINX plugin
- Installing and configuring
- Adding WAF rules
- Troubleshooting

Technical requirements

This chapter requires a clear understanding of how a web server works. Complete knowledge of DNS HTTP(S) and TLS protocols is also essential.

Publishing websites to the world

Nowadays, our modern internet is, essentially, based on web applications. It is rare to see a modern app that is built to run installed on a computer, and even the smartphone-based apps are, for the most part, a responsive version of the website. While managing an OPNsense firewall, you will probably have to deal with websites and web applications. As a modern next-generation firewall solution, OPNsense can provide enough features to keep a website safely online, protecting it against threats. In the following sections, we will explore the NGINX plugin, which does an outstanding job while publishing web server applications and websites protected by OPNsense.

About the NGINX plugin

In the old days, a firewall was just a packet filtering system, and to publish a web server service to the internet, simply adding a NAT rule was enough. With the evolution of the internet, more sophisticated web applications were raised, but the attacks followed at the same pace, becoming more harmful. Good firewall solutions added features such as IDS and IPS to increase the protection level of applications and the users behind them. Still, web applications require more detailed filters to protect them against the threats of bad actors than packet filtering and a network IPS.

A solution to help web servers and applications become better protected emerged: HTTP reverse proxies. Similar to a web proxy, the reverse proxy stands between the users and the web servers, but in reverse, that is, the users are outside the local network and the web servers are inside.

The following diagram illustrates how a reverse proxy works:

Figure 18.1 – A reverse proxy topology example

Some well-known open source reverse proxies are Apache HTTP Daemon, NGINX, and HAProxy. Deciso, the company behind OPNsense, recently launched an Apache-based plugin that is available for OPNsense Business subscribers. So, supposing you started reading this chapter as a reference to set up a reverse proxy or even a WAF, I suggest you take some time to research the available options for OPNsense so that you can choose the better choice for your project needs.

This chapter will focus on NGINX, as OPNsense has a robust plugin based on it. This plugin is freely available, so you can use a regular OPNsense installation to install and configure it.

NGINX

NGINX (pronounced engine X) is a powerful web server used widely by big websites on the internet. It can also be deployed as a reverse proxy and load balancer. Other than *HTTP*, NGINX also supports the *SMTP*, *IMAP*, and *POP3* email protocols for email proxying. The project claims to be faster than its rival, Apache.

> **Note**
> At the time of writing, the SMTP, IMAP, and POP3 protocols are currently not supported by the plugin.

In OPNsense, the NGINX plugin was created by Mr. Franz Fabian, an active project contributor – and, proudly, one of the technical revisors of this book!

As a complete plugin with many features, NGINX could be complex to configure. Still, a complete plugin can be used as a reverse proxy, and it eases a lot of the configuration process compared to doing everything using the CLI. The protection level it delivers is pretty decent, and you can do almost all of the usual action only using the webGUI, except if you are an NGINX hacker!

Plugin concepts

Before moving on to the installation and configuration, we need to learn about some essential concepts that the NGINX plugin uses.

Upstream server

Usually, this is the web server host. The upstream server is the host that will serve the website or application to the world or the network.

Upstream

The upstream could be a cluster of servers (or upstream servers). This concept could be used in a similar manner to a load balancer, in which multiple servers (and ports) can be allocated to a single upstream (with a minimum of a single node).

Location

The location is a configuration that maps an HTTP resource path to some configuration. This could be a load balancer target, but it can also be a local directory to serve or force HTTPS.

HTTP server

An HTTP server represents a server configuration block that defines the hostnames, TLS certificate, and more. It will specify the listening port (HTTP/HTTPS) to serve requests, usually to the internet.

These are the basic concepts we will need to understand to configure the NGINX plugin in OPNsense. As I mentioned earlier, the plugin has complete features, and talking about them would require maybe two or three dedicated chapters.

In this chapter, we will focus on how to do the basic configuration of the NGINX plugin with WAF capabilities. Now that we have the basics of the plugin covered, let's move on to the practical part, where we will explore how to install and configure it.

Installing and configuring the NGINX plugin

To install the NGINX plugin, follow these steps:

1. Go to the **System | Firmware | Plugins** tab:

System: Firmware

Figure 18.2 – Adding the NGINX plugin

2. To add the plugin, find it and click on the + button.

    Before enabling the NGINX service, we need to adjust the webGUI *configuration to avoid any port conflict between NGINX and the Lighttpd* (the process that serves the webGUI).

3. To change the webGUI connection port, go to **System | Settings | Administration**.

4. Change the **TCP port** option from 443 to another port such as 8443, *for example*.

5. Check the **HTTP Redirect Disable web GUI redirect rule** option. This will free *TCP port 80 (HTTP)*:

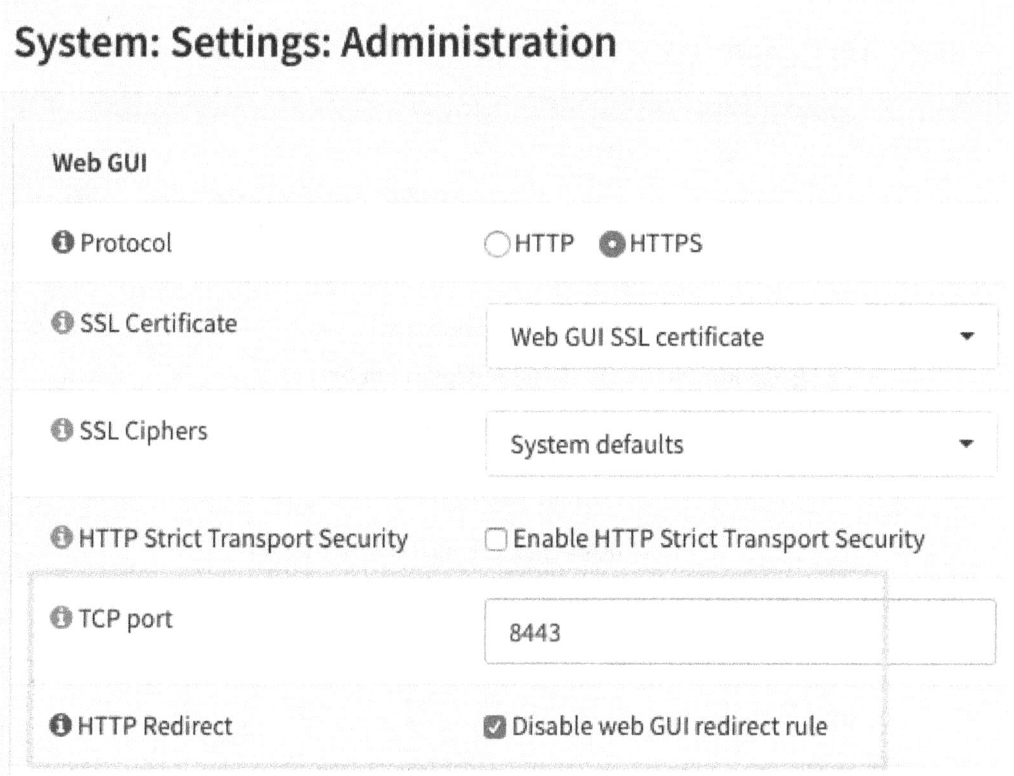

System: Settings: Administration

Web GUI

ⓘ Protocol   ◯ HTTP   ◉ HTTPS

ⓘ SSL Certificate | Web GUI SSL certificate   ▼ |

ⓘ SSL Ciphers | System defaults   ▼ |

ⓘ HTTP Strict Transport Security   ☐ Enable HTTP Strict Transport Security

ⓘ TCP port | 8443 |

ⓘ HTTP Redirect   ☑ Disable web GUI redirect rule

Figure 18.3 – The webGUI configurations details

> **Note**
> The following steps will change the ports of the webGUI. Ensure that you have the firewall rules to allow access from the webGUI to the new TCP port configuration before applying the configurations.

6. After configuring these options, click on the **Save** button and access the newly configured webGUI port. You can also access the newly configured webGUI port by changing it on the URL.

7. After being installed, the plugin configuration page can be located at **Services | NGINX | Configuration**.

8. Let's start the configuration by creating an upstream server entry. In the **Upstream Server** page, we will set the web server that will serve requests and has the web page/application hosted in it:

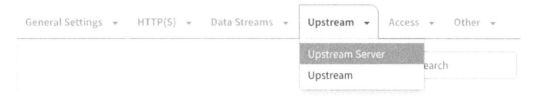

Services: Nginx: Configuration
General Settings
HTTP(S)
Data Streams
Upstream
Access
Other
Upstream Server
Upstream
earch

Figure 18.4 – The Upstream Server submenu

9. Click on the **Upstream** menu and then the **Upstream Server** submenu. To add a new server, click on the + button:

+

Figure 18.5 – The add button detail

10. On the **Edit Upstream** server page, the following fields are available:

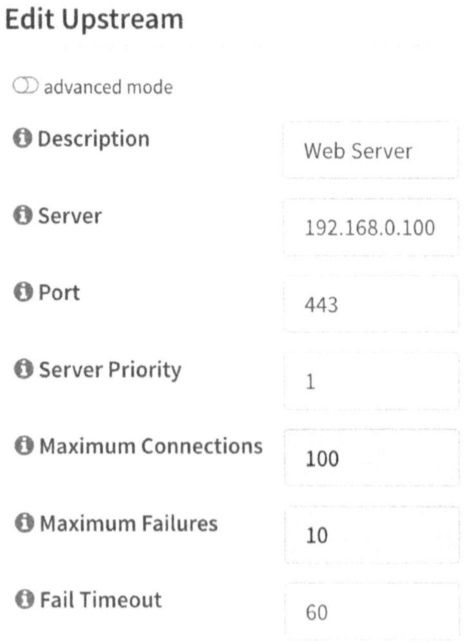
Edit Upstream
advanced mode
Description
Web Server
Server
192.168.0.100
Port
443
Server Priority
1
Maximum Connections
100
Maximum Failures
10
Fail Timeout
60

Figure 18.6 – The upstream server edit page

Let's look at each field in a little more detail:

- **Description**: Here, you can enter the description of the web server entry.

- **Server**: Here, you can enter the web server's IP address or internal hostname (note that NGINX must be able to resolve it).

- **Port**: Here, you can enter the web server's port, usually 80 (HTTP) or 443 (HTTPS).

- **Server Priority**: This defines the priority for this server entry. While using a backend upstream with multiple web servers, this priority will determine which servers will be receiving requests preferentially.

- **Maximum Connections**: This defines the limit of simultaneous requests that this server will handle.

- **Maximum Failures**: This refers to the maximum number of connection failures before considering this server offline for the backend.

- **Fail Timeout**: This refers to the maximum amount of time, in *seconds*, that the backend will wait for the server to reply before considering it offline.

Click on the **Save** button to finish the configuration. You should see it listed as follows:

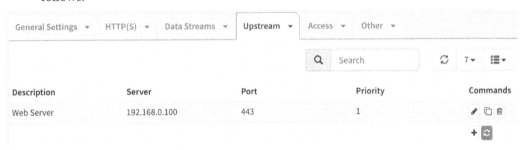

Figure 18.7 – The configured upstream server

11. After configuring the upstream server, we need to link it to an upstream. Click on the **Upstream** *menu* and then the **Upstream** *submenu* to create a new upstream by clicking on the + button.

On the **Edit Upstream** backend page, the following fields are available:

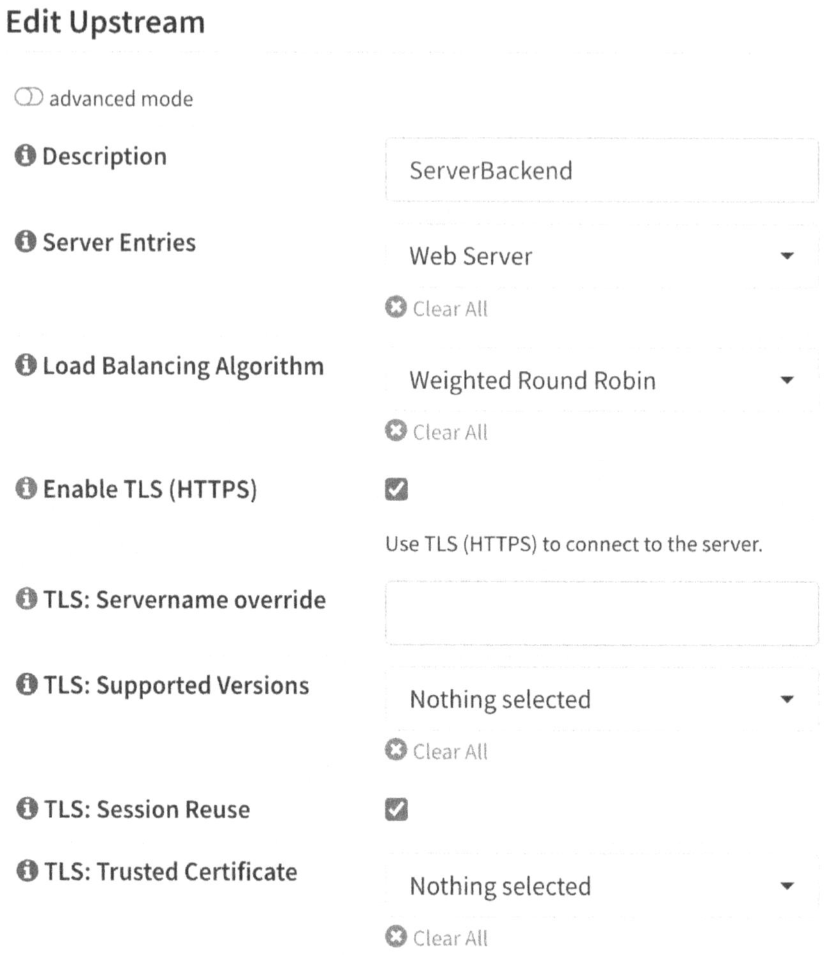

Edit Upstream

◯ advanced mode

ⓘ Description

> ServerBackend

ⓘ Server Entries

> Web Server   ▾
>
> ⊗ Clear All

ⓘ Load Balancing Algorithm

> Weighted Round Robin   ▾
>
> ⊗ Clear All

ⓘ Enable TLS (HTTPS)

> ☑
>
> Use TLS (HTTPS) to connect to the server.

ⓘ TLS: Servername override

> [   ]

ⓘ TLS: Supported Versions

> Nothing selected   ▾
>
> ⊗ Clear All

ⓘ TLS: Session Reuse

> ☑

ⓘ TLS: Trusted Certificate

> Nothing selected   ▾
>
> ⊗ Clear All

Figure 18.8 – Editing the upstream backend

The following options are available in the **Edit Upstream** dialog:

- **Description**: Enter a description for this backend.

- **Server Entries**: Select the previously added upstream server.

- **Load Balancing Algorithm**: Select the **Weighted Round Robin** option.

- **Enable TLS (HTTPS)**: Mark this option to enable HTTPS connection to the upstream server.

- While using a web server with a self-signed certificate, it is necessary to click on the **advanced mode** option to disable the certificate validation:

Edit Upstream

advanced mode

Figure 18.9 – The advanced mode button detail

12. Leave the **TLS: Verify Certificate** option unchecked:

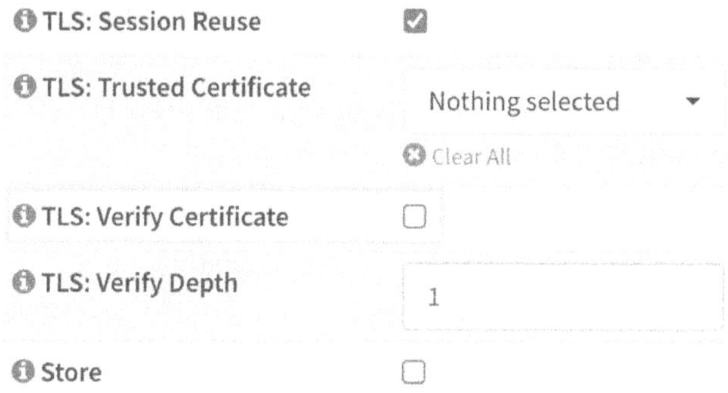

ⓘ TLS: Session Reuse	☑
ⓘ TLS: Trusted Certificate	Nothing selected ▾
	✖ Clear All
ⓘ TLS: Verify Certificate	☐
ⓘ TLS: Verify Depth	1
ⓘ Store	☐

Figure 18.10 – The advanced mode TLS: Verify Certificate option detail

13. To finish adding the new backend, click on the **Save** button.

14. After configuring the backend, we need to add a new location. To add a new location, go to the **HTTP(S)** menu and then the **Location** submenu.

15. Click on the + button to add a new location.

    On the **Edit Location** page, fill in the following fields:

    - **Description**: Enter a description for this location.

    - **URL Pattern**: Fill this with the / (slash character). It will instruct NGINX to set the URL pattern will match to any path published by the web server.

    - **Upstream Servers**: Select the previously created upstream backend.

16. If you want to redirect HTTP requests to an HTTPS port, check the **Force HTTPS** option.

17. Finish the configuration by clicking on the **Save** button. The newly configured location will be listed as follows:

Description	URL Pattern	URL Path Pr...	Match Type	WAF Status	Force HTTPS	Commands
WebServer root	/		none	disabled	0	✏ ⧉ 🗑
						+ ↻

Figure 18.11 – Added location example

18. To add a new HTTP server, go to the **HTTP(S)** menu and then the **HTTP Server** submenu. Click on the + button to add a new HTTP server.

    On the **Edit HTTP Server** page, set the following configuration options:

    - **HTTP Listen Address**: This will set which port will listen to *HTTP* requests. You can leave it with the default values – 80(IPv4) [::]:80 (IPv6).

    - **HTTPS Listen Address**: This will set which port will listen to *HTTPS* encrypted requests. You can leave it with the default values – 443 (IPv4) [::]:443 (IPv6).

    - **Server Name**: This refers to the FQDN hostname for this HTTP server; *for example*, www.example.com.

    - **Locations**: Select the previously created location.

    - **TLS Certificate**: Select the TLS certificate that will be used for HTTPS requests. Usually, it would be necessary to install a valid certificate in OPNsense for this option to be available. For testing purposes, you can select the **Web GUI TLS certificate** option.

    > **Note**
    >
    > To use valid certificates at no cost, you can install the os-acme-client that provides the *Let's Encrypt* backend, which is an open certificate authority created by Internet Security Research Group. Installing the mentioned plugin provides the functionality of creating valid certificates for free.
    >
    > Another approach is to create a self-signed certificate.

19. Finish the configuration by clicking on the **Save** button.

20. Now the NGINX service can be started without any issues. Go back to the **Services: NGINX: Configuration** menu and enable the service:

Figure 18.12 – Enabling the NGINX service

21. Check the **Enable NGINX** option, and click on the **Apply** button.

The following is a screenshot of the NGINX service's running state on the webGUI:

Services: Nginx: Configuration

Figure 18.13 – NGINX service running

We will need to simulate the added www.example.com address by adding it to the DNS resolution for our local machine to test. The simplest way to do that is to add it to our operating system's /etc/hosts file.

The path of this file on *Windows* is c:\windows\system32\drivers\etc\hosts.

On a *Unix-like* OS (including *Mac* and *Linux*) you can find this file at /etc/hosts.

22. Edit the file using a user with administrative privileges, and add the following line:

```
192.168.3.3        www.example.com
```

Use one of your OPNsense-configured IP addresses. In this example, I will use IP 192.168.3.3, which is *my OPNsense VM WAN's IP address.*

23. In *step 5*, we have used IP address 192.168.0.100 as the *upstream server*. If you have a web server in your local network to test, you can change it to its IP address. Otherwise, we can set it to an external website IP address, such as www.opnsense.org.

24. To find the website IP address, navigate to **Interfaces | Diagnostics | DNS Lookup** on the webGUI and type in the website address (www.opnsense.org) to find its IP address.

As I'm writing this chapter, the OPNsense website has an IP address of 178.162.131.118. So, I will go back to *step 5* and change it on the **Upstream Server** page from 192.168.0.100 to 178.162.131.118, to *test using an external web server* instead of a local one.

Every time you need to apply changes on the NGINX plugin, after saving the changes, you must click on the button shown in the following screenshot:

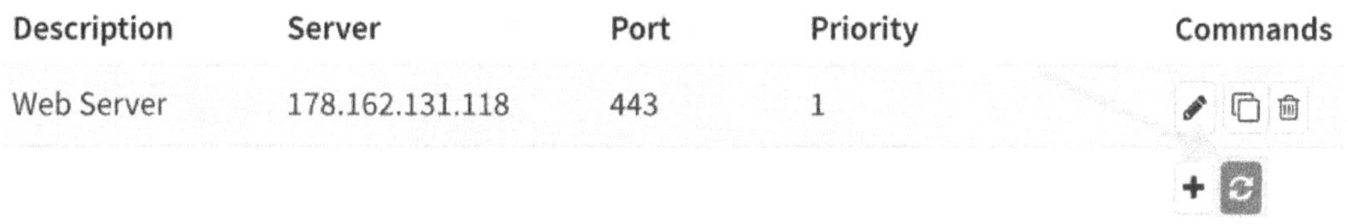

Description	Server	Port	Priority	Commands
Web Server	178.162.131.118	443	1	✎ ⧉ 🗑
				＋ ↻

Figure 18.14 – The NGINX plugin's apply changes button

25. Click on the apply changes button.

26. Open an anonymous browser window and type in the following URL: `www.example.com`. You should see OPNsense's website! Congratulations; you just configured your OPNsense as a reverse proxy!

With this, you have finished the NGINX basic configuration. Now, we can move on to add some website protection to our NGINX plugin's configuration.

Adding WAF rules

The NGINX plugin implements a WAF with the help of the NAXSI (NGINX Anti XSS & SQL Injection) module. This module works with predefined rules that match 99% of known patterns found in website vulnerabilities. The NAXSI module was created and maintained by NBS System, a French security company (ref: `https://www.nbs-system.com/`):

1. To add the NAXSI rules to the NGINX plugin, go to the **HTTP(S)** menu, followed by the **Naxsi WAF Policy** submenu. When the rules haven't been installed, the following button will be visible:

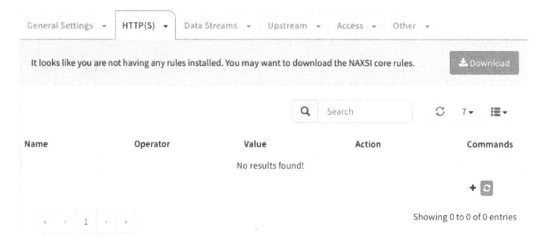

Figure 18.15 – The NAXSI download rules button

2. Click on the **Download** button to install the rules. You will be prompted with a dialog box about the NAXSI rules licensing. You must click on the **Accept and Download** button to agree with the license terms (the GPLv3 license):

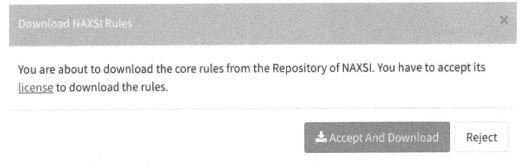

Figure 18.16 – The NAXSI license terms dialog

After it has been downloaded, the rules will be listed as follows:

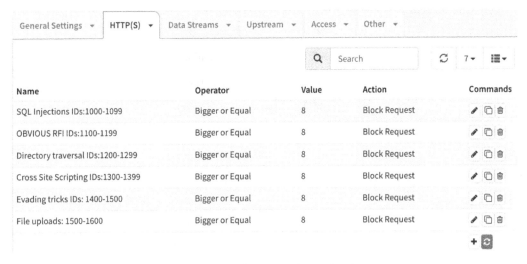

Figure 18.17 – The installed NAXSI rules

3. To enable the NAXSI rules and WAF capabilities, go back to **HTTP(S) | Location** and edit the existing location:

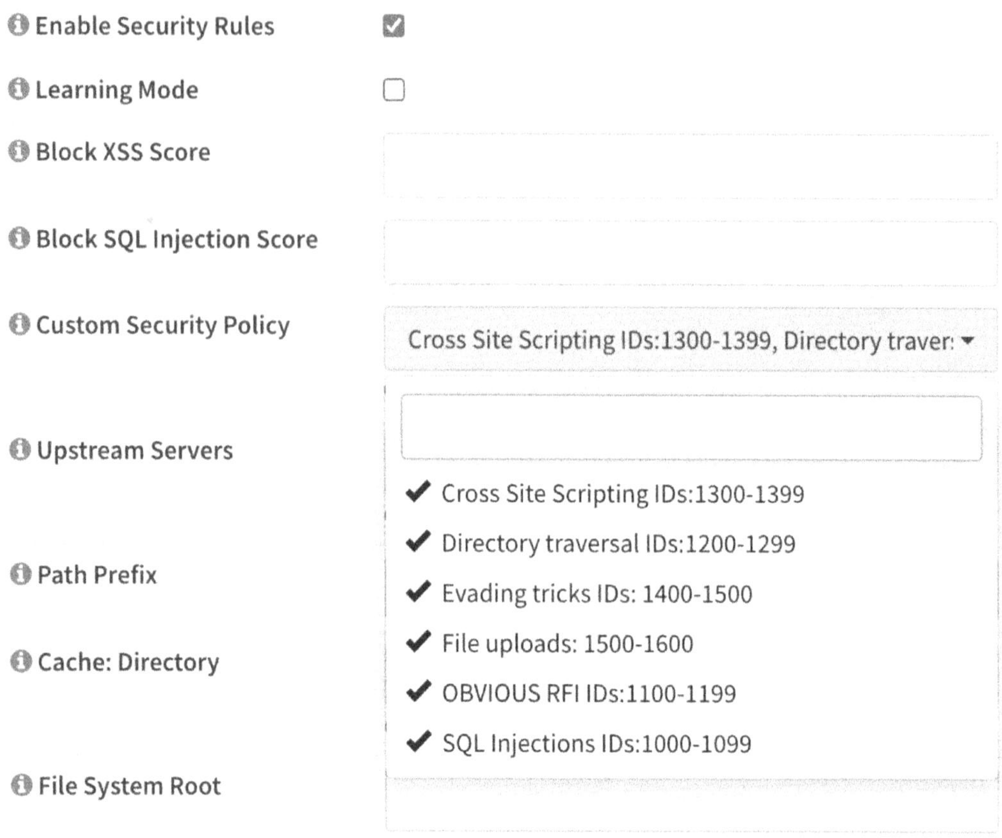

ⓘ Enable Security Rules   ☑

ⓘ Learning Mode   ☐

ⓘ Block XSS Score

ⓘ Block SQL Injection Score

ⓘ Custom Security Policy

Cross Site Scripting IDs:1300-1399, Directory traver: ▼

ⓘ Upstream Servers

✔ Cross Site Scripting IDs:1300-1399

✔ Directory traversal IDs:1200-1299

ⓘ Path Prefix

✔ Evading tricks IDs: 1400-1500

✔ File uploads: 1500-1600

ⓘ Cache: Directory

✔ OBVIOUS RFI IDs:1100-1199

✔ SQL Injections IDs:1000-1099

ⓘ File System Root

Figure 18.18 – Editing the existing location

4. Check the **Enable Security Rules** option, which enables the WAF in the NGINX plugin.

5. The **Learning Mode** option is useful as it allows you to enable the WAF but without blocking. It will just log the requests that should be blocked. In this way, you can analyze the logs and adjust the rules before blocking any requests. For our example, *leave it unchecked*.

6. In the **Custom Security Policy** field, select all the existing NAXSI group rules.

7. Click on **Save**, and then click on the **apply changes** button.

8. Now you can try to simulate an attack, such as a SQL injection, for example, and see the WAF doing its job:

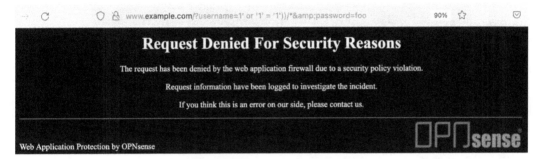

Figure 18.19 – Testing the WAF rules

Here is the tested URL: `http://www.example.com/?username=1'%20or%20 '1'%20=%20'1'))/*&amp;password=foo`.

> **Note**
>
> You can check out some attack examples and how to use them in the complete OWASP guide at `https://owasp.org/www-project-web-security-testing-guide/latest/`.

Congratulations! Now your OPNsense has one more security feature: a WAF!

Next, let's look at practical tools you can use for troubleshooting the NGINX plugin.

Troubleshooting

As a complex system, a reverse proxy or a WAF can lead you to troubleshooting scenarios that require a lot of logs reading along with some web server and application knowledge. Here, we will explore a few tools that might help you to solve a quest.

Testing for configuration issues

Sometimes, even with all of the help and automations that the webGUI plugin frontend has, some configuration issues could appear, making the NGINX service unable to start. To test the NGINX configuration, you can log in to the OPNsense CLI and run the following command:

```
root@OPNsense-1:~ # nginx -t
nginx: the configuration file /usr/local/etc/nginx/nginx.conf
syntax is ok
nginx: configuration file /usr/local/etc/nginx/nginx.conf test
is successful
```

For a more complete testing and configuration output, you can also run `nginx -t`.

Logs reading

In the webGUI, we find the **NGINX | Logs** menu. Inside this page, it is possible to check *every created HTTP server log file* and the NGINX service log file:

- **HTTP Access Logs**: This will show the HTTP requests to an existing HTTP server.
- **HTTP Error Logs**: In this menu, you will find the errors related to the requests. *The NAXSI blocks can also be checked in this log file.*
- **Global Error Log**: This is the service log file. If something goes wrong with the NGINX daemon, you will find information about it in this log file.

As I mentioned earlier, a WAF or a reverse proxy is a complex system. It could present many issues depending on the complexity of the configuration, the number of hosted web applications, integrations, and more. Eventually, troubleshooting could demand web server logs, the web developers that are involved, DNS checks, firewall rules, and IPS checks.

> **Note**
> Remember to add a firewall rule to allow incoming traffic for NGINX's configured ports.

Summary

At this point, you can affirm that OPNsense has a robust security stack and can even act as a WAF in the frontline of a cloud or network infrastructure. This chapter taught you how to install and configure the NGINX plugin, and you also learned how to enable its WAF features and protect the infrastructure of web servers. In the following chapter, we will jump into the CLI world and see how it can extend your tools and knowledge by executing commands in OPNsense.

19
Command-Line Interface

This chapter will explore the shell **command-line interface (CLI)**. We will explore some of the most relevant FreeBSD commands to manage the system, networking, and firewalling. We will also learn how to customize some parts of the system and use commands to improve information extraction from logs. At the end of this chapter, you will be capable of running commands on the OPNsense CLI for diagnostics and troubleshooting purposes.

We will explore the following topics:

- Directory structure
- Managing the configd backend daemon
- Useful system commands
- Advanced customization
- Filtering log files

Technical requirements

To follow this chapter, you need to have basic Unix-like shell and filesystem knowledge. A running OPNsense firewall might help try the commands. All commands presented in this chapter are executed as the root user, so pay attention before pressing the *Enter* key after each command.

Note

Never run the presented commands in production environments without supervision if you don't know exactly what you are doing

---

Before starting to run CLI commands, let´s first learn more about the OPNsense filesystem.

Directory structure

If you are not familiar with the *FreeBSD* filesystem layout, I recommend you read this manual page: `https://www.freebsd.org/cgi/man.cgi?query=hier&sektion=7`. You may find some stimulating directory descriptions on this page, so it is worth reading.

On OPNsense, the most relevant directories reside on the `/usr/local` path:

```
root@bluebox:/usr/local # ls
Makefile            libdata         openssl         share
bin             etc         libexec         opnsense
var
cloudfence      include     man             sbin
wizard
datastore       lib                         www
```

The subdirectories we will explore inside this path (`/usr/local`) are as follows:

- `/usr/local/etc`: This directory contains the configuration files used by OPNsense daemons. Most configuration files generated using the webGUI will reside inside this directory. Occasionally, you might want to check some service's configuration by looking at its configuration file(s).

- `/usr/local/www`: The legacy webGUI **Hypertext Preprocessor** (**PHP**) pages are stored in this directory.

- `/usr/local/sbin`: OPNsense's binaries, such as the command to sound the boot beep (`opnsense-beep`) and `opnsense-shell`, are stored inside this directory, for example.

- `/usr/local/opnsense`: Inside this directory, we will find the OPNsense **Model-View-Controller** (**MVC**) framework, which the webGUI relies on to manage the whole system, for example. The MVC is inside the `mvc` subdirectory. Other relevant subdirectories inside `/usr/local/opnsense` are as follows:

  - `scripts`: The auxiliary scripts called by the plugins and other OPNSense framework components are stored inside this directory.

  - `service/conf/actions.d`: Inside this subdirectory are the configuration files used by the `configctl` command, which manages daemons used by OPNsense. We will explore `configctl` in a dedicated topic in this chapter.

Other directories on the filesystem's / root that we will mention here are as follows:

- `/boot`: This is the directory that contains the FreeBSD files used in the bootstrap. You should rarely need to change a file inside it, but on some special occasions in advanced troubleshooting or development, you might need to interact with files inside this directory.

- `/conf`: This is where the OPNsense configuration file resides. The `config.xml` file is the webGUI's generated **Extensible Markup Language** (**XML**) configuration file.

- `/var`: Some daemons store variable information inside this directory. If you need to check some daemon's temporary information, you will find it inside the `/var` directory.

- `/var/log`: This directory is often used while troubleshooting or in a debugging process where the system's log files are stored inside it; I venture to say that it is the most accessed directory in the CLI for this reason.

- `/tmp`: Last but not least, the system's temporary directory. As the name suggests, this directory is used by the daemon and the operating system to store most temporary information. A specific example in OPNsense is the generated firewall rules stored in a temporary file called `rules.debug`.

Now that we have a basic map to explore the CLI world, let's move on with OPNsense's backend command.

> **Note**
>
> As OPNsense supports RAM disk, and both /var and /tmp directories can be configured using it, all content is recreated after a reboot in these directories.

Managing the backend daemons

The OPNsense architecture is composed of a *frontend* and a *backend*. In the next chapter, we will explore OPNsense's architecture to learn how to use its **Application Programming Interface** (**API**). For now, let's say that the *backend* is the part that interacts with the system daemons, and eventually, you may want to change some daemon states by stopping, starting, or restarting using it. The service in charge of interacting with daemon states and also writing its configuration file is configd. To manage the configd service with the CLI, we can use the configctl command:

```
root@bluebox:/tmp # configctl -h
usage: configd_ctl.py [-h] [-m] [-e] [-d] [-q] [-t T] command
[command ...]

positional arguments:
  command       command(s) to execute

optional arguments:
  -h, --help  show this help message and exit
  -m          execute multiple arguments at once
  -e          use as event handler, execute command on
receiving input
  -d          detach the execution of the command and return
immediately
  -q          run quietly by muting standard output
  -t T        threshold between events, wait this interval
before executing commands,
              combine input into single events
```

> **Note**
>
> The configctl -h output will show configd_ctl.py as the command to run, but you can ignore it and remain using just configctl instead.

The first thing you might check before using the `configctl` command is whether the `configd` service is up and running:

```
root@bluebox:/tmp # service configd status
configd is running as pid 143.
```

> **Note**
>
> Unix-like systems use a **process identification number** (**PID**). This number is not fixed, so the PID of your running OPNsense `configd` process will probably differ from the preceding example.

OK! It is running, but if for some reason the `configd` service stops, you can restart it by running the following:

```
root@bluebox:/tmp # service configd restart
Stopping configd...done
Starting configd.
```

Now, let's list the available options while using `configctl`:

```
root@bluebox:/tmp # configctl configd actions
service reload delay [  ]
service reload all [  ]
webgui restart [ Restart web GUI service ]
unbound dumpcache [  ]
unbound dumpinfra [  ]
unbound stats [  ]
unbound listinsecure [  ]
unbound listlocalzones [  ]
unbound listlocaldata [  ]
...
```

The list may vary depending on the OPNsense installed plugins. Each plugin may install a configuration file inside the `/usr/local/opnsense/service/conf/actions.d/` directory, as we explored in the last section. These files are invoked by the `configctl` command to execute the listed actions shown by the preceding command.

Some examples of the `configctl` usage include the following:

- `configctl webgui restart`: Will restart the webGUI process
- `configctl interface list arp`: Will list the **Address Resolution Protocol (ARP)** table
- `configctl captiveportal list_clients`: Will list the current captive portal's connected users

There are dozens of useful commands you can try using `configctl`. Take some time and test them on your own!

In the following section, we will explore other useful system commands through the CLI.

Useful system commands

As a Unix-like system, OPNsense has a lot of useful commands. In the following, we will explore some of these commands.

Let's dive into a practical scenario.

Assume that you lost access to the webGUI and want to find which process is using **Transmission Control Protocol (TCP)** port `443` on OPNsense. We can find running processes with listening ports on FreeBSD using the `sockstat` command:

```
root@bluebox:/usr/local/etc # sockstat -41 -p 443
USER      COMMAND    PID    FD PROTO   LOCAL ADDRESS
FOREIGN ADDRESS
root      nginx    7065   5  tcp4    *:443                    *:*
```

In this example, the `nginx` process is using port `443` (just the IPv4 address specified in the `-41` parameter), which caused you to lose webGUI access due to some configuration issue.

> **Note**
>
> We are assuming that you followed the previous chapter, so we might see the `nginx` process running on port `443` instead of `lighttpd` (OPNsense's default configuration).

You might want to stop the `nginx` service and start the webGUI again to solve this problem.

To stop the running process, we can use the `service` command:

```
root@bluebox:/usr/local/etc #  service  nginx stop
```

Let's now restart the webGUI. As you already learned, you can use `configctl` to interact with OPNsense daemons, so what we need to do is find the right command:

```
root@bluebox:/tmp # configctl configd actions | grep -i webgui
webgui restart [ Restart web GUI service ]
```

Bingo! Now, Let's get the webGUI back up and running:

```
configctl webgui restart
```

Confirming that it is back, we see the following:

```
root@bluebox:/tmp # sockstat -41 -p 443
USER      COMMAND      PID    FD PROTO  LOCAL ADDRESS
FOREIGN ADDRESS
root      lighttpd   87811 5  tcp4    *:443                    *:*
```

Well done! With this, the webGUI will be back online. To solve the port conflict issue, changing the port number on the webGUI or nginx configurations will be necessary. As it was just an example, we will not detail this process here.

If I listed all the examples using my personal experience of possible commands, we would probably need an extra book!

But, let's list some useful commands that you can count on with the OPNsense CLI:

- **Network commands**:

  - List routing table: `netstat -nr` or `configctl interface routes list`
  - List interfaces: `ifconfig -a` or `configctl interface list ifconfig`

> **Note**
>
> Some `configctl` commands output in JSON format, which can be very confusing to read in the CLI. To help format the JSON output, you can count on the `jq` utility. You can install it using `pkg install jq`.
>
> An example of formatting a JSON command output using `jq`: `configctl interface list ifconfig | jq`.

  - Test connectivity: `ping <IP address>`

- Trace a route path: `traceroute <IP address>`
- DNS diagnostics: `drill <domain>`
- Traffic usage per interface: `systat -ifstat -match <network interface name>` (*for example:* `igb0`)

- **Firewall commands**:

  - List firewalls states: `pftop` or `pfctl -ss`
  - List active rules: `pfctl -sr` or `cat /tmp/rules.debug`
  - List active NAT rules: `pfctl -sn` or `cat /tmp/rules.debug | grep nat`
  - Show statistics: `pftctl -si`
  - Show active aliases (tables): `pfctl -s Tables`

- **System commands**:

  - Show system usage statics: `top -aSH`
  - Get system kernel and hardware details: `sysctl -a`
  - Managing system packages: `pkg <info|install|del|add>`

As you can see, there are a lot of commands that help us to manage our OPNsense on the CLI.

It might sometimes be necessary to customize configuration on OPNsense that you can't do using the webGUI. Still, you must take care while customizing the configuration and avoid doing that outside the webGUI where possible. *Proceed with maximum caution!*

Advanced customization

The OPNsense webGUI is a powerful configuration tool, and you rarely need to change or customize the OPNsense configuration while using the CLI. Still, it can happen, and it is essential to know which tools we can count on for doing this task.

The `config.xml` file is the webGUI's generated XML configuration file, it is where the OPNsense configuration file resides.

Customizing the XML configuration file

All configurations generated by webGUI through the OPNsense framework are saved in `/conf/config.xml`.

> **Note**
>
> Before modifying the `config.xml` file, *always* make a backup copy! Editing this file can crash your OPNsense installation, so take caution!

The following is an example where the manual editing of the `config.xml` file could help.

This is an example showing the substitution of OPNsense hardware. Let's suppose that you need to replace the OPNsense hardware, and the new one uses a different network card, which means changing the **Network Interface Card** (**NIC**) name on FreeBSD. If you try to back up the configuration from the old hardware and just restore it on the new one, it might not get the network up and running without a bit of adjustment. Let's see how to change it.

This example shows changing from `emX` to `igbX` network cards:

```
root@bluebox:/conf # grep -E "<if>em[0-9]</if>" config.xml
        <if>em1</if>
        <if>em2</if>
        <if>em0</if>
```

The preceding command lists the configured network devices inside the `config.xml` file.

To change it, edit the `config.xml` file with the `vi` editor (making a copy before):

```
cp /conf/config.xml /tmp/config.xml
vi /tmp/config.xml
```

Inside the file, look for the XML block, for example `<if>em1</if>`, and change it to the new device name.

We can change it from this:

```
<interfaces>
    <wan>
        <if>em1</if>
```

To this:

```
<interfaces>
    <wan>
        <if>igb1</if>
```

Repeat the preceding steps for each network interface. Now, you can restore the new config.xml file to the new hardware to have the network up and running when it is done.

> **Note**
> Always check the XML file syntax before trying to upload it to the OPNsense webGUI to restore it. Several utilities might help to do that. An example is xmllint, which is available on OPNsense.

The root password reset

If you lose or forget the root password, you must reset it to get OPNsense access back. To do that, you need to access the OPNsense console and reboot it. During the boot process, the following steps must be taken to reset the root password:

1. Select option 2: **Boot Single User**:

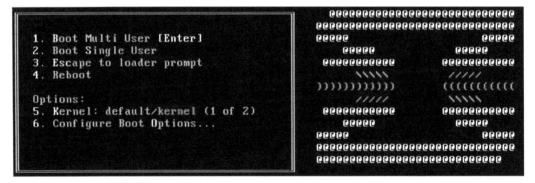

Figure 19.1 – OPNsense boot loader menu

2. After the boot process, OPNsense will show the command prompt (in *single-user mode*), but as this mode mounts the filesystem as *read-only*, it will be necessary to put it in *read-write* mode before changing the root password:

```
/sbin/mount -o rw /
```

3. With the filesystem in read-write mode, now it is possible to change the root password:

```
/usr/local/sbin/opnsense-shell password
The root user login behavior will be restored to its
defaults.

Do you want to proceed? [y/N]: y

Type a new password:
Confirm new password:

The root user has been reset successfully.
```

4. Finally, reboot OPNsense:

```
/sbin/reboot
```

After the reboot, you should log in with the newly defined root password.

Filtering log files

As we have explored in this book, one of the most valuable resources for diagnosing issues or troubleshooting is the log files. The log files are always *telling the truth* and learning about some CLI tools can help us filter data inside the logs and save a lot of time!

In the following, we will explore how to filter logs on OPNsense.

On the CLI, we will change our current directory to the /var/logs path, as we have learned that the logs are stored in this path.

Filtering logs

To have a continuous output of a log file, we can use the tail -f command:

```
root@bluebox:/var/log # tail -f system/system_`date +%Y%m%d`
```

> **Note**
> The `date +%Y%m%d` command will set the filename to the current date.

grep is one of the most helpful tools to filter data inside log files. We can use it to extract information more clearly:

```
root@bluebox:/var/log # grep --color -i error system/
system_20220305.log
```

In the preceding example, we filter the word error inside the /var/log/system/ system_20220305.log file.

--color colorfully outputs the filtered expression, and -i is for the filter in a case-insensitive case.

Some logs can be trickier than others to filter, for example, while using the nginx plugin with **NAXSI** rules.

This is an example of an nginx error log file:

```
022/03/07 00:01:01 [error] 66395#100210: *257649 NAXSI_
FMT: ip=203.0.110.100&server=example.cloudfence.com.
br&uri=/xmlrpc.php&vers=1.3&total_processed=37547&total_
blocked=4900&config=block&cscore0=$LIBINJECTION_XSS&score0=8&
zone0=BODY|NAME&id0=18&var_name0=%3C%3Fxml%20version, client:
203.0.110.100, server: example.cloudfence.com.br, request:
"POST /xmlrpc.php HTTP/1.1", host: "example.cloudfence.com.br"
```

This logline can appear non-human readable at first glance, but if you use some CLI tricks, it could help a lot:

```
head -1 example.com.br.error.log | awk -F'&' '{print "Server:
"$2 "\nRequest: \n" $12 "\nMatch rule ID:" $11 "\n"}'
Server: server=example.cloudfence.com.br
Request:
var_name0=%3C%3Fxml%20version, client: 203.0.110.100, server:
example.cloudfence.com.br, request: "POST /xmlrpc.php
HTTP/1.1", host: "example.cloudfence.com.br"
Match rule ID:id0=18
```

Using the awk command, it is possible to break the line into fields and output a filtered and cleaner output. This is one of many possibilities that we can use CLI filtering logs for. You can change the head -1 parameter to fit the number of lines you want to list.

The command can be explained as follows:

- -F'&': This defines the field delimiter.
- '{print "Server: "**$2** "\nRequest: \n" **$12** "\nMatch rule ID:" **$11** "\n"}': This line will print the *2nd*, *12th*, and *11th* fields, considering the & limiter.

Another example is filtering a Squid access.log file with a colorful output depending on the log message:

1. For this, we will create a little shell script:

   ```
   cat > /tmp/colorful-access_log.sh
   ```

2. Paste these contents to the script file:

   ```
   tail -f /var/log/squid/access.log | awk '
                       {matched=0}
                       /INFO:/    {matched=1; print "\033[0;37m"
   $0 "\033[0m"}    # WHITE
                       /NOTICE:/  {matched=1; print "\033[0;36m"
   $0 "\033[0m"}    # CYAN
                       /WARNING:/ {matched=1; print "\033[0;34m"
   $0 "\033[0m"}    # BLUE
                       /NONE\/200/ {matched=1; print
   "\033[0;34m" $0 "\033[0m"}    # BLUE
                       /ERROR:/   {matched=1; print "\033[0;31m"
   $0 "\033[0m"}    # RED
                       /TCP_DENIED/   {matched=1; print
   "\033[0;35m" $0 "\033[0m"}    # PURPLE
                       matched==0                 {print "\033[0;33m"
   $0 "\033[0m"}    # YELLOW
                       '
   ```

3. Finish by pressing *Enter + Ctrl + D*.

   The new script should be saved in /tmp/colorful-access_log.sh.

4. Change the script permissions by running the following command:

   ```
   chmod +x /tmp/colorful-access_log.sh
   ```

To test, run the following command:

```
/tmp/colorful-access_log.sh
```

You should see different line colors depending on the filtered message in the Squid `access.log` file.

As we have explored, we can do extensive filtering and ease the log outputs with a few commands such as `tail`, `grep`, and `awk`.

Summary

This chapter has explored the CLI world with its multiple possibilities. From now on, you can take advantage of the OPNsense CLI (with caution) and run advanced commands and customizations on it. In the next chapter, we will explore the API calls that will extend the power of the CLI over HTTP(S) calls.

20

API – Application Programming Interface

In this chapter, we will learn how the OPNsense **application programming interface** (**API**) works and how to use it externally by looking at some examples. By the end of this chapter, you will understand the OPNsense API structure and how to create custom calls using it.

In this chapter, we will cover the following topics:

- Concepts
- Setting up API keys
- API calls

Technical requirements

For this chapter, you will need basic knowledge of using the `curl` command on the CLI and a basic understanding of how HTTP protocol methods work.

Concepts

APIs are a way that software can talk with other software and integrate and automate routines. The **REpresentational State Transfer** (**REST**), also known as the RESTful API, is a web API that uses HTTP requests. For example, it allows you to make calls to a web application to display information or execute commands on the web server's backend. The OPNsense framework supports APIs for most webGUI features to interact with the backend controller as a modern web application.

The following diagram shows OPNsense's framework architecture, which has been extracted from the official documentation:

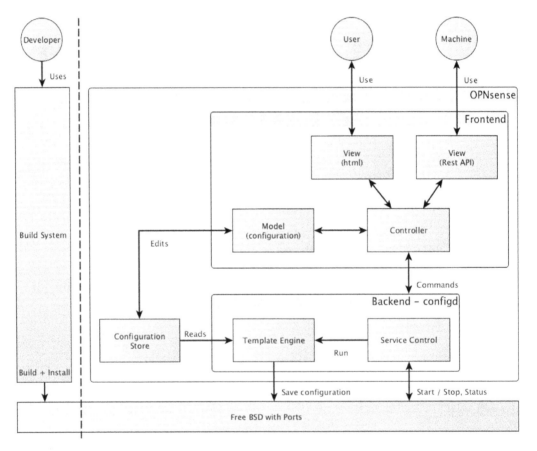

Figure 20.1 – OPNsense architecture extracted from https://docs.opnsense.org/development/architecture.html

As we can see, OPNsense's architecture is composed of a frontend, where the webGUI resides, and a backend, where the configd service resides. Every time we use OPNsense's API, we are utilizing this architecture. One of the advantages of this model is that the webGUI doesn't execute commands directly on the **operating system (OS)**, for example. Executing commands from web applications directly in the OS can lead to misconfigurations, service interruptions, or even security vulnerabilities.

Once OPNsense has an API, it is possible to integrate it with other software, such as security tools, administrative consoles, other online services, and more. Some example code for this can be found in Cloudfence's GitHub repository, which integrates Wazuh with OPNsense. You can check it out here: `https://github.com/cloudfence/opnsense-wazuh`.

The OPNsense API's structure is as follows:

```
https://opnsense.local/api/<module>/<controller>/<command>/
[<param1>/[<param2>/...]]
```

The API works with the POST and GET HTTP methods. As a rule of thumb, the GET method is used to display information, while the POST method is used to send data to change settings.

For example, if we want to get the existing firewall aliases on OPNsense, we can make a GET call using our browser (we must be logged into the webGUI):

```
https://<OPNsense_IP_Address>/api/firewall/alias/get
```

The preceding example uses the following OPNsense API structure:

- Module: `firewall`
- Controller: `alias`
- Command: `get`

> **Important Note**
>
> For a complete OPNsense API reference, please go to `https://docs.opnsense.org/development/api.html`.

Executing the preceding example on an OPNsense firewall outputs data in JSON format in your browser, as follows:

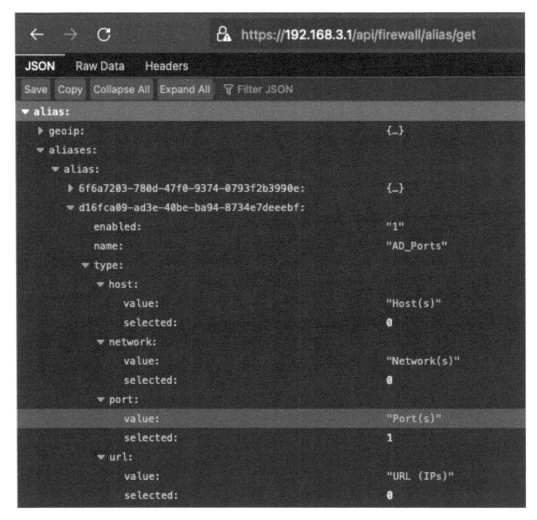

Figure 20.2 – API GET output example using the Firefox web browser

As you can see, the API call's JSON output shows the firewall aliases in a structured manner, displaying its contents.

Now, let's learn how to set up the API keys for external usage of the API and look at some other examples.

Setting up API keys

In the previous example, we made an API call internally using the webGUI. To make an external call, we need to configure the API's authentication.

API authentication in OPNsense can be created by associating a *key* and a *secret* with an existing user. To create an API key and secret for the root user, follow these steps:

1. On the webGUI, go to **System | Access | Users** and edit the root user:

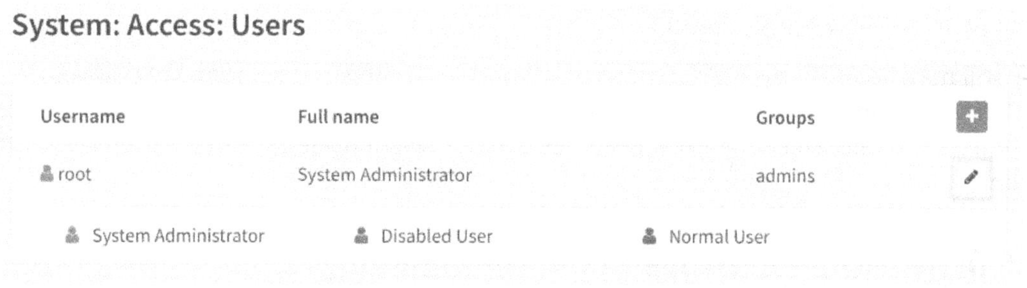

System: Access: Users

Username	Full name	Groups	
👤 root	System Administrator	admins	✏

👤 System Administrator	👤 Disabled User	👤 Normal User

Figure 20.3 – Editing the root user

2. On the user's edit page, click the + button under **API keys**:

Figure 20.4 – Adding a new API key/secret pair to the root user

3. This will save a text file that contains the API's key and secret:

```
key=xc8Odoms2r+a45Z9UOXG8drL5MHl+PkeUOrCTbj9abxZ7SD1FUJZ
vp8s23uGB5eygqQDr15wTv7f/yJm
secret=RzYoJ/uw+8btPCbXVljqijDLRIywXpOUg2xmeashsl/l/+XhR
D1DHaNVu0N3yq2xV1BJKdhoi1txcsnm
```

This file contains two lines. The first line specifies the API's *key*, while the second specifies the API's *secret*.

These credentials will allow you to remotely execute remote commands on OPNsense using the API, so keep them in a safe place!

Now, let's make some API calls using these credentials.

API calls

The simplest way to test API calls is by using the `curl` tool. You can install it on Linux, FreeBSD, macOS, and Windows. On Windows 10/11, it is installed by default. Choose your preferred OS and start testing!

We will need access to the OPNsense webGUI port for testing, so I recommend that you do the tests from the LAN network. Avoid keeping webGUI and SSH access open to the internet so that you don't use it from the WAN. It would be best if you used a VPN instead.

GET method example

We will use the following `curl` parameters in this example:

> **Important Note**
>
> We are using the `curl` and jq parameters for Unix-like environments in the following steps. If you want to use it on another shell, such as Windows Powershell, the parameters and outputs may vary.

- `-k`: Allows insecure server connections when you're using TLS. Do not use this in a production environment! Consider using a trusted certificate instead.
- `-u`: `<user:password>` This is the server's *user* and *password*. Here, we must specify the API *key* as the *user* and the API *secret* as the *password*.
- `-s`: Silent mode – prevents curl *noise* output.

Let's repeat the previous example of getting the firewall aliases but using external API access:

```
curl -s -k -u "xc8Odoms2r+a45Z9UOXG8drL5MHl+
PkeUOrCTbj9abxZ7SD1FUJZvp8s23uGB5eygqQDrl5wTv
7f/yJm":"RzYoJ/uw+8btPCbXV1jqijDLRIywXpOUg2
xmeashsl/1/+XhRD1DHaNVu0N3yq2xV1BJKdhoiltxcsnm"
https://192.168.3.1/api/firewall/alias/get
```

The command's output will look like this:

curl -k -u https://192.168.3.1/api/firewall/alias/get | more

{"alias":{"geoip":{"url":"http:\/\/192.168.254.30\/cloudsoc\/geoip\/geoip.zip"},"aliases":{"alias":{"6f6a7203-780d-47f0-9374-0793f2b3990e":{"enabled":"0","name":"GeoIP_China","type":{"host":{"value":"Host(s)","selected":0},"network":{"value":"Network(s)","selected":0},"port":{"value":"Port(s)","selected":0},"url":{"value":"URL (IPs)","selected":0},"urltable":{"value":"URL Table (IPs)","selected":0},"geoip":{"value":"GeoIP","selected":1},"networkgroup":{"value":"Network group","selected":0},"mac":{"value":"MAC address","selected":0},"external":{"value":"External (advanced)","selected":0}},"proto":{"IPv4":{"value":"IPv4","selected":0},"IPv6":{"value":"IPv6","selected":0}},"counters":"1","updatefreq":"","content":{"CN":{"value":"CN","selected":1}},"description":"GeoIP China"},"d16fca09-ad3e-40be-ba94-8734e7deeebf":{"enabled":"1","name":"AD_Ports","type":{"host":{"value":"Host(s)","selected":0},"network":{"value":"Network(s)","selected":0},"port":{"value":"Port(s)","selected":1},"url":{"value":"URL (IPs)","selected":0},"urltable":{"value":"URL Table (IPs)","selected":0},"geoip":{"value":"GeoIP","selected":0},"networkgroup":{"value":"Network group","selected":0},"mac":{"value":"MAC address","selected":0},"external":{"value":"External (advanced)","selected":0}},"proto":{"IPv4":{"value":"IPv4","selected":0},"IPv6":{"value":"IPv6","selected":0}},"counters":"0","updatefreq":"","content":{"53":{"value":"53","selected":1},"88":{"value":"88","selected":1},"123":{"value":"123","selected":1},"135":{"value":"135","selected":1},"137":{"value":"137","selected":1},"389":{"value":"389","selected":1},"445":{"value":"445","selected":1},"3268":{"value":"3268","selected":1},"3389":{"value":"3389","selected":1},"49667:49674":{"value":"49667:49674","selected":1}},"description":"Active Directory Ports"},"5a3053f7-7f8c-4a31-b923-e4de7c60734f":{"enabled":"1","name":"RFC_1918","type":{"host":{"value":"Host(s)","selected":0},"network":{"value":"Network(s)","selected":1},"port":{"value":"Port(s)","selected":0},"url":{"value":"URL (IPs)","selected":0},"urltable":{"value":"URL Table (IPs)","selected":0},"geoip":{"value":"GeoIP","selected":0},"networkgroup":{"value":"Network group","selected":0},"mac":{"value":"MAC address","selected":0},"external":{"value":"External (advanced)","selected":0}},"proto":{"IPv4":{"value":"IPv4","selected":0},"IPv6":{"value":"IPv6","selected":0}},"counters":"0","updatefreq":"","content":{"172.16.0.0\/12":{"value":"172.16.0.0\/12","selected":1},"192.168.0.0\/16":{"value":"192.168.0.0\/16","selected":1},"10.0.0.0\/8":{"value":"10.0.0.0\/8","selected":1}},"description":"Redes Locais"},"13e8f81f-6fb5-45fb-a1da-732469dee1d7":{"enabled":"1","name":"Mail_Ports","type":{"host":{"value":"Host(s)","selected":0},"network":{"value":"Network(s)","selected":0},"port":{"value":"Port(s)","selec

Figure 20.5 – The curl command's output

To help format this output, we can use the `jq` tool. It can be installed on Linux, FreeBSD, Windows, and macOS. Check it out at `https://stedolan.github.io/jq/`:

```
curl -s -k -u "xc8Odoms2r+a45Z9UOXG8drL5MHl+PkeUOrCTbj9
abxZ7SD1FUJZvp8s23uGB5eygqQDrl5wTv7f/yJm":"RzYoJ/uw+8bt
PCbXVljqijDLRIywXpOUg2xmeashsl/l/+XhRD1DHaNVu0N3yq2xVlB
JKdhoiltxcsnm" https://192.168.3.1/api/firewall/alias/get | jq
```

Just add | `jq` to the end of the previous `curl` command line.

This will output a more structured (and colorful) JSON:

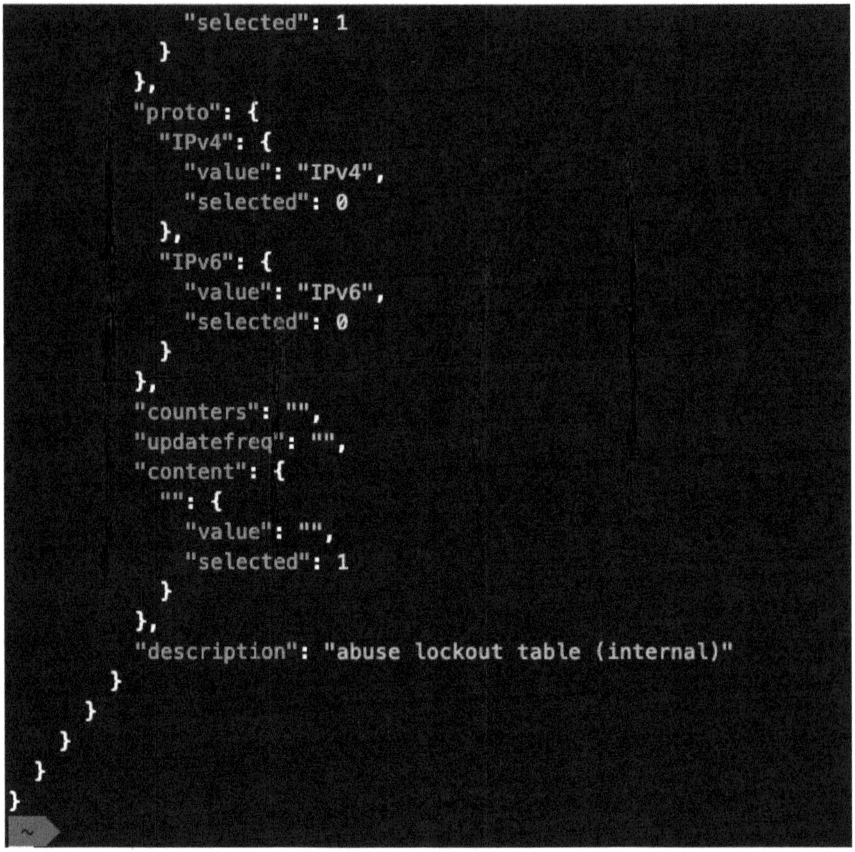

```
                "selected": 1
            }
        },
        "proto": {
          "IPv4": {
            "value": "IPv4",
            "selected": 0
          },
          "IPv6": {
            "value": "IPv6",
            "selected": 0
          }
        },
        "counters": "",
        "updatefreq": "",
        "content": {
          "": {
            "value": "",
            "selected": 1
          }
        },
        "description": "abuse lockout table (internal)"
      }
    }
  }
}
```

Figure 20.6 – curl output combined with the jq tool

So, as we can see, it is also possible to extract data from the webGUI using the API with tools such as curl and jq. This is a pretty straightforward example, but if we think about all the possibilities of integrating, managing, and automating tasks using API calls, we will realize that the sky is the limit while using the OPNsense API.

Now, let's learn how to use the API to modify firewall alias content.

POST method example

Now, let's learn how to modify firewall alias content using an API call. To change the content of an alias using the API, we must use the *POST HTTP* method.

For example, let's create a new alias on the webGUI by using the **External** type, which is the recommended type for using API calls.

Go back to the webGUI and create a new firewall alias (by going to **Firewall | Aliases**):

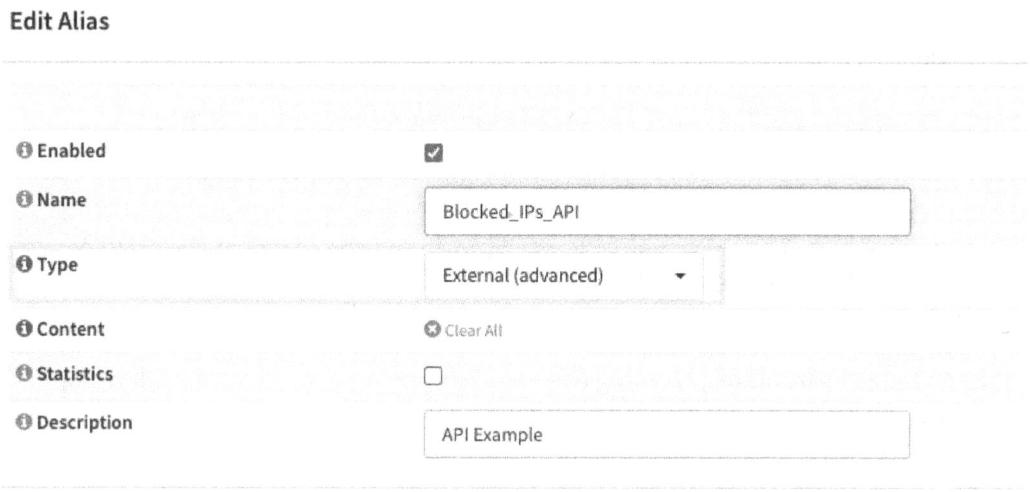

Edit Alias

ⓘ Enabled	☑
ⓘ Name	Blocked_IPs_API
ⓘ Type	External (advanced) ▼
ⓘ Content	⊗ Clear All
ⓘ Statistics	☐
ⓘ Description	API Example

Figure 20.7 – Creating an external firewall alias

In this example, we will assume that this alias will be used in a firewall rule to automatically block offenders' IP addresses that an API call will add. To finish, click the **Apply** button.

For this example, we will use the following API structure:

- Module: `firewall`
- Controller: `alias_util`
- Commands: `list` and `add`

Now, let's make an API call using curl and jq to format the JSON output:

1. First, let's list the newly created alias content using an API call:

```
curl -s -k -u "xc8Odoms2r+a45Z9UOXG8drL5MHl+PkeUOrCTbj9
abxZ7SD1FUJZvp8s23uGB5eygqQDrl5wTv7f/yJm":"RzYoJ/uw+8bt
PCbXVljqijDLRIywXpOUg2xmeashsl/l/+XhRD1DHaNVu0N3yq2xV1B
JKdhoiltxcsnm" https://192.168.3.1/api/firewall/alias_ut
il/list/Blocked_IPs_API | jq
{
  "total": 0,
  "rowCount": -1,
  "current": 1,
```

```
        "rows": []
    }
```

Now, let's add an IP address to it by using the POST method:

```
curl -s -k -u "xc8Odoms2r+a45Z9UOXG8drL5MH1+PkeUOrCTbj9
abxZ7SD1FUJZvp8s23uGB5eygqQDrl5wTv7f/yJm":"RzYoJ/uw+8bt
PCbXVljqijDLRIywXpOUg2xmeashs1/1/+XhRD1DHaNVu0N3yq2xV1B
JKdhoiltxcsnm" https://192.168.3.1/api/firewall/alias_ut
il/add/Blocked_IPs_API -XPOST -d '{"address": "203.0.110.10"}'
 -H "Content-Type: application/json"
{"status":"done"}%
```

As the highlighted part of the preceding command shows, we have added the following arguments:

- add/Blocked_IPs_API: the URL part specifies the add command and the alias's name as a parameter.

- -XPOST: This argument tells curl that it is a POST method call.

- -d '{"address": "203.0.110.10"}': This is the POST call payload that will send the IP address we want to add to the alias.

- -H "Content-Type: application/json": This specifies that our payload is using JSON format in the request header.

As we can see, the result of the API call was done:

```
{"status":"done"}%
```

Now, let's repeat the GET call to list the alias content:

```
curl -s -k -u "xc8Odoms2r+a45Z9UOXG8drL5MH1+PkeUOrCTbj9abxZ7
SD1FUJZvp8s23uGB5eygqQDrl5wTv7f/yJm":"RzYoJ/uw+8btPCbXVljqij
DLRIywXpOUg2xmeashs1/1/+XhRD1DHaNVu0N3yq2xV1BJKdhoiltxcsnm"
https://192.168.3.1/api/firewall/alias_util/list/Blocked_IPs
_API | jq
{
  "total": 1,
  "rowCount": -1,
  "current": 1,
  "rows": [
    {
```

```
        "ip": "203.0.110.10"
    }
  ]
}
```

As we can see, the IP address was added to the alias!

Another way to check the alias content is by using the `pfctl` command on the CLI:

```
root@bluebox:~ # pfctl -t Blocked_IPs_API -T show
    203.0.110.10
```

That's it – it's confirmed! Our API call that uses the POST method has worked!

You can improve this example by writing a script using programming languages such as shell scripting, Python, PHP, and others.

Summary

In this chapter, you learned about the OPNsense API and how to use it to integrate with other tools. Now, you can try to create some code and extend OPNsense's functions by using its API calls with other tools. With that, we have come to the end of this book! Thank you for reading it, and I hope this book has contributed to your OPNsense knowledge.

Index

Packt.com

Subscribe to our online digital library for full access to over 7,000 books and videos, as well as industry leading tools to help you plan your personal development and advance your career. For more information, please visit our website.

Why subscribe?

- Spend less time learning and more time coding with practical eBooks and Videos from over 4,000 industry professionals

- Improve your learning with Skill Plans built especially for you

- Get a free eBook or video every month

- Fully searchable for easy access to vital information

- Copy and paste, print, and bookmark content

Did you know that Packt offers eBook versions of every book published, with PDF and ePub files available? You can upgrade to the eBook version at packt.com and as a print book customer, you are entitled to a discount on the eBook copy. Get in touch with us at customercare@packtpub.com for more details.

At www.packt.com, you can also read a collection of free technical articles, sign up for a range of free newsletters, and receive exclusive discounts and offers on Packt books and eBooks.

Other Books You May Enjoy

If you enjoyed this book, you may be interested in these other books by Packt:

Mastering Palo Alto Networks - Second Edition

Tom Piens

ISBN: 9781803241418

- Discover the core technologies and see how to maximize your potential in your network
- Identify best practices and important considerations when configuring a security policy
- Connect to a freshly deployed/installed appliance or VM via a web interface or command-line interface
- Get your firewall up and running with a rudimentary but rigid configuration
- Gain insight into encrypted sessions by setting up SSL decryption
- Configure the GlobalProtect VPN for remote workers as well as site-to-site VPN
- Leverage the latest in SD-WAN to connect remote locations
- Explore Cortex XSOAR to automate event response

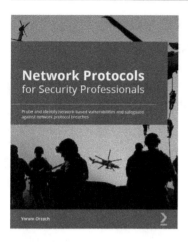

Network Protocols for Security Professionals

Yoram Orzach

ISBN: 9781789953480

- Understand security breaches, weaknesses, and protection techniques
- Attack and defend wired as well as wireless networks
- Discover how to attack and defend LAN, IP, and TCP/UDP-based vulnerabilities
- Focus on encryption, authorization, and authentication principles
- Gain insights into implementing security protocols the right way
- Use tools and scripts to perform attacks on network devices
- Wield Python, PyShark, and other scripting tools for packet analysis
- Identify attacks on web servers to secure web and email services

Packt is searching for authors like you

If you're interested in becoming an author for Packt, please visit `authors.packtpub.com` and apply today. We have worked with thousands of developers and tech professionals, just like you, to help them share their insight with the global tech community. You can make a general application, apply for a specific hot topic that we are recruiting an author for, or submit your own idea.

Share Your Thoughts

Now you've finished *OPNsense Beginner to Professional*, we'd love to hear your thoughts! Scan the QR code below to go straight to the Amazon review page for this book and share your feedback or leave a review on the site that you purchased it from.

https://packt.link/r/1-801-81687-5

Your review is important to us and the tech community and will help us make sure we're delivering excellent quality content.

www.ingramcontent.com/pod-product-compliance
Lightning Source LLC
Chambersburg PA
CBHW081457050326
40690CB00015B/2834